Second Edition

Maths Progress
Core Textbook

Series editors: Dr Naomi Norman and Katherine Pate

1

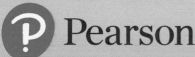

Pearson

Published by Pearson Education Limited, 80 Strand, London, WC2R 0RL.

www.pearsonschoolsandfecolleges.co.uk

Text © Pearson Education Limited 2019
Project managed and edited by Just Content Ltd
Typeset by PDQ Digital Media Solutions Ltd
Original illustrations © Pearson Education Limited 2019
Cover illustration by Robert Samuel Hanson

The rights of Nick Asker, Jack Barraclough, Sharon Bolger, Gwenllian Burns, Greg Byrd, Lynn Byrd, Andrew Edmondson, Bobbie Johns, Catherine Murphy, Naomi Norman, Mary Pardoe, Katherine Pate, Harry Smith and Angela Wheeler to be identified as authors of this work have been asserted by them in accordance with the Copyright, Designs and Patents Act 1988.

First published 2019

22 21 20 19
10 9 8 7 6 5 4 3 2

British Library Cataloguing in Publication Data
A catalogue record for this book is available from the British Library.

ISBN 978 1 292 28005 9

Printed in the UK by Bell & Bain Ltd, Glasgow

Note from the publisher
Pearson has robust editorial processes, including answer and fact checks, to ensure the accuracy of the content in this publication, and every effort is made to ensure this publication is free of errors. We are, however, only human, and occasionally errors do occur. Pearson is not liable for any misunderstandings that arise as a result of errors in this publication, but it is our priority to ensure that the content is accurate. If you spot an error, please do contact us at resourcescorrections@pearson.com so we can make sure it is corrected.

Contents

Maths Progress Second Edition

Confidence at the heart

Maths Progress Second Edition is built around a unique pedagogy that has been created by leading mathematics educational researchers and Key Stage 3 teachers in the UK. The result is an innovative structure, based around 10 key principles designed to nurture confidence and raise achievement.

Pedagogy – our 10 key principles

- Fluency
- Problem-solving
- Reflection
- Mathematical reasoning
- Progression
- Linking
- Multiplicative reasoning
- Modelling
- Concrete–Pictorial–Abstract (CPA)
- Relevance

This edition of Maths Progress has been updated based on feedback from thousands of teachers and students.

The Core Curriculum

Textbooks with tried-and-tested differentiation

Core Textbooks *For your whole cohort*

Based on a single, well-paced curriculum with built-in differentiation, fluency, problem-solving and reasoning so you can use them with your whole class. They follow the unique unit structure that's been shown to boost confidence and support every student's progress.

Support Books

Strengthening skills and knowledge

Provide extra scaffolding and support on key concepts for each lesson in the Core Textbook, giving students the mathematical foundations they need to progress with confidence.

Depth Books

Extending skills and knowledge

Deepen students' understanding of key concepts, and build problem-solving skills for each lesson in the Core Textbook so students can explore key concepts to their fullest.

Welcome to Maths Progress Second Edition Core Textbooks!

Building confidence

Pearson's unique unit structure has been shown to build confidence. Here's how it works.

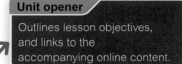

Master

1 Students are helped to **master** fundamental knowledge and skills over a series of lessons.

Check

2 Before moving on with the rest of the unit, students **check** their understanding in a short formative assessment, and give an indication of their confidence level.

Master

Learn fundamental knowledge and skills over a series of lessons.

Unit opener
Outlines lesson objectives, and links to the accompanying online content.

Key point Explains key concepts and definitions where students need them.

Hints
Guide students to help build problem-solving strategies throughout the course.

Warm up
Lessons begin with accessible questions designed to recap prior knowledge, and develop students' mathematical fluency in the facts and skills they will soon be using.

Challenge
Rich, problem-solving questions to help students apply what they've learned in the lesson and think differently.

Worked example
Provides guidance around examples of key concepts with images, bar models and other pictorial representations where needed.

Reflect Metacognitive questions that ask students to examine their thinking and understanding.

Check up

At the end of the Master lessons, students check their understanding with a short, formative Check up test, to help decide whether to Strengthen or Extend their learning.

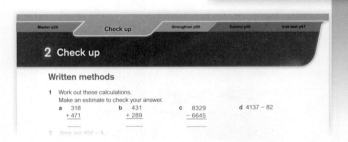

In areas where students have yet to develop a solid understanding or do not feel confident, they can choose to **strengthen** their learning.

3 Students decide on their personalised route through the rest of the unit.

Strengthen

Extend

Test

In areas where students performed well in the assessment and also feel confident, they can choose to **extend** their learning.

4 Finally, students do a **test** to determine their progression across the unit.

Strengthen

Students can choose the topics that they need more practice on. There are lots of hints and supporting questions to help.

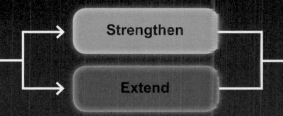

Extend

Students can apply and develop the maths they know in different situations.

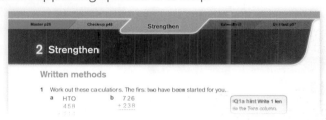

Test

Students can show everything they have learned and check their progress using the end-of-unit test.

Students can use the Support and Depth Books at any point throughout the unit. They're designed to give the right level of support and additional problem-solving content to help strengthen students' understanding of key concepts.

Progress with confidence!

This innovative Key Stage 3 Mathematics course builds on the first edition KS3 Maths Progress (2014) course, drawing on input from thousands of teachers and students, and a 2-year study into the effectiveness of the course. All of this has come together with the latest cutting-edge approaches to shape Maths Progress Second Edition.

Take a look at the other parts of the series

*Active*Learn service

The *Active*Learn service enhances the course by bringing together your planning, teaching and assessment tools, as well as giving students access to additional resources to support their learning. Use the interactive Scheme of Work, linked to all the teacher and student resources, to create a personalised learning experience both in and outside the classroom.

What's in *Active*Learn for Maths Progress?

- ☑ **Front-of-class student books** with links to PowerPoints, videos, animations and homework activities

- ☑ **96 new KS3 assessments and online markbooks,** including end-of-unit, end-of-term and end-of-year tests

- ☑ **Over 500 editable and printable homework worksheets,** linked to each lesson and differentiated for Support, Core and Depth

- ☑ **Online, auto-marked homework activities**

- ☑ **Interactive Scheme of Work** makes re-ordering the course easy by bringing everything together into one curriculum for all students with links to Core, Support and Depth resources, and teacher guidance

- ☑ **Student access to videos, homework and online textbooks**

*Active*Learn Progress & Assess

The Progress & Assess service is part of the full *Active*Learn service, or can be bought as a separate subscription. It includes assessments that have been designed to ensure all students have the opportunity to show what they have learned through:

- a 2-tier assessment model
- approximately 60% common questions from Core in each tier
- separate calculator and non-calculator sections
- online markbooks for tracking and reporting
- mapping to indicative 9–1 grades

New *Assessment Builder*

Create your own classroom assessments from the bank of Maths Progress assessment questions by selecting questions on the skills and topics you have covered. Map the results of your custom assessments to indicative 9–1 grades using the custom online markbooks. *Assessment Builder* is available to purchase as an add-on to the *Active*Learn service or Progress & Assess subscriptions.

Purposeful Practice Books

Over 3750 questions using minimal variation that:

- ☑ build in small steps to consolidate knowledge and boost confidence
- ☑ focus on strengthening skills and strategies, such as problem-solving
- ☑ help every student put their learning into practice in different ways
- ☑ give students strong preparation for progressing to GCSE study

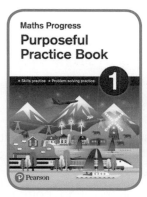
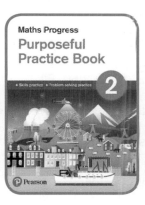
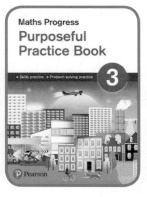

1 Analysing and displaying data

1.1 Mode, median and range

• Find the mode, median and range for a set of data

Active Learn
Homework

Warm up

1 Fluency What is the difference between
  a 11 and 7   **b** 40 and 6

2 Fluency Which number is halfway between 8 and 12?

3 Look at this set of data. 7, 2, 4, 2, 7, 8, 1, 1, 7
  a Which number occurs most often in this set?
  b Write the numbers in order, from smallest to largest.
  c Which is the middle number in the ordered list?

> **Key point**   **Data** is a set of information. Each piece of information is called a **value**.
> The **range** is the difference between the smallest and largest values. The larger the range, the more spread out the values.

4 Order the values from smallest to largest. Work out their **range**.
  a 7, 9, 3, 12, 10   **b** 8, 3, 15, 6, 2, 12, 6
  c 70, 20, 20, 40, 100   **d** 21, 50, 17
  e 250, 150, 400, 300, 350, 250, 200   **f** 0.7, 0.3, 0.1, 0.9, 0.4

5 Some students spent these amounts in a café.

  £2.50, £3, £1, £4, £0.50, £3, £1, £1, £2.50, £1

What is the **range** of the data?

> **Key point**   The **mode** is the most common value. It is also called the **modal** value.

6 Write down the **mode** for each set of data.
  a TV, phone, phone, computer, iPad, TV, TV, phone, iPad, phone, iPad, computer, phone
  b 4   7   2   2   4   2   7   2   4   7
  c 4   7   7   2   4   7   7   2   4   7
  d 0.4   0.7   0.2   0.4   0.4   0.7   0.2
  e 40   20   40   40   70   20
  f 0   20   20   0   0

7 a Work out the range for each set of values **a–f** in Q6, where possible.

b Reasoning Which set of data in Q6 does not have a range? Why not?

8 Twenty Year 7 students recorded the number of times in a week that they visited Wikipedia for information.

10	7	4	5	6	5	9	7	6	8
7	7	5	5	6	8	8	7	10	8

Find the **modal** number of visits.

9 Find the two modes of each data set.

a 5   6   1   5   6   5   6   2

b 0.8 0.6 0.5 0.8 0.6 0.7

c 0   1   2   1   1   2   2

10 Problem-solving

a Write down a set of data that has two modes.

b Write down a set of data that has no mode.

11 Here are the number of press-ups done by some Year 7 students.

12, 8, 5, 7, 2, 3, 6, 5, 6, 2, 5

a Work out the range.

b Find the mode.

12 Work out the range and mode for each data set.

a 6 cm, 4 cm, 8 cm, 8 cm, 5 cm, 4 cm, 8 cm, 7 cm

b 25 g, 27 g, 20 g, 22 g, 20 g, 23 g, 23 g, 22 g, 26 g, 20 g

c 100 mm, 300 mm, 500 mm, 400 mm, 200 mm, 300 mm, 500 mm, 100 mm, 300 mm

d 65°, 50°, 55°, 60°, 55°, 70°, 50°, 70°, 50°, 55°, 50°

Key point   The **median** is the middle value when the data is written in order.

Worked example

Find the **median** of 8, 3, 15, 6, 2, 12, 6.

2   3   6   6   8   12   15 ——— Write the values in order from smallest to largest.

median = 6 ——— Count in to the middle.

median = 6

---

13 Write each set of data in order, from smallest to largest. Then find the median.

a 5 cm, 2 cm, 7 cm, 4 cm, 4 cm

b 17, 12, 16, 15, 15, 18, 11, 16, 12

c 20 g, 70 g, 50 g, 30 g, 75 g, 15 g, 55 g

14 Find the median for each set of test results.

a 4 7 7 8 9 10
b 4 7 8 9 10 10
c 4 5 6 6 7 9
d 4 4 4 6 7 9
e 3 3 4 7 7 9
f 3 3 4 10 10 10

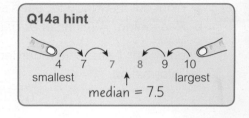

Q14a hint

4   7   7   8   9   10
smallest   largest
median = 7.5

15 Work out the median for each data set.

a 8, 3, 2, 2, 5, 9

b 6, 10, 7, 15, 8, 17, 11, 9

c 14 km, 7 km, 18 km, 12 km, 16 km, 11 km, 13 km, 15 km

d 16 ml, 10 ml, 10 ml, 12 ml, 18 ml, 20 ml

e Reasoning What fraction of the values are less than the median?

16 The numbers of children in the families of some Year 7 students are

4   3   2   2   1   2   3   4   3
2   1   2   1   6   2   3   3   4

a Find the median.

b Reasoning Does the median have to be a value in the data set?

17 Problem-solving The data shows the numbers of hours some students spent on the internet on Sunday. One value is missing.

4   0   3   4   2   1   1   ?

The median is 2.5. What could the missing value be?

Q17 hint Try out different values.

18 These are the prices paid for video games on an internet auction site.

£4, £5, £7, £12, £13, £14, £19, £21, £22

a Work out the range.   **b** Find the median.

19 Find the mode, range and median for each of these sets of data.

a 9, 3, 5, 5, 7, 4, 6, 4, 8, 5, 6   **b** 30, 10, 50, 30, 30, 100, 70, 40

c 700, 300, 200, 500, 200, 400   **d** 0.1, 0.7, 0.3, 0, 0.3

e 1, 1, 1, 10, 10   **f** 10, 10, 50, 50, 50

Challenge

1 For a set of data,
- can the mode and median be the same
- can the mode be greater than the median
- can the range be less than the mode?

Write down a simple set of data to show each answer.

2 The range of a set of data is 0. What can you say about the median and the mode?

Reflect In this lesson, you ordered numbers to work out the median.

What other maths skills did you use to work out the median?

What maths skills did you use for other topics in this lesson?

Copy and complete this sentence to help you list them all:

I used ____ to work out the ____.

1.2 Displaying data

- Find information from tables and diagrams
- Display data using tally charts, tables, bar charts and bar-line charts

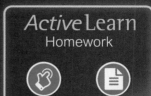

Active Learn
Homework

Warm up

1 Fluency Starting at 0, count on in steps of
   a 2   **b** 10

2 Write down the number that each arrow points to on the scale.

3 The table shows the number of faults on cars that were tested for their MOT.
   a How many cars had 2 faults?
   b How many cars had more than 2 faults?
   c How many cars were tested altogether?
   d What was the most common number of faults?

Number of faults	Number of cars
0	8
1	12
2	10
3	8
4	6

> **Key point** A **pictogram** uses pictures to show data. The **key** shows what each picture represents.

4 This **pictogram** shows the numbers of texts Kayleigh sent.
   a How many texts do these represent?
      i ◒   **ii** ◷
   b Copy and complete the table to show the number of texts she sent each day.

Day	Sat	Sun	Mon	Tue	Wed
Texts			2		

   c On which day did she send the most texts?
   d How many texts did she send altogether?

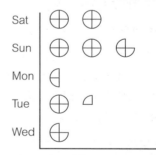

Numbers of texts sent

Key: ⊕ represents 4 texts

5 This **pictogram** shows the numbers of messages Gary sent.
   a How many messages did he send each day?
   b He sent 55 messages on Saturday and 70 messages on Sunday.
      Draw diagrams to show this information.
   c How many messages did he send altogether on these five days?

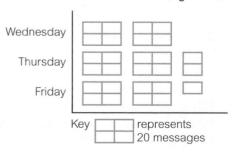

Number of messages sent

Key ⊞ represents 20 messages

Unit 1 Analysing and displaying data   4

Key point   A **bar chart** uses bars of equal width to show data.

6   Marcia drew this **bar chart** to show her classmates' favourite cold drinks.

   a   Which drink is the most popular?

   b   How many students like juice best?

   c   How many more students chose fizzy drinks than juice?

   d   How many students are there altogether?

   e   **Reasoning**   How can you work out the mode from a bar chart?

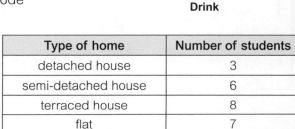

7   The table shows the types of home in which some Year 7 students live.

Type of home	Number of students
detached house	3
semi-detached house	6
terraced house	8
flat	7

   a   How many students live in a house?

   b   Copy and complete the bar chart.

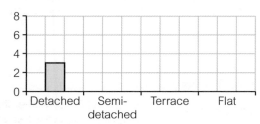

> **Q7b hint** Make the bars the same width. Put spaces between the bars. Label the vertical axis. Give your chart a title.

   c   What is the mode?

Key point   A **bar-line chart** is like a bar chart but uses lines instead of bars.

8   The **bar-line chart** shows Year 9 girls' shoe sizes.

   a   How many of these girls have a shoe size of 7?

   b   Write down the modal shoe size.

   c   Work out the range of the girls' shoe sizes.

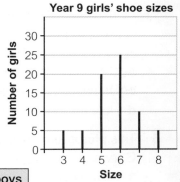

9   The table shows Year 9 boys' shoe sizes.

   a   What is the modal shoe size?

   b   What is the range of the shoe sizes?

   c   Draw a bar-line chart for the data.

   d   Work out the total number of boys in Year 9.

Shoe size	Number of boys
5	5
6	10
7	10
8	20
9	10
10	5

> **Q9c hint** Use a similar grid to Q8.

10 The **tally chart** shows how some Year 7
students travel to school.

a How many students walk to school?

b Sixteen students travel by car.
Draw the tally marks for this data.

c What is the modal method of travel?

d Write down the **frequency** of each
method of travel.

Method of travel	Tally	Frequency
walk	ⅲⅉ ⅲⅉ ⅲⅉ ⅲⅉ ⅰ	
car		16
bicycle	ⅲ	
bus	ⅲⅉ ⅲⅉ ⅲⅉ	
other	ⅱ	

11 A class of Year 7 students chose their favourite social website.
Facebook (F)   Instagram (I)   Snapchat (S)   YouTube (Y)   Twitter (T)

F, T, F, Y, S, F, I, T, F, F, Y, I, F, T, Y, T, F,
F, T, Y, F, F, I, I, S, Y, T, F, Y, T, F

Make a tally chart for the data.

12 Yolanda counted the food items in 20 lunch boxes.

6, 7, 5, 6, 5, 9, 6, 4, 5, 5, 7, 6, 5, 6, 4, 7, 6, 5, 6, 4

a Copy and complete the **frequency table**.

Number of food items	4	5	6	7	8	9
Frequency						

> **Q12b hint** Remember that
> even if a data value has a
> frequency of 0, it should still
> be shown on the bar chart.

b Draw a bar chart for the data.

c What is the modal number of food items?

d What is the range?

13 **Reasoning**   Jeremy and Ashley have collected
shoe size data for their class.
Jeremy says, 'The mode is 12.'
Ashley says, 'The mode is 3.'
Who is correct? Explain why.

Shoe size	2	3	4	5
Frequency	4	12	7	4

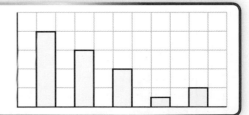

1.3 Grouping data

Active Learn
Homework

- Interpret simple charts for grouped data
- Find the modal class for grouped data

Warm up

1 Fluency What values are included in the group '20–29 pens'?

2 Count the tally marks.

     a 𝄃𝄃𝄃 𝄃𝄃𝄃 𝄃𝄃𝄃   **b** 𝄃𝄃𝄃 ||| **c** 𝄃𝄃𝄃 𝄃𝄃𝄃 𝄃𝄃𝄃 𝄃𝄃𝄃 ||

Key point

Data is sometimes organised into **groups** or **classes**, such as 1–5, 6–10, 11–15, ...
The **modal class** is the one with the highest frequency.

---

3 Eduardo measured the pulse rates of some classmates, in beats per minute.
He crossed out each value as he made a tally mark for it in a chart.

     81, 96, 90, 97, 78, 100, 88, 91, 90, 84, 96, 85, 84, 89, 80, 102, 95, 89, 109, 89

     a Complete the Tally and Frequency columns in the **grouped tally chart**.

     b What is the **modal class**?

     c How many students had pulse rate of 90 or more?

Pulse rate	Tally	Frequency
70–79	|	
80–89	|	
90–99	|||	
100–109		

4 A PE teacher asked a Year 7 class to do as many
star jumps as they could in 30 seconds.
The **grouped frequency table** shows the results.

     a What is the modal class?

     b How many students are in the class?

     c Reasoning Three more students each did 33 star jumps.
Has the modal class changed? If so, how?

Star jumps	Frequency
20–24	4
25–29	11
30–34	9
35–39	4
40–44	1

5 Sui Main asked her classmates how many
coins they had on them. Here are her results.

Students' coins

Number of coins	Frequency
0–2	6
3–5	
6–8	
9–11	
12–14	

a Copy and complete the frequency table.

b What is the modal class?

c **Problem-solving** How many classmates had nine or more coins?

d **Reasoning** Can you tell from Sui Main's frequency table how many classmates had eight coins?

6 Pulen carried out a survey of the number of books that people in his class owned. Here are his results.

Number of books	Frequency
0–4	11
5–9	9
10–14	4
15–19	2
20–24	3
25–29	2

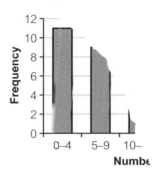

a What is the modal class?

b Copy the axes and complete the bar chart for the data. Give your chart a title.

7 An engineer's toolbox contains drills with these sizes (in mm).

    7   10   4   12   2   5   4   9   3   18   6   15

The engineer records the sizes in this **grouped frequency table.**

Drill size (mm)	Tally	Frequency
1–5		
6–10		
11–15		
16–20		

a Copy the table. Tally the drill sizes in the table. Fill in the frequency column.

b Draw a bar chart for the data.

c What is the modal class?

Key point   For data that comes from measuring, such as height, there are no gaps between the bars of a bar chart.

Worked example

Abigail timed, to the nearest second, how long some students took to solve a puzzle.

Draw a bar chart for the data.

Time (seconds)	Frequency
10–19	10
20–29	25
30–39	20
40–49	10
50–59	5
60–69	5

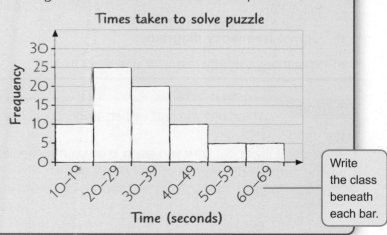

Write the class beneath each bar.

8 The frequency table shows the heights of students in a Year 7 class, measured to the nearest centimetre.

Height (cm)	Frequency
130–139	5
140–149	10
150–159	13
160–169	2

a Draw a bar chart for the data.

b What is the modal class?

c Reasoning Can you tell the height of the tallest student?

9 Problem-solving / Reasoning The bar chart shows the volumes of liquid drunk by some Year 7 students in a day.

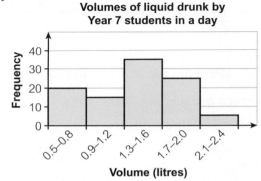

It is recommended that Year 7 students should drink about 1.2 litres a day.

a Are these Year 7 students drinking enough? Give a reason for your answer.

b How could grouping the data differently help you to answer the question? What groups would you use?

Challenge

a Find a 6-sided pencil. Use a pen to label two sides with A, two with B and two with C.

b Roll the pencil 30 times.

c Record the letter on the top side in this tally chart.

d Draw a **frequency diagram**.

e Was there a modal result? If yes, what was it?

Result	Tally	Frequency
A		
B		
C		
Total		

Reflect   Which did you find easier: working out the modal class from a table or from a bar chart?

Think carefully about how you learn in all your subjects. Do you understand things better when there is a diagram?

1.4 Averages and comparing data

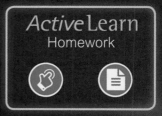

Active Learn
Homework

- Calculate the mean of a set of data
- Compare sets of data using their ranges and averages

Warm up

1 **Fluency** What is the total of 5, 3 and 7?

2 Work out 48 ÷ 8.

3 5 members of a paintball team have these numbers of paintballs.
  8, 4, 9, 3, 1
  a How many paintballs do they have altogether?
  b They share the paintballs equally. How many does each player have?

Key point The **mean** of a set of values is the total of the set of values divided by the number of values.

Worked example

Find the **mean** of 2, 2, 6, 4, 1.

total = 2 + 2 + 6 + 4 + 1 = 15

mean = total ÷ number of values

   = 15 ÷ 5 = 3

mean = 3

Add up the values.

total = 15

| 2 | 2 | 6 | 4 | 1 | 5 values |

There are 5 values so divide the total by 5.

| 3 | 3 | 3 | 3 | 3 | 15 ÷ 5 = 3 |

4 Work out the mean for each set of numbers.
  a 6, 10   b 2, 6, 10   c 2, 6, 10, 10   d 0, 2, 6, 10, 10
  e **Reasoning** Can you ignore values that are 0 when working out the mean?

5 Work out the mean for each set of values.
  a 1, 2, 5, 6   b 4, 6, 8, 10, 12
  c 2, 1, 6, 3, 3, 5, 2, 4, 1, 3   d 20p, 20p, 30p, 30p, 50p
  e 5 m, 2 m, 8 m, 2 m, 10 m, 6 m
  f **Reasoning** Look at your answers. Does the mean have to be one of the data values?

Q5 hint Remember to press the = key on your calculator after adding up the values.

6 A high jump champion made these high jumps in a competition.
  189 cm, 193 cm, 197 cm, 200 cm, 203 cm
  Use your calculator to work out the mean height she jumped.

The **average** of a set of data gives a typical value for the data.
The mode, median and mean are different ways of describing the average of a set of data.

7   Three players scored these points in a table tennis tournament
   Manjit   7, 11, 8, 13, 11, 5, 3, 12, 8, 7, 11, 9
   Tony   40, 20, 60, 50, 30, 20, 60
   Sebastian   8, 19, 7, 23, 9, 15
   For each player, work out
   a the mode   **b** the median   **c** the mean   **d** the range

8   A school tested the performance of some BMX bicycles.
   The results show the times to complete a course.

Time (min)	17	14	16	17	18	14	18	23	16

   a Work out the range.
   b **Reasoning**   Which averages can you find for this data? Work out these values.

Key point   To **compare** two sets of data, find and compare an average (the mode, median or mean) and the range.

Worked example

Daniel's five long jumps were 3.7 m, 3.4 m, 4.1 m, 3.8 m, 4.1 m
Paul's five long jumps were 3.9 m, 4.3 m, 3.2 m, 4.2 m, 3.1 m
Compare their jumps.

Daniel: range = 4.1 − 3.4 = 0.7 m
        total = 3.7 + 3.4 + 4.1 + 3.8 + 4.1 = 19.1 m
        mean = 19.1 ÷ 5 = 3.82 m

Paul: range = 4.3 − 3.1 = 1.2 m
        total = 3.9 + 4.3 + 3.2 + 4.2 + 3.1 = 18.7 m
        mean = 18.7 ÷ 5 = 3.74 m

Paul's jumps are less consistent than Daniel's because his jumps have a greater range.

> Compare the ranges. The closer the results are to one another, the more **consistent** they are. The smaller the range, the more consistent the values.

Daniel jumps further on average than Paul because his jumps have a greater mean.

> Compare the means.

9   Competitors in highland games have to throw a heavy object over a bar.
   These are the heights of the bars for successful throws, in feet, for two teams.
        Team 1: 15, 13, 14, 12, 13, 12, 11, 13
        Team 2: 11, 16, 12, 11, 14, 11, 12, 10
   a **i** Work out the range for each team.
       ii Which team has the bigger range?
   b **i** Write down the mode for each team.
       ii Which team has the higher mode?

10 Reasoning The bar charts show the sports that Kieran's friends and Robin's friends enjoy most.

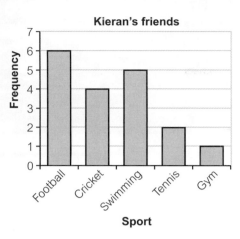

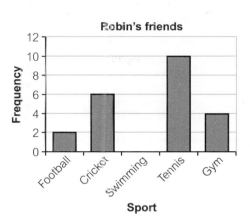

a Compare the types of sport that Kieran's and Robin's friends like least.

b Compare the types of sport that Kieran's and Robins friends like most. Which average are you using to compare?

11 The table shows the top scores in a horse riding event at the Olympics.

Men	76	74	81	71	75	72	76	74
Women	79	82	76	83	74	78	78	75

a Work out the range for the men and the range for the women.

b Work out the median for the men and the median for the women.

c **Reasoning** Use the ranges and medians to compare the scores of the men and the women.

12 A school tested the performance of some BMX bikes made from aluminium and some made from steel. The results show the times (in minutes) to complete a test course.

Steel	9.4	9.7	9.9	9.2	10.1	9.6	9.6	10.2	9.6
Aluminium	8.9	9.6	10.3	9.4	10.5	9.0	9.7	9.6	9.9

a Find the mean and range for each type of BMX bike.

b **Reasoning** Use the means and ranges to compare the times for the two types.

Challenge The table shows the heights (in centimetres) of the top ten male athletes in two sports at the 2016 Olympics.

Javelin	191	175	183	188	199	189	186	190	191	186
Shot put	203	180	185	198	188	204	199	178	190	190

a Are the heights of the athletes different in the two sports? Explain the method you used.

b Give two more sports where the contestants might be physically different.

c Investigate your answer using information from websites.

Reflect You often see the word 'average' in headlines. For example, 'Average screen size of TVs grows again'. 'Average' in everyday language could be the mean, median or mode. Write notes in your own words on the differences between these.

1.5 Line graphs and more bar charts

- Understand and draw line graphs
- Understand and draw dual and compound bar charts

*Active*Learn
Homework

Warm up

1   **Fluency**   What labels should every bar chart have?

2   Write down the values shown with letters.

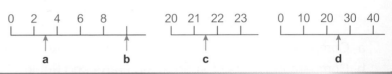

Key point

A **line graph** shows how quantities change.

3   The graph shows the average daily temperature in Leeds.

   a   Copy this sentence and fill in the gaps.
     The graph shows the average temperature every _____ for a _____ in December.

   b   What was the average temperature on
     i   Saturday   **ii**   Sunday   **iii**   Tuesday?

   c   On what days was the average temperature 5 °C?

   d   What was the **minimum** average temperature?

   e   What was the **maximum** average temperature?

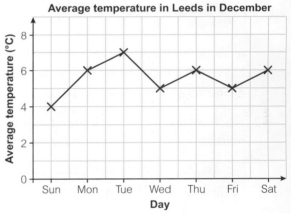

4   The **line graph** shows how Rikki's temperature changed during the morning of one day.

   a   What was Rikki's maximum temperature?

   b   When was Rikki's temperature 36.5 °C?

   c   Normal body temperature is 37 °C. How many of Rikki's readings are above this?

   d   Between which times was Rikki's temperature decreasing?

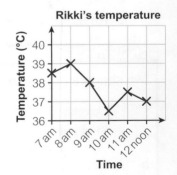

5   A supermarket aims to collect 100 kg of coins for charity. The line graph shows how much was donated each month.

   a   When was the least mass of coins donated in a month?

   b   Which of these describes the mass of coins donated in the first three months: increasing, decreasing, or staying the same?

   c   In which months were more than 15 kg of coins donated?

   d   **Problem-solving**   When did the supermarket reach its target of 100 kg?

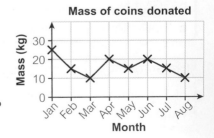

6   The table shows the numbers of hours that Michelle spent on Facebook each day.

Day	Fri	Sat	Sun	Mon	Tue	Wed	Thu
Time (hours)	2	5	4	0	1	2	2

Draw a line graph for the data.

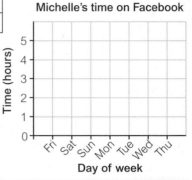

Key point   A **dual bar chart** compares two sets of data

7   The **dual bar chart** shows the numbers of genuine and junk emails
    that Aroti received each day.

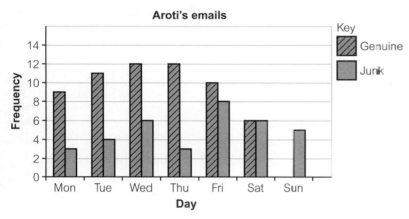

a   How many junk emails did she receive on Tuesday?
b   Which day did she receive equal numbers of junk and genuine emails?
c   Why has Sunday only got one bar?
d   **Reasoning**   Aroti says that about half of the emails she receives
    are junk. Is she correct? Explain your answer.

8   115 girls and 115 boys from Year 7 were asked what they like reading
    the most.

	Fiction	Graphic novels	Non-fiction	Magazines
Boys	25	35	30	25
Girls	40	25	20	30

a   Copy and complete the dual bar chart for the data.
b   For these students, do the boys and girls like reading different things?

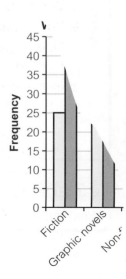

9   The **compound bar chart** shows some exam results for a school in 2018.

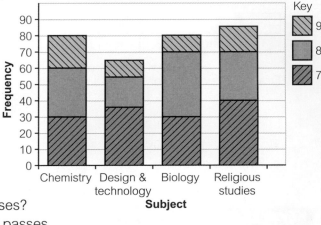

School exam results, 2018

a   Look at the bar for chemistry. How many students

   i   took chemistry

   ii   got a grade 7 in chemistry?

b   How many students got

   i   a grade 8 in religious studies

   ii   a grade 8 or 9 in biology?

c   In which subjects did

   i   30 students get a grade 7

   ii   students get the most grade 8 passes?

d   **Problem-solving**   How many grade 9 passes were there altogether in these subjects?

10   Here is some information about the photos Gareth uploaded.

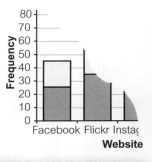

	Facebook	Flickr	Instagram	Snapchat
Photos of Gareth	25	35	40	15
Other photos	20	35	25	50

a   Copy and complete the compound bar chart for the data.

b   **Problem-solving**   Were there more photos of Gareth than other photos? Explain your answer.

Challenge   People in different countries were asked how often, during the last 30 days, they had talked on their mobile phones while driving. The chart shows the results.
Write a news story about the data shown in the graph. Don't forget to give it a headline.

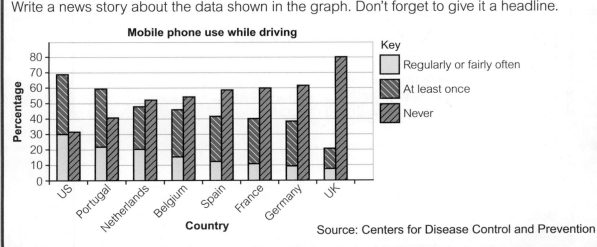

Source: Centers for Disease Control and Prevention

Reflect   Make a list of the questions from this lesson that you found easiest to answer. What made them easier for you?

1 Check up

Averages and range

1 Use your calculator to find the mean for this set of lengths.

2.4 m, 3.6 m, 1.9 m, 5.2 m, 2.9 m

2 Look at this set of data.   6, 5, 8, 0, 2, 2, 12

    a Work out the range.   **b** Write down the mode.

    c Find the median.   **d** Calculate the mean.

Charts and tables

3 Year 7 students chose their favourite big cats.

tiger, lion, lion, cheetah, tiger, leopard, lion, tiger,
lion, jaguar, tiger, cheetah, lion, leopard, lion

    a Copy and complete the tally chart.

Big cat	Tally	Frequency
lion		
tiger		
cheetah		
leopard		
jaguar		

    b What is the mode?

4 The bar-line chart shows the numbers of children in the
families of some Year 7 students.

    a How many families have 3 children?

    b What is the mode?

    c Why is there no bar for 5 children?

    d How many families have fewer than 3 children?

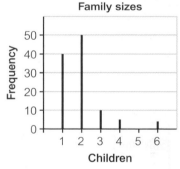

Family sizes

5 The line graph shows the classroom temperature
every 2 hours.

    a What was the temperature at 3 pm?

    b When was the temperature 14 °C?

    c Describe what happened to the temperature
between 7 am and 11 am.

    d What was the maximum temperature?

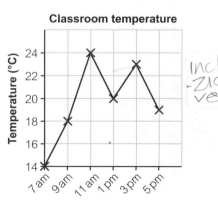

Classroom temperature

Comparing data

6　100 Year 7 students and 100 Year 8 students were asked which topics in maths they liked best.
The bar chart shows the results.

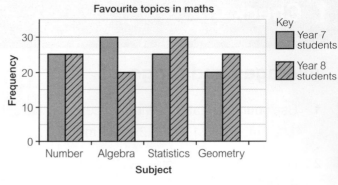

Favourite topics in maths

a Which was the Year 8 students' favourite topic?

b Which was the Year 7 students' favourite topic?

c Which topic did equal numbers of Year 7 and Year 8 students like best?

d How many more Year 8 students than Year 7 students liked Statistics best?

e How many students chose Number as their favourite topic?

7　The bar chart shows the medals won by four countries at the 2016 Olympics.

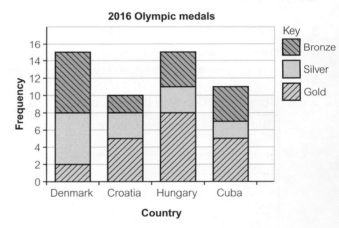

2016 Olympic medals

a How many gold medals did Cuba win?

b Which of these countries won the most bronze medals?

c Which of these countries won 6 silver medals?

d Which two of these countries won the same number of medals?

8　The table shows the distances (in kilometres) that some Year 7 students travel to school.

Oakbridge School	1.2	0.8	2.2	1.1	1.5	2.6	0.9	3.0	2.1	1.6
St John's School	3.0	0.5	6.1	1.5	1.1	5.2	9.9	2.8	8.5	1.4

a Calculate the mean travel distance for each school.

b Calculate the range for each school.

c Use the means and ranges to compare the distances travelled to the two schools.

Challenge

a The numbers 4, 2, 7, ☐ have a mode of 2. What is the missing number?

b The numbers 4, 2, 5, ☐ have a mean of 4. What is the missing number?

c The numbers 20, 70, 10, ☐ have a range of 80. What is the missing number?

d The numbers 8, 4, 6, ☐ have a median of 5. Write down a possible value for the missing number.

Reflect

How sure are you of your answers? Were you mostly

😞 Just guessing　😐 Feeling doubtful　🙂 Confident

What next? Use your results to decide whether to strengthen or extend your learning.

1 Strengthen

Averages and range

1 a These snacks were bought from a vending machine.

crisps, drink, crisps, chocolate, drink, crisps, chocolate, biscuits

Find the mode.

> **Q1a hint** Mode = most common

b The amounts paid for the snacks are

50p, 80p, 50p, 60p, 80p, 50p, 60p, 40p

Write down the mode.

2 This set of data has two modes.

5, 5, 5, 6, 7, 8, 8, 8

Write down both modes.

3 a The diagram shows some crayons arranged in order of size. Write down the length of the middle crayon.

b Another set of crayons have these lengths.

5 cm, 5 cm, 8 cm, 9 cm, 9 cm

Write down the median.

4 cm   6 cm   7 cm   8 cm   12 cm

> **Q3b hint** Median is the middle value when they are in order.

4 The diagram shows some pencils arranged in order of size.

a i Write down the lengths of the middle two pencils.

ii Find the median length of the pencils.

b Another set of pencils have these lengths.

10 cm, 12 cm, 12 cm, 16 cm, 17 cm, 17 cm

Find the median.

9 cm   9 cm   12 cm   14 cm   14 cm   16 cm

> **Q4a ii hint** The median is halfway between the two middle lengths. Draw a number line to find halfway between two numbers.
>
> 12 cm   □   14 cm

5 Work out the median of these values.

7, 0, 6, 4, 4, 2, 7, 5, 1

> **Q5 hint** Write the numbers in order first.

6 Work out the median for these basketball scores for one team.

14, 30, 21, 9, 25, 18, 39, 26

7 Some Year 7 students counted the number of items in their pencil cases.

4, 2, 11, 5, 8, 13, 6

a Which is the smallest value?   **b** Which is the largest value?

c Work out the range.

8   **a** Some Year 7 students did as many press-ups as they could manage.

8, 3, 15, 6, 2, 12, 6

Work out the range.

b Some pianists measured the length of their little fingers.

5.7 cm, 6.2 cm, 5.1 cm, 6.6 cm, 5.5 cm, 5.2 cm

Use a calculator to work out the range.

Q8a hint The **range** is a number. Subtract the smallest value from the largest value.

9 Follow these steps to arrange these counters in 4 equal piles.
  a Collect them together. $1 + 3 + 6 + 2 = \square$
  b Share them into 4 equal piles. $\square \div 4 = \square$

10   Look at these values.   12, 4, 6, 0, 8
  a Add up the values.
  b How many values are there?
  c Work out the mean.

Q10c hint Divide the total by the number of values.
$\square \div \square = \square$
The answer is a decimal.

11 Work out the mean of these values.
  a 4, 2, 1, 5   **b** 8, 5, 0, 2, 8, 9, 1, 7
  c 4, 2, 5, 8, 6   **d** 13, 28, 0, 24, 19, 14, 18, 24, 16, 14

Charts and tables

1 The bar-line chart shows the mass of bowling balls in a national competition.

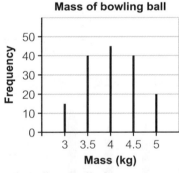

Mass of bowling ball

a How many bowling balls have a mass of 3 kg?
b Which mass of bowling ball did 45 players use?
c Write down the mass with the tallest bar-line to find the **modal** mass.

Q1a hint The height of the bar-line for 3 kg lies halfway between 10 and 20.

Q1c hint The **modal** mass is the most common mass. It is another way of saying the mode.

2 Scarlett won these prizes in a video game.

$50, $20, $20, $100, $50, $100, $20, $20, $20, $50

a Copy the tally chart and frequency table.

Prize	Tally	Frequency (number of prizes)
$20		
$50		
$100		

b Copy the list of prizes. Cross off the first prize in the list.
Make a tally mark in the table. Do this for all the prizes.

c Count up the tally marks for each prize.
Write the number in the Frequency column.

d Which prize was the mode?

3 Alex recorded the time he spent talking on his
mobile phone each day for a month.

Time (minutes)	Frequency
10–19	15
20–29	9
30–39	6
40–49	0
50–59	1

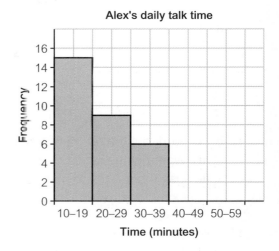
Alex's daily talk time

a What is the modal class?

b Copy and complete the bar chart.

4 The table shows the amount of money in a charity box
at the beginning of each month.

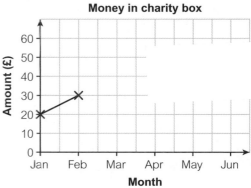
Money in charity box

Month	Jan	Feb	Mar	Apr	May	Jun
Amount (£)	20	30	0	15	15	50

a Copy the axes and the first two points.

b Plot the points for March, April, May and June
using crosses.

c Join the points with straight lines.

d Between March and June, did the amount increase or decrease?

Comparing data

1 The table shows the number of adults and
children visiting a sports centre during
the summer.

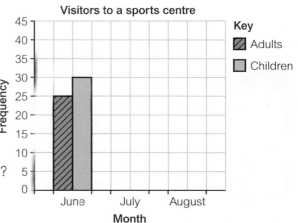
Visitors to a sports centre

Key
▨ Adults
▨ Children

	Adults	Children
June	25	30
July	30	40
August	20	40

a How many adults visited the centre in June?

b Copy and complete the dual bar chart.
Draw the bar for adults in July first.

c In which month was there the largest difference
between adults and children?

2 Two students are neighbours. They compared their travel times (in minutes) to school by bus and by car.

Bus	15	22	19	18	21	19	20
Car	7	28	15	21	17	29	9

a **i** Work out the median travel time by bus.

  ii Work out the median travel time by car.

  iii Use the medians to compare the travel times by bus and by car.
  Complete the sentence using the words quicker or slower, less or greater.
  Travelling by car is _____ than by bus on average because the
  median time by bus is _____ than by car.

b **i** Work out the range of travel times by bus.

  ii Work out the range of travel times by car.

  iii Use the ranges to compare the travel times by bus and by car.
  Complete the sentence using the words more, less, or greater.
  Travel times by car vary _____ than by bus because the range of car
  travel times is _____ than the range of bus travel times.

3 The chart shows the amount of time two students, Atifa and Joanne, spent talking, texting and playing games on their mobile phones one day.

a How long did Joanne spend talking?

b How long did Atifa spend texting?

c Who spent the most time playing games?

d How many minutes altogether did Atifa spend talking, texting and gaming?

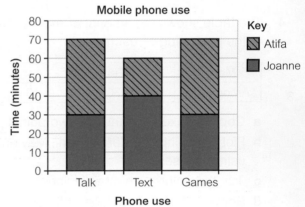

Q3 hint First look at the key.
What does the top box represent?
What does the bottom box represent?

Challenge

a Copy this tally chart and use it to record the number of letters in each word of this sentence.

b Repeat part **a** using a long sentence from a different book.

c Compare the two sentences using the mode.

Letters	Tally	Frequency
1–2		
3–4		
5–6		
7–8		

Reflect   Write down the names of all the different charts you have learned about in this unit.
Beside each chart, write *one* thing that makes it different from the others.
Write *two* things that are the same for all charts.

1 Extend

1 a The maximum wind speeds (in km/h) of the Atlantic hurricanes in 2012 were:

140, 155, 175, 130, 165, 130, 185, 150, 150, 185

   i Work out the median wind speed.
   ii Work out the range of the wind speeds.

 b The maximum wind speeds (in km/h) of the Atlantic hurricanes in 2017 were:

140, 175, 215, 285, 250, 165, 185, 230, 150, 185

   i Work out the median wind speed.
   ii Work out the range of the wind speeds.

 c Reasoning   Use the medians and ranges to compare the wind speeds of hurricanes in 2012 and 2017. Write two sentences.

2 Reasoning   Dave wanted to estimate the average height of a Year 7 student. He measured the heights of five of his friends and calculated their mean to be 142 cm.
 a Explain why Dave's friends might be different from the average for all Year 7 students.
 b Explain why his mean might not be a good model for the average height of Year 7 students.
 c How can his model be improved?

3 The table shows the challenge award badges given to scouts of three troops.

	Creative	Community	Expedition	Fitness
Trascombe	4	11	7	7
Drax Valley	10	8	0	6
1st Sandon	2	5	4	12

 a What was the most popular award overall?
 b Draw a compound bar chart for the data. Draw one bar for each scout troop.
 c Which troop got the most awards?

4 Wigan and Bradford rugby teams each played 35 games.
 The charts show the points they scored.

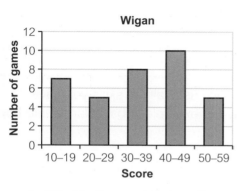
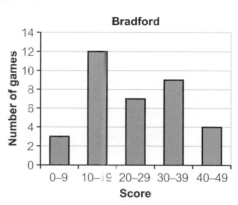

 a i What is the modal class for each team?
    ii Which team performed better? Give a reason for your answer.
 b Reasoning / Problem-solving   Can you tell which team had the greater range of scores? Give a reason for your answer.

5 Problem-solving / Reasoning Asifa recorded the goals scored by her hockey team. She calculated the mean score.

🏑 0 5 3 1   mean = 3

 a One of the scores got rubbed out accidentally. Work out the missing score.

 b How can you use the mean of a set of values to find their total?

6 The table shows the emergency response times for ambulances in a town during May.

Time (minutes)	Frequency
3–5	12
6–8	16
9–11	10
12–14	7
15–17	3
18–20	1

 a What is the modal class?

 b **Reasoning** The Government's target is for ambulances to respond within 8 minutes.

  i How many ambulances achieved this target?

  ii How many did not?

 c Draw a bar chart for the data.

 d Here are the first few emergency response times for June.

  4, 7, 5, 6, 10, 7, 8, 95

  i What is unusual about this data? Give a possible reason for this.

  ii Work out the mode, median and mean.

  iii **Reasoning** Which one of these three averages best represents the data?

Challenge   Answer each of the following questions and explain the method you use.

 a The mode of three whole numbers ☐, ☐, ☐ is 2. Their range is 3.
   Find the three numbers.

 b The median of the four whole numbers 3, ☐, ☐, 7 is 5.5.
   Find two possible missing numbers.

 c Find five whole numbers whose mode is 1, median is 2 and mean is 3.

 d The mean of three whole numbers is 4. Work out the total of the numbers.

 e The mean of the three whole numbers 2, 3, ☐ is 5. Find the missing number.

 f The mean of the three whole numbers ☐, ☐, 3 is 0. Find two possible missing numbers.

Reflect   The Government's Office for National Statistics uses all the skills you have learned in this unit to interpret data about Britain today (for example, jobs, salaries and how we live).

What other organisations might use these skills?

What do you think they might use them for?

1 Unit test

1 The pictogram shows the favourite treats of some dogs.

a How many dogs like biscuits best?

b How many dogs like bones best?

c 11 dogs like rawhide best. Draw symbols to show this.

d How many dogs are there altogether?

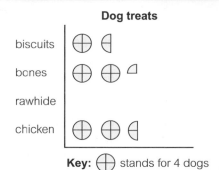

Dog treats

Key: ⊕ stands for 4 dogs

2 a The chart shows the accidents each month on a busy road before a speed camera was fitted.

　　i In how many months were there exactly three accidents?

　　ii In how many months were there more than three accidents?

　　iii Write down the modal number of accidents.

　　iv Work out the range.

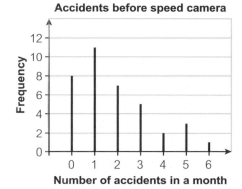

Accidents before speed camera

b The chart shows the accidents each month after a speed camera was fitted.

　　i In how many months were there no accidents?

　　ii Write down the mode.

　　iii Work out the range.

c Use the modes and ranges to write two sentences comparing the numbers of accidents before and after the speed camera was fitted.

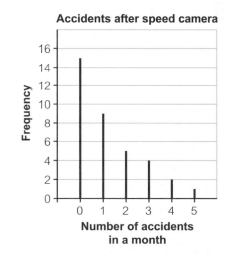

Accidents after speed camera

3 A doctor recorded the number of breaths some patients took in 1 minute.

a How many patients took 10 to 12 breaths?

b Jasmine took 17 breaths in 1 minute. Which class contains this value?

c Find the modal class.

d How many patients took more than 15 breaths in a minute?

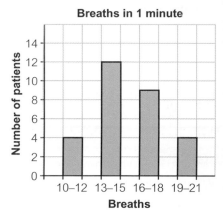

Breaths in 1 minute

Unit 1 Analysing and displaying data　24

4 Look at this set of data.

8, 0, 5, 2, 5

a Write down the mode.
b Work out the range.
c Work out the median.
d Work out the mean.

5 The chart shows the volume of water drunk by a rabbit each day over one week.

a How much water did the rabbit drink on Sunday?
b When did the rabbit drink 6 cl of water?
c What was the maximum amount of water that the rabbit drank in a day?

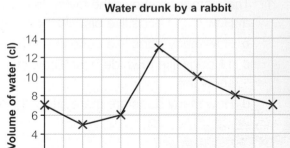

Water drunk by a rabbit

6 Andy used electronic tagging to record the journeys of five cats. The table shows the total distances travelled (in km) over a week.

	Tizzy	Rolf	Mittens	Zoe	Patch
Daytime	1	6	3	4	3
Night	4	2	8	6	5

a Draw a compound bar chart for the data.
b Do these cats travel further at night or during the day? Explain your answer.

7 Here are the masses of the boxes in a removal van.

32 kg, 15 kg, 51 kg, 22 kg, 30 kg, 20 kg, 44 kg, 18 kg, 27 kg, 33 kg, 38 kg, 42 kg, 33 kg, 27 kg, 19 kg, 41 kg, 30 kg, 25 kg, 36 kg, 25 kg

a Work out the median.
b Copy and complete the tally chart for the data.

Mass (kg)	Tally	Frequency
15–24		

c Jason says that the mean is 62 kg. Without calculating the mean, explain how you know he is wrong.
d Calculate the correct mean.

Challenge The numbers of North Atlantic hurricanes from 2007 to 2017 were

6, 8, 3, 12, 7, 10, 2, 6, 4, 7, 10

a Work out
  i the median   ii the mean   iii the range
b In 2005 there were 15 hurricanes, and in 2006 there were 5 hurricanes. If these hurricanes were included in the data, how would your answers to part **a** be affected?

Reflect What new ideas and skills have you learned in this unit?
When could these skills be useful in mathematics?
Could they be useful in other school subjects?

2 Number skills

2.1 Mental maths

- Use the priority of operations, including brackets
- Use multiplication facts up to 10 × 10 and the laws of arithmetic to do mental multiplication and division
- Multiply by multiples of 10, 100 and 1000

Active Learn
Homework

Warm up

1   **Fluency**   Do the calculations in each pair have the same or different answers? Explain why.
    a 6 + 2 and 2 + 6   **b** 6 − 2 and 2 − 6   **c** 6 × 2 and 2 × 6   **d** 6 ÷ 2 and 2 ÷ 6

2   Work out
    a 7 × 10   **b** 100 × 9   **c** 50 × 100   **d** 10 × 25
    e 4 × 1000   **f** 1000 × 18   **g** 300 ÷ 100   **h** 540 ÷ 10
    i 7200 ÷ 100   **j** 8000 ÷ 1000

3   Work out the missing numbers.
    a □ × 10 = 70   **b** 10 × □ = 80   **c** □ × 100 = 300   **d** 1000 × □ = 12 000

4   **a** Work out
        i 2 × 3 × 4   **ii** 2 × 4 × 3   **iii** 3 × 2 × 4
        iv 6 × 4   **v** 8 × 3
    b **Reasoning**   Does changing the order of the multiplication change your answer?

5   Work out
    a 8 × 2 × 5   **b** 7 × 5 × 2
    c 3 × 6 × 3   **d** 9 × 4 × 2

    Q5a hint It is easier to do 2 × 5 first, so change the order of the multiplication.
    8 × 2 × 5 = 2 × 5 × 8.

Worked example

a   Work out 6 × 20

6 × 20 = 6 × 2 × 10 ———— Rewrite 20 as 2 × 10

    = 12 × 10

    = 120

b   Work out 30 × 7

30 × 7 = 3 × 10 × 7

    = 3 × 7 × 10 ———— Change the order to make the multiplication easier.

    = 21 × 10

    = 210

6 Work out

a 4×20 b 4×30

c 4×300 d 4×3000

e 50×6 f 50×60

g 50×600 h 50×6000

i 12×30 j 400×12

k 22×30 l 11×60

> **Key point**   **Partitioning** splits the bigger number to make some easier multiplications.

Worked example

Work out 27×6

$\boxed{27} \times 6 = \boxed{20} \times 6 + \boxed{7} \times 6$ Split 27 into 20 + 7.

$\qquad = \boxed{2} \times 6 \times \boxed{10} + 7 \times 6$ Split 20 into 2×10.
It is easier to do $2 \times 6 = 12$ first and then 12×10.

$\qquad = 120 + 42 = 162$

---

7 Work out

a 32×6 b 32×8

c 4×63 d 5×63

e 7×72 f 72×4

> **Key point**   You must use the **priority of operations** to do calculations. Use **BIDMAS**:
> → **B**rackets
> → **I**ndices (powers)
> → **D**ivision and **M**ultiplication
> → **A**ddition and **S**ubtraction

Worked example

Work out $6 + 3 \times 10$

$6 + 3 \times 10 = 6 + 30$ Work out the multiplication first.

$\qquad\qquad = 36$

---

8 Work out

a $6 + 4 \times 2$ b $4 + 6 \times 2$

c $2 \times 4 + 6$ d $2 + 4 \times 6$

e $10 - 4 \times 2$ f $10 - 2 \times 4$

g $10 \times 4 - 2$ h $2 \times 10 - 4$

Q8 hint Multiplication and Division *before* Addition and Subtraction.

9 **Problem-solving / Reasoning** Use each of the numbers 2, 5, 10 once only to write three different calculations with the answer

a 20

b 0

10 Work out

   a $6 + 8 \div 2$ **b** $6 \div 2 + 8$

   c $8 \div 2 + 6$ **d** $8 + 6 \div 2$

   e $24 \div 6 - 2$ **f** $24 \div 2 + 6$

   g $30 \div 5 - 5$ **h** $30 - 5 \div 5$

> **Key point**   When you have only × and ÷, or only + and −, then work from left to right.

11 Work out

   a **i** $12 + 54 + 18$ **ii** $54 + 18 + 12$

   b **i** $35 - 12 + 14$ **ii** $12 + 14 - 35$

   c **i** $42 - 11 - 15$ **ii** $42 - 15 - 11$

12 Work out

   a $3 \times 4 \div 2$ **b** $18 \div 3 \times 4$

   c $20 \div 5 \div 4$ **d** $30 \div 5 \times 2$

   e $30 \div 2 \times 5$ **f** $30 \times 10 \div 5$

13 Work out

   a $3 \times 2 + 5$ **b** $3 \times (2 + 5)$

   c $7 \times 6 - 2$ **d** $7 \times (6 - 2)$ **e** $(7 - 6) \times 2$

   f $(12 - 4) \times (5 + 1)$ **g** $(12 - 4) \div (5 - 1)$

> **Q13 hint** Brackets first.

14 **Problem-solving** $4 + 2 \times 8 - 6 \div 2$

   Put sets of brackets in different places in the calculation to find as many answers as possible.

15 **Problem-solving** An expedition group need a total of 22 AA batteries for all their torches. They buy four packs of two AA batteries and three packs of four AA batteries. Have they bought enough batteries?

> **Challenge**
>
> **a** Use your calculator to work out
>
>   **i** $4 + 8 \times 2 - 7$ **ii** $2 \times 8 - 6 \div 2$ **iii** $5 \times (3 + 6)$
>
>   **iv** $(4 - 2) + 4 \times 7$ **v** $10 \times 3 \div 5 + (2 - 1)$
>
> **b** Compare your answers with another person's. Are they the same?
>
> **c** Does your calculator use BIDMAS? How do you know?

> **Reflect**   Write the new mathematical terms and words you have learned in this lesson.
> Use your own words to write a definition for each of them.
> For each definition, make up a calculation to show what it means.

2.2 Addition and subtraction

- Make an estimate to check an answer
- Use inverse operations to check an answer
- Use a written method to add and subtract whole numbers of any size
- Round whole numbers to the nearest 10 000, 100 000 and 1 000 000

Active Learn
Homework

Warm up

1 **a** **Fluency** Find the sum of 29 and 32.
   b **Fluency** Find the difference between 78 and 13.

2 Round each number to the nearest 10 and to the nearest 100.
   a 92   **b** 538   **c** 145
   d 499   **e** 1549   **f** 1550

3 Round each number to the nearest 1000.
   a 4900   **b** 3260   **c** 6095
   d 1458   **e** 16 326   **f** 16 623

Key point   An **approximation** is a number that is not exact. It is close enough for it to be useful though. Use approximations to **estimate** the answer to calculations.
≈ means 'approximately equal to'.

4 Use **approximation** to **estimate** these sums.
   a $48 + 57 ≈ 50 + 60 =$
   b $57 + 58 ≈$
   c $87 + 101 ≈$
   d $123 + 48 ≈$
   e Work out the exact answers to Q4 parts **a** and **d**. How close are your estimates?

5 **Reasoning**   Two of the calculations below are wrong.
   Use **estimation** to work out which two.
   What mistakes have been made?

   A   247
      + 329
      ‾‾‾‾‾
       566

   B   1375
      + 2148
      ‾‾‾‾‾‾
       3523

   C   482
      + 271
      ‾‾‾‾‾
      6153

   D   1482
      + 6530
      ‾‾‾‾‾‾
       8012

6 **Problem-solving** Website A has 326 hits on Monday. On the same day, Website B has 118 more hits than Website A.
How many hits does Website B have?

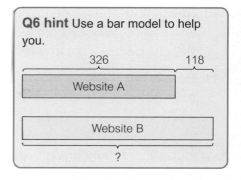

Q6 hint Use a bar model to help you.

7 Work out these calculations. Make an estimate first to check your answer.
   a 389 + 46   b 1752 + 179
   c 247 + 2008   d 1426 + 145 + 63

Q7 hint Line up 100s with 100s, 10s with 10s, ones with ones.

8 **Problem-solving** A garage sells 86 more red cars than blue cars. The garage sells 1048 blue cars.
How many red and blue cars do they sell?

Q8 hint Use a bar model to help you.

Key point You can check a subtraction calculation using the inverse operation of addition.

Worked example

Use the column method to work out 392 − 165.

$\begin{array}{r} 392 \\ -165 \\ \hline \end{array}$ — Write the larger number on top. Line up 100s with 100s, 10s with 10s, ones with ones.

Start with the ones column.
You can't subtract 5 from 2 because this gives a negative answer.

$\begin{array}{r} 3\,^8\!9\,^1\!2 \\ -165 \\ \hline 7 \end{array}$ — Take a ten from the 9 tens to make 8 tens and 12 ones.
12 − 5 = 7

$\begin{array}{r} 3\,^8\!9\,^1\!2 \\ -165 \\ \hline 227 \end{array}$ — Now look at the tens column and the hundreds column.

Check:
$\begin{array}{r} 227 \\ +165 \\ \hline 392 \\ {}_{1} \end{array}$

9 Complete these calculations using the column method.
   a 438 − 347   b 264 − 139
   c 381 − 193   d 436 − 257
   e 175 − 58   f 231 − 86
   g 2845 − 380   h 1763 − 97

Q9c hint

$\begin{array}{r} ^2\!3\,^{17}\!8\,^1\!1 \\ -\,1\,9\,3 \\ \hline 8 \end{array}$

You can't subtract 9 from 7. Take a hundred from the 3 hundreds to make 2 hundreds and 17 tens.

10 **Reasoning** Samia needs to subtract 37 from 54.
She can't subtract 7 from 4, so she subtracts 4 from 7.
Will she get the right answer? Explain how you know.

11 **Reasoning** What is the value of the digit 4 in each number?
    a 24 307   **b** 42 307   **c** 423 007   **d** 1 423 007
    e 4 000 000   **f** 1 240 000   **g** 3 204 000   **h** 6 421 000

Key point For rounding to the nearest 10 000

- **5000 and above rounds up**
- **4999 and below rounds down**

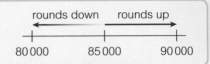

Worked example

Round 376 462 to the nearest 10 000.

376 462 to the nearest 10 000 is 380 000

Look at the digit in the thousands column.

---

12 Round each number to the nearest 10 000.
    a 86 319   **b** 756 319   **c** 1 756 319
    d 13 928   **e** 19 328   **f** 20 999

13 Round each number to the nearest 100 000.
    a 327 016   **b** 630 172   **c** 279 124
    d 4 667 001   **e** 5 437 999   **f** 98 423

> **Q13 hint** To round to the nearest 100 000, look at the digit in the ten thousands column.

14 Round these numbers to the nearest 1 000 000
    a 1 623 883   **b** 4 555 306   **c** 5 463 880

15 **Problem-solving** Charity A raises £25 654.
Charity B raises £848 less than Charity A.
How much money do the charities raise in total?

Challenge The table gives the numbers of children and adults attending some concerts.

Concert	Children	Adults
A	3 624	35 876
B	327 419	48 826
C	79 902	816 121

a Work out the total number of people at each concert.
b Estimate the number of people at all three concerts, to the nearest 10 000.
c Work out the difference between the total number of adults and the total number of children.

Reflect This lesson used some bar models for problem-solving. Would you use bar models to help you solve mathematics problems in future? Explain why, or why not.

2.3 Multiplication

Active Learn
Homework

- Use an estimate to check an answer to a multiplication
- Use a written method to multiply whole numbers

Warm up

1 Fluency Find the missing numbers.
 a $9 \times 8 = \square$ **b** $\square \times 7 = 49$ **c** $8 \times \square = 80$ **d** $\square \times \square = 21$

2 Work out
 a 50×10 **b** 100×5 **c** 30×20

3 Round each number to the nearest 100.
 a 218   **b** 658   **c** 80   **d** 999

> **Key point** You can use estimation to check the answer to a multiplication calculation.
> Round the numbers and then multiply.

4 Use approximation to estimate these multiplications.
 a 49×7 **b** 182×4 **c** 93×608

> **Key point** In the **column method** you write the numbers in the calculation in their place
> value columns like this:
> $$\begin{array}{r} 242 \\ \times \quad 3 \\ \hline \end{array}$$

Worked example

Work out 625×3 using the **column method**.

$$\begin{array}{r} 625 \\ \times \quad 3 \\ \hline 5 \\ {\scriptstyle 1} \end{array}$$

Start in the ones column. $5 \times 3 = 15$.
That's 5 ones and **1** ten.

$$\begin{array}{r} 625 \\ \times \quad 3 \\ \hline 75 \\ {\scriptstyle 1} \end{array}$$

In the tens column:
2 tens × 3 = 6 tens.
6 tens + **1** ten = 7 tens.

$$\begin{array}{r} 625 \\ \times \quad 3 \\ \hline 1875 \\ {\scriptstyle 1} \end{array}$$

In the hundreds column:
6 hundreds × 3 = 18 hundreds.

Check: $600 \times 3 = 1800$, which is close to 1875.

5 Work out
   a 121 × 4   **b** 2131 × 3   **c** 124 × 3   **d** 2081 × 4
   e 624 × 5   **f** 239 × 6   **g** 3714 × 8   **h** 4623 × 9

6 **Reasoning** Two of the calculations below are wrong.
   Use **approximation** to work out which two.
   What mistakes have been made?

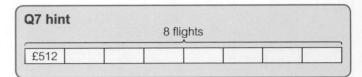

	A 242		**B** 375		**C** 564
	× 3		× 8		× 4
	6126		3000		2046

7 A flight costs £512.
   How much do 8 flights cost?

> **Q7 hint**
>   8 flights
> | £512 | | | | | | | |

8 **Problem-solving** A bookcase has 8 shelves.

> **Q8 hint** Use bar models to help you.

   a Each shelf holds 115 books.
      How many books are there in total?

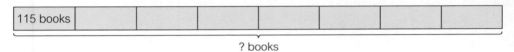

115 books							
   ? books

   b There are 342 fiction books. How many books are non-fiction?

   342 fiction   ? non-fiction

9 Ann earns £2813 per month. How much does she earn in 6 months?

> **Key point** **Long multiplication** is a written method to multiply by numbers with two or more digits.

> **Worked example**
>
> Work out 34 × 29
>
> ```
>     3 4
> ×   2 9
>   3 0 6      First work out 34 × 9.
> + 6 8 0      Now work out 34 × 20.
>   9 8 6      Add to give the final answer.
> ```
> Check: 30 × 30 = 900, which is close to 986

10 Work out
   a 32 × 15   **b** 46 × 54   **c** 62 × 39
   d 132 × 15   **e** 243 × 26   **f** 327 × 41

11 Malik travels by train to London 17 times a year.
Each return train ticket costs £65.
How much does he spend in total on train tickets in a year?

12 Problem-solving Sami has 265 followers on Twitter.
Tom has 34 times as many followers as Sami.
How many more people follow Tom than follow Sami?

Key point If money received is greater than money spent, then you make a **profit**.
If money spent is greater than money received, then you make a **loss**

loss

money received
money spent

profit

money received
money spent

---

13 Reasoning A T-shirt seller buys 384 T-shirts at £14 each.
She sells 208 of the T-shirts at £22 each.
Has she made a profit or a loss? How much?

Challenge

Arrange the digits 3, 5, 6, 8 as a $\times \underline{}^{\,*\,*}$ calculation.

For example:

```
    35
×   68
_____

_____
_____
```

a Find the answer.

Arrange the digits in a different $\times \underline{}^{\,*\,*}$ calculation.

```
_____
_____
```

b What is the smallest possible answer? What is the largest possible answer?

Reflect Look at your working for part **a** of the Challenge. Did you use a strategy to make sure you found all the possible answers?
If you answered 'Yes', what strategy did you use, and why?
If you answered 'No', what strategy could you have used, and why?
Could you have used a different strategy? Would this have been better?

2.4 Division

* Use a written method to divide whole numbers
* Use inverse operations to check an answer

Active Learn
Homework

Warm up

1 Fluency Find the missing numbers.

a $4 \times 8 = \square$ **b** $32 \div 8 = \square$ **c** $32 \div 4 = \square$

d $9 \times 7 = \square$ **e** $63 \div 9 = \square$ **f** $63 \div 7 = \square$

2 a How many times does 3 go into 9?

   b How many times does 3 go into 8? What is the remainder?

   c How many times does 7 go into 20? What is the remainder?

3 Work out

   a $4\overline{)88}$ **b** $4\overline{)96}$ **c** $6\overline{)96}$ **d** $7\overline{)91}$

Worked example

Work out $132 \div 4$ using short division.

$$4\overline{)13\,2}\quad\frac{3\,...}{}$$

Look at the digits in 132, starting on the left.
4 doesn't go into 1, so look at 13.
4 goes into 13 three times so write a 3 in the tens column.

$$4\overline{)13\,^12}\quad\frac{3\,...}{}$$

The difference between 13 and 4×3 is 1
so write the remainder ten in the ones column, to make 12.

$$4\overline{)13\,^12}\quad\frac{3\,3}{}$$

Check:
$$\begin{array}{r} 3\,3 \\ \times \quad 4 \\ \hline 1\,3\,2 \\ {}_{1} \end{array}$$

4 goes into 12 three times. So write 3 in the units column.

4 Work out

   a $3\overline{)93}$ **b** $3\overline{)936}$

   c $4\overline{)84}$ **d** $4\overline{)844}$

   e $6\overline{)84}$ **f** $6\overline{)846}$

   g $6\overline{)852}$ **h** $6\overline{)8526}$

Key point   You can check a division calculation using the inverse operation of multiplication.

5 Reasoning Two of the calculations below are wrong.

A $\quad\dfrac{119}{3\overline{)627}}$

B $\quad\dfrac{204}{4\overline{)816}}$

C $\quad\dfrac{31\text{-}}{6\overline{)2076}}$

Q5a hint

Check:

$\begin{array}{r} 119 \\ \times \quad 3 \\ \hline \end{array}$

  a Use the inverse operation of multiplication to work out which two are wrong.

  b Work out the correct answers.

6 Use a written method to work out these divisions.

  a $3\overline{)369}$
  b $189 \div 3$
  c $378 \div 3$
  d $192 \div 3$

  e $532 \div 4$
  f $415 \div 5$
  g $2345 - 5$
  h $1449 \div 7$

7 Sara receives £135 for 9 hours' work. What is her hourly rate?

8 How many teams of 4 can be made from 256 people?

Worked example

Work out $134 \div 4$

The difference between 14 and 4×3 is 2.
So write remainder 2.

$\begin{array}{r} 3\ 3 \text{ remainder } 2 \\ 4\overline{)13\,{}^14} \end{array}$

---

9 Work out these divisions with remainders.

  a $3\overline{)94}$
  b $128 \div 5$
  c $247 - 4$

  d $586 \div 8$
  e $3421 \div 7$
  f $7016 \div 6$

Key point **Long division** is a written method to divide by numbers with two or more digits. It breaks down the calculation into smaller steps than short division.

Worked example

Work out $448 \div 16$

Short division:

$\begin{array}{r} 2\ \dots \\ 16\overline{)4\,4\,{}^{12}8} \end{array}$

Try multiples of 16:
$1 \times 16 = 16$
$2 \times 16 = 32$

The difference between 44 and 32 is 12.

Long division:

$\begin{array}{r} 2\ \dots \\ 16\overline{)448} \\ -\ 32 \\ \hline 128 \end{array}$ $\qquad$ 2×16

$\begin{array}{r} 2\ 8 \\ 16\overline{)4\,4\,{}^{12}8} \end{array}$

Try multiples of 16:
$10 \times 16 = 160$ (too big)
$5 \times 16 = 80$ (too small)
$8 \times 16 = 128$

$\begin{array}{r} 2\ 8 \\ 16\overline{)448} \\ -\ 32 \\ \hline 128 \\ -\ 128 \\ \hline O \end{array}$ $\qquad$ 8×16

Check: $\begin{array}{r} 2\ 8 \\ \times\quad 1\ 6 \\ \hline 1\ 6\ 8 \\ 2\ 8\ O \\ \hline 4\ 4\ 8 \end{array}$

Remainder 0

10 Work out

  a $22\overline{)396}$ **b** $14\overline{)406}$ **c** $17\overline{)1938}$

  d $7250 \div 25$ **e** $4950 \div 33$ **f** $1092 \div 52$

11 Work out these divisions with remainders.

  a $310 \div 14$ **b** $365 \div 15$ **c** $685 \div 19$ **d** $1032 \div 21$

12 Problem-solving A gardener has 322 tulip bulbs.
  She wants to plant rows of 15 bulbs.
  a How many full rows can she plant?
  b How many bulbs will she have left over?

13 Problem-solving There are 676 students in Year 7.
  A coach takes 52 people.
  How many coaches are needed for everyone in Year 7?

14 Problem-solving Micah takes out a
  business loan of £1000.
  She makes an initial repayment of £136.
  She then makes 12 equal repayments.
  How much is each repayment?

> **Q14 hint**
>
> £1000
>
> [| £136]
>
> [| | | | | | | | | | | |]

15 Problem-solving Lewis buys a sofa for £950.
  He makes an initial payment of £198.
  He then makes 8 equal repayments.
  How much is each repayment?

2.5 Money and time

- Round money to the nearest pound or penny
- Interpret the display on a calculator in different contexts
- Use a calculator to solve problems involving money and time

Active Learn
Homework

Warm up

1   **Fluency**   How many pennies in £1?

2   **a**   Match each decimal to the correct fraction.

0.25	0.5	0.75

$\frac{3}{4}$	$\frac{1}{4}$	$\frac{1}{2}$

   b   Copy and complete.

     i $\frac{1}{4}$ hour = ☐ minutes

     ii $\frac{1}{2}$ hour = ☐ minutes

     iii $\frac{3}{4}$ hour = ☐ minutes

3   Match each p amount to the correct £ amount.

500p	50p	5p

£5.00	£0.05	£0.50

4   Write these p in £.
   a   80p = £0.☐☐   **b**   75p   **c**   7p
   d   125p   **e**   520p   **f**   502p

Key point

To **round** an amount to the nearest pound, look at the pence.

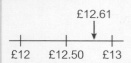

£12.61

+————+————+————+
£12   £12.50   £13

£12.61 rounds up to £13

5   Round each amount to the nearest pound.
   a   £4.80   **b**   £10.29   **c**   £38.55
   d   £0.62   **e**   £1040.89   **f**   £0.4⁻

6   **Reasoning**   Jin has £4.38, Luke has £7.62 and Ian has £9.11.
   Round each amount to the nearest whole number and then add the rounded amounts.
   Do they have enough for a £20 taxi fare?

7 James sells a bicycle online.
He pays £1.45 as a registration fee, £2.85 for the advert and a selling fee of £1.95.
What are his total fees to the nearest pound?

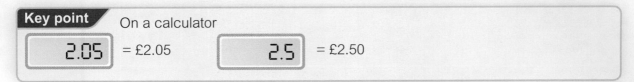

Key point On a calculator

2.05 = £2.05   2.5 = £2.50

8 Work out
  a £0.90 × 3   **b** £1.25 × 6   **c** £1.41 × 5
  d £19.20 ÷ 6   **e** £21.14 ÷ 7   **f** £0.72 ÷ 12

9 **Reasoning** Suzanne has 240 5p coins.
She works out 240 × 5 = 1200.
Ben works out 240 × 0.05 = 12.
Which answer is in pounds and which is in pence?

10 A company has an advert on a billboard for 5 weeks.
The total cost is £399.50.
What is the weekly cost of the advert?

> **Q10 hint** Estimate the answer first so that you can check the answer given by the calculator.

11 **Problem-solving** Shakira advertises her laptop for sale online.
The table shows advert prices.
  a How much does it cost to advertise a laptop worth £300 for 14 days?
  b How much more would it cost to advertise a laptop valued at £750 for 14 days?

Selling price	7 days online advert
Up to £50	£3.10
£50.01– £200	£5.20
£200.01– £500	£7.50
£500.01 and above	£9.90

Key point Change is the money you get back after paying for something with more money than it costs.

12 Sarah buys two train tickets for £7.85 each. She pays with a £20 note.
How much change does she get?

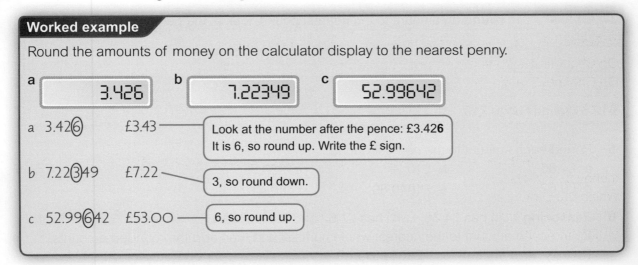

Worked example

Round the amounts of money on the calculator display to the nearest penny.

a   3.426   b   7.22349   c   52.99642

a 3.42⑥   £3.43 ── Look at the number after the pence: £3.42**6**
                    It is 6, so round up. Write the £ sign.

b 7.22③49   £7.22 ── 3, so round down.

c 52.99⑥42 £53.00 ── 6, so round up.

13 Round these calculator values to the nearest penny (2 decimal places).

a | 13.231
b | 27.258
c | 23.8953

d | 89.0906
e | 72.999
f | 99.997

14 Work out the answer then round to the nearest penny.
- **a** £2679 ÷ 8
- **b** £1.75 ÷ 4
- **c** £5.98 − 5
- **d** £6254 ÷ 9
- **e** £4690 ÷ 16
- **f** £238 ÷ 21

15 A restaurant bill for 3 people comes to £45.73.
- **a** How much does each person pay?
- **b** **Reasoning** Why does rounding to the nearest pound or penny not always work?

16 The calculator screens show the answers to questions on time.

| 0.75 | 0.5 | 0.25 |
| A | B | C |

Which screen shows an answer of
- **a** 15 minutes
- **b** 30 minutes
- **c** 45 minutes?

17 A drama is $2\frac{1}{2}$ hours long.
It is divided into ten episodes, which are all the same length.
How many minutes long is each episode?

> 2.5 ÷ 10
>
> 0.25

18 Work out
- **a** 0.5 hours × 10 = ☐ hours
- **b** 0.75 hours × 10 = ☐ hours ☐ minutes
- **c** 0.25 hours × 10 = ☐ hours ☐ minutes
- **d** 0.25 hours × 3 = ☐ minutes
- **e** 31 hours ÷ 2 = ☐ hours ☐ minutes
- **f** 50 hours ÷ 8 = ☐ hours ☐ minutes
- **g** 15 hours ÷ 4 = ☐ hours ☐ minutes
- **h** 21 hours ÷ 10 = ☐ hours ☐ minutes

19 **Problem-solving** Sara works for the same number of hours each day.
In 4 days, she works 21 hours.
How many hours does she work per day?
Write your answer in hours and minutes.

Challenge On Channel 1, a 10-second advert costs £200.
On Channel 2, a 20-second advert costs £388.50.
On Channel 3, a 30-second advert costs £559.56.

Which advert is best value for money?

Reflect Write down two new things you have learned about using a calculator for money calculations.
Which questions used this new knowledge?
When you answered these questions, did you make any mistakes? If so, check that you understand where you went wrong.

2.6 Negative numbers

- Order positive and negative numbers
- Add and subtract positive and negative numbers
- Begin to multiply with negative numbers

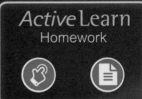

Active Learn
Homework

Warm up

1 Fluency Count down from 5 to −5.

2 This temperature scale shows positive and negative temperatures in degrees Celsius (°C).
   a What was the temperature on
     i Monday   **ii** Tuesday
     iii Wednesday   **iv** Thursday?
   b Is the temperature getting warmer or colder from Monday to Thursday?

3 The temperature now is 14 °C. It cools down by 5 degrees.
What is the new temperature?

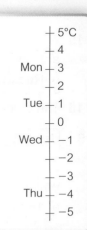

Worked example

The temperature is −2 °C. It gets 5 °C warmer.
What is the new temperature?

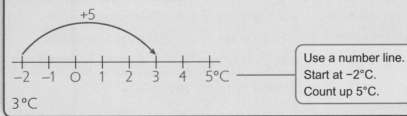

Use a number line.
Start at −2 °C.
Count up 5 °C.

3 °C

4 Find the new temperatures.
   a The temperature is 3 °C. It increases by 4 °C.
   b The temperature is −3 °C. It goes up by 4 °C.
   c The temperature is −5 °C. It rises by 9 °C.

> **Q4 hint** 'Rise', 'increase' and 'go up' all mean the same thing.

5 **a** Copy this temperature scale.

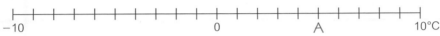

Mark these temperatures on it.
A 5 °C   **B** −4 °C   **C** 2 °C   **D** 10 °C   **E** −6 °C   **F** −9 °C   **G** 8 °C
The first one has been done for you.
   b Which is the lowest temperature?   **c** Which is the largest number?

The symbol > means greater than. The symbol < means less than.

6 Write the correct symbol, < or >, between each pair of temperatures.
   a 3°C ... 9°C   **b** 8°C ... 2°C   **c** −3°C ... 5°C
   d 6°C ... −2°C   **e** −4°C ... −3°C   **f** −2°C ... −7°C

7 Write the correct symbol, < or >, between each pair of numbers.
   a 4 ... 7   **b** 5 ... 2   **c** 9 ... −1
   d −6 ... 3   **e** −3 ... −8   **f** −7 ... −2

8 Here are the goal differences of some football teams.
   Arrange each set of numbers in order of size, smallest first.

   Q8 hint Use a number line to help.

   a 7, 8, 19, −2, −5, −14, −9
   b −4, −6, −2, −11, −6, 7, −13
   c −21, −30, 24, 63, −12, −41, 23, −15

9 Use a number line to work these out.
   a −3 − 1   **b** −3 − 2   **c** −3 − 3   **d** −3 − 4   **e** −4 − 3
   f −4 − 4   **g** −4 − 5   **h** −5 − 4   **i** −5 − 5   **j** −5 − 6

   ```
   ├─┼─┼─┼─┼─┼─┼─┼─┼─┼─┼─┼─┼─┼─┼─┼─┼─┼─┼─┼─┤
  −10 −9 −8 −7 −6 −5 −4 −3 −2 −1  0  1  2  3  4  5  6  7  8  9 10
   ```

10 Work out
   a 3 − 4   **b** 2 − 4   **c** 1 − 4   **d** −4 + 3   **e** −4 + 4
   f −4 + 5   **g** −4 + 6   **h** −5 + 6   **i** −6 + 8   **j** −8 + 6

11 Work out
   a 3 × 2   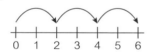   **b** 3 × −2

   c 4 × 1   **d** 4 × −1   **e** 2 × 5   **f** 2 × −5
   g 2 × 3   **h** 2 × −3   **i** 6 × 8   **j** 6 × −8

12 Work out
   a 3 × −4   **b** 7 × −2   **c** 5 × −8   **d** 9 × −3   **e** 6 × −1

Challenge   The answer to a calculation is −8.

a Write three different calculations that give the answer −8.

b Now write three different calculations using a different operation, choosing from +, −, × in each one.

Reflect   Write this calculation: −1 + 2 − 3

Without looking at this book, imagine a number line and work out the answer to the calculation. Now check your answer by drawing a number line.

Did imagining a number line help you? If so, how?
If not, do you have a different way of imagining negative numbers?

2.7 Factors, multiples and primes

- Find all the factor pairs for any whole number
- Identify common factors, the highest common factor and the lowest common multiple
- Recognise prime numbers

Active Learn
Homework

Warm up

1 Fluency

| 10 | 12 | 15 | 20 | 22 | 24 |

Which of the numbers in the box are in the
a 2
b 5
c 10 multiplication tables?

2 Fluency Work out
a 6×9
b 4×5
c 7×4
d 6×8
e 9×8
f 7×12

3 Fluency Work out
a $24 \div 8$
b $35 \div 7$
c $56 \div 7$
d $48 \div 12$
e $42 \div 6$
f $63 \div 9$

4 Write the first five multiples of 7.

Key point A **multiple** of a number is in that number's multiplication table.

5 Copy these numbers.

18   25   27   30   32   37   45

a Ring the multiples of 10, using a coloured pencil.
b Ring the multiples of 5, in a second colour.
c Ring the multiples of 2, in a third colour.
d Ring the multiples of 9, in a fourth colour.
e Copy and complete these sentences.
   i 45 is a multiple of ☐ and ☐.
   ii 18 is a multiple of ☐ and ☐.
   iii 30 is a multiple of ☐, ☐ and ☐.
f Reasoning Which number has no ring round it? Why?

6 Reasoning Choose a number to make these statements true.
a 12 is a multiple of ☐
b 30 is a multiple of ☐
c 16 is a multiple of ☐
d ☐ is a multiple of 5
e ☐ is a multiple of 9
f ☐ is a multiple of 25

7 Reasoning Which is the odd one out in each of these lists?
Explain your reason using the word 'multiple'.
a 10, 20, 25, 30, 40, 50
b 4, 8, 12, 15, 20, 24
c 10, 25, 50, 75, 125
d 18, 36, 42, 45, 81
e 15, 21, 30, 35, 50
f 14, 25, 32, 48, 58

8 a List the first eight multiples of
   i 4   **ii** 6
 b From your lists, which numbers are multiples of 4 *and* 6?

9 Look at these numbers:
   2, 3, 4, 6, 8, 9, 10, 12, 14, 15, 16, 18

 a Which of the numbers are multiples of 2?
 b Which are multiples of 3?
 c Which are multiples of 2 and 3?
 d Copy this **Venn diagram**.
   Write each of the numbers in the correct section.

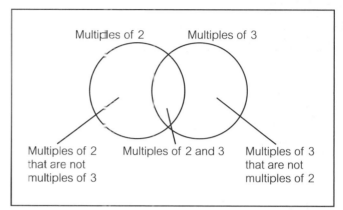

10 Complete this Venn diagram for multiples of 4 and 5 up to (and including) 40.

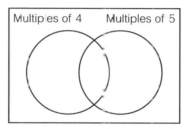

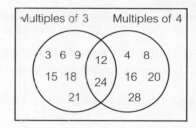

11 a Write the first ten multiples of each of these numbers.
   i 3   **ii** 8
 b What is the lowest common multiple of 3 and 8?

12 Find the lowest common multiple of each of these pairs of numbers.
   a 8 and 14   **b** 12 and 20   **c** 15 and 25

13 Copy and complete all the factor pairs for 24.
       1 and 24
       2 and 12
       3 and ...
       ... and ...

14 a List all the factors of

  i 6   **ii** 15

  b From your lists, which number is a factor of 6 *and* 15?

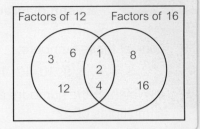
15 a Copy and complete the Venn diagram to show the factors of 4 and 18.

  b List the common factors of 4 and 18.

  c What is the highest common factor of 4 and 18?

16 a Draw a Venn diagram or write lists to find the factors of 8 and 12.

  b What are the common factors of 8 and 12?

  c What is the highest common factor of 8 and 12?

17 Find the highest common factor of each of these pairs of numbers.

  a 9 and 15   **b** 18 and 27   **c** 16 and 24

  d 6 and 24   **e** 15 and 25   **f** 8 and 24

18 a Write the prime numbers between 1 and 20.

  b Reasoning   Are all prime numbers odd? Give a reason for your answer.

19 a List the factors of each of these numbers.

  i 23   **ii** 35   **iii** 29

  b Which of the numbers in part **a** are **prime** numbers?

20 Problem-solving   a   Write two numbers that have highest common factor 30.

  b Write two numbers that have lowest common multiple 30.

Challenge   Write numbers to fit each description.

  a A factor of both 12 and 15   **b** A multiple of both 2 and 9

  c Two prime numbers between 50 and 60   **d** An odd-and-even factor pair of 30

Reflect   Write your own short definition for each of these mathematical words:

  highest   lowest   common   factor   multiple

Now use your definitions to write (in your own words) the meaning of:

  highest common factor   lowest common multiple

2.8 Square numbers

- Recognise square numbers
- Use a calculator to find squares and square roots
- Use the priority of operations, including powers
- Use index form for powers
- Do mental calculations with squares and square roots

Active Learn
Homework

Warm up

1 **Fluency** Work out
 a 4×4 b 9×9 c 7×7 d $2 \times 2 \times 2$

2 What number comes next?
 a 1, 3, 5, 7, …   b 2, 6, 10, 14, …

3 Work out
 a $2 + 6 \times 6$ b $10 - 4 \times 2$ c $6 + 12 \div 3$

> **Key point**   **Square numbers** make a square pattern of dots.
> To find the square of a number, you multiply it by itself.

4 These patterns of dots show the first three **square numbers**.
 a Draw the dot pattern for the 4th square number.
 b Copy and complete this calculation for the
   4th square number.
   $4 \times \square = \square$
 c How many rows of dots are there in the 5th pattern? How many columns?
 d How many dots are there in the 10th pattern?
 e What is the value of the 10th square number?

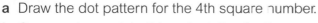

1×1 2×2 3×3
1   4   9

> **Key point**   You can write 3×3 as 3^2. You read this as '3 squared'.
> The '2' in 3^2 is called the power or **index**. The plural of index is **indices**.

5 Work out
 a 6^2 b 8^2 c 9^2 d 11^2

> **Key point**   You can use the $\boxed{x^2}$ key on your calculator to work out a square.

6 Work out
 a 20^2 b 15^2 c 100^2 d 13^2

Key point

A **square root** is a number that is multiplied by itself to produce a given number.

Finding the square root is the **inverse** of squaring.

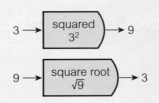

7 Find the **square root** of each of these numbers.
   a 36   **b** 25   **c** 64   **d** 100

Key point

You can use the $\sqrt{}$ key on your calculator to find a square root.

8 Find
   a $\sqrt{225}$ **b** $\sqrt{361}$ **c** $\sqrt{400}$ **d** $\sqrt{10\,000}$

9 **Problem-solving**   Which of the numbers in the box are square numbers?

| 100 | 8 | 6 | 15 | 49 | 50 | 36 | 4 | 2 | 81 |

10 **Reasoning**   Can 41 dots make a square pattern? Explain your answer.

> **Q11 hint** You must use the **priority of operations** to do calculations. Use **BIDMAS**:
> → **B**rackets
> → **I**ndices (powers)
> → **D**ivision and **M**ultiplication
> → **A**ddition and **S**ubtraction

11 Work out
   a $2^2 \times 4$ **b** $2^2 + 4$ **c** $4 + 2^2 \times 5$ **d** $4^2 \div 2$
   e $4^2 \div 2 + 6$ **f** $50 - 5^2$ **g** $50 - 5^2 \times 2$

12 Work out
   a $6 - 2^2$ **b** $(6 - 2)^2$ **c** $(6 - 2)^2 \times 8$ **d** $(6 - 2^2) \times 8$
   e $(8^2 + 3) - 5$ **f** $15 \times (3^2 - 9)$ **g** $10 \times (50 - 5^2) + 4$

13 Copy and complete these calculations.
   a $4 \times \square = 4^2$ **b** $2 \times 2 \times \square = 2^3$ **c** $4 \times 4 \times 4 = 4^\square$
   d $5^3 = 5 \times \square \times \square$ **e** $10 \times 10 = 10^\square$ **f** $10^3 = \square \times \square \times \square$

Challenge

These patterns of dots show the first three triangle numbers.

a Continue the pattern to work out the next three triangle numbers.
b Write down the first six triangle numbers.
c What is being added on each time?
d Continue the pattern to work out the 7th and 8th triangle numbers.
e Is 44 a triangle number? Explain your answer.
f How does the way the dot pattern grows match the triangle numbers?
g Add two consecutive triangle numbers: $1 + 3$, $3 + 6$, and so on. What type of numbers are the answers?

1   3   6

Reflect

Think about the *square* of 9 and the *square root* of 9. Which is 3 and which is 81? Make sure you know the difference between these two mathematical terms. Write down a hint in your own words to help you remember which is which

2 Check up

Written methods

1   Work out these calculations.
    Make an estimate to check your answer.

    a   318
            + 471
            ‾‾‾‾

    b   431
            + 289
            ‾‾‾‾

    c   8329
            − 6645
            ‾‾‾‾

    d 4137 − 82

2   Work out 404 ÷ 4.
    Show how you checked your answer.

3   Work out these calculations.
    Make an estimate first.

    a 2513 × 4   **b** 28 × 34   **c** 419 ÷ 19

Mental work

4   **a** Round 28 417 to the nearest 10 000.
    b Round 562 104 to the nearest 10 000.

5   Work out
    a 8 × 30   **b** 60 × 70   **c** 20 × 5000

6   Work out 23 × 9.

7   **a** In Edinburgh the temperature was −3 °C.
        In Liverpool the temperature was 5 degrees warmer.
        What was the temperature in Liverpool?
    b In Cardiff the temperature was 4 °C.
        In Llanberis the temperature was −5 °C.
        How many degrees warmer was it in Cardiff than in Llanberis?

8   Write the correct sign, < or >, between each pair of numbers.
    a 3 ... −1
    b −5 ... 2
    c −4 ... −6

9
16	2	4	19
9	18	23	12

    Which of the numbers in the box are
    a factors of 36   **b** multiples of 8
    c prime   **d** square numbers?

10 Work out
    a −2 + 5   **b** −6 + 3   **c** 1 − 5   **d** −5 − 2

11 a i Find the common factors of 12 and 14.

ii What is the highest common factor of 12 and 14?

b Find the lowest common multiple of 5 and 7.

12 Work out

a $2 + 5 \times 4$

b $6 \times (2 + 7)$

c $25 - 5 + 4$

d $6 \times 0 + 8$

e $(18 - 9) \times 4$

f $32 \div 8 \div 4$

g $4^2 + 5$

h $(2 + 3)^2 - 4$

i $64 \div 16 \div 2$

Problem-solving

13 Phillipe spent £285 on a shopping trip.
Joe spent £189 more than Phillipe.
How much did Joe spend?

14 Layla wants to buy three items.
The items are priced at £6.49, £10.90 and 82p.
Layla has £20.
Use estimation to decide whether Layla has enough to buy all three items.

15 Zoe and Stewart have a total of 632 followers on Twitter.
Zoe has three times as many followers as Stewart.
How many followers does Stewart have?

16 Anika works for £5.72 per hour.
How much will she be paid for working 15 hours?

17 A piano teacher teaches 8 lessons in 6 hours.
The lessons are all the same length.
How long is one lesson, in minutes?

Challenge

1

| 2 | 3 | 5 | 6 |

Use each of the numbers in the box once and any combination of +, −, ×, ÷ and brackets to make 60.

2 When a number is subtracted from another number, the answer is −1.
What could the two numbers be?

3 When you reverse the digits of 13 you get 31.
Both 13 and 31 are prime numbers.
Find another pair of prime numbers that have this property.

Reflect How sure are you of your answers? Were you mostly

☹ **Just guessing** 😐 **Feeling doubtful** 🙂 **Confident**

What next? Use your results to decide whether to strengthen or extend your learning.

2 Strengthen

Written methods

1 Work out these calculations. The first two have been started for you.

a
```
 HTO
 4 5 8
+ 2 1 4
───────
     2
    ₁
```

b
```
  7 2 6
+ 2 3 8
───────
```

> **Q1a hint** Write 1 ten in the Tens column.

c 348 + 491   **d** 223 + 585   **e** 2438 + 192

f 164 + 52   **g** 75 + 139   **h** 642 + 4389

> **Q1f hint** Line up the Hundreds, Tens and Ones.

2 Work out these calculations. The first two have been started for you.

a
```
  H T O
  5 ³4̸ ¹2
− 1 2 3
────────
      9
```

b
```
  9 7 4
− 5 2 6
───────
```

> **Q2a hint** Use 1 ten from the Tens column to make 12 ones.

c 346 − 182   **d** 925 − 671   **e** 518 − 236

f 764 − 493   **g** 3495 − 1523   **h** 6822 − 351

3 Work out

a 375 − 188   **b** 942 − 366   **c** 638 − 479   **d** 127 − 73

e 164 − 47   **f** 212 − 65   **g** 1726 − 52   **h** 1637 − 848

4 Use short multiplication to work out

a 154 × 3   **b** 237 × 4

c 6218 × 2   **d** 4126 × 7

> **Q4a hint** 154 = 100 + 50 + 4
> What is 4 × 3?
> What is 50 × 3?
> What is 100 × 3?
> ```
>   HTO
> ×   O
> ─────
>   HTO
> ```

5 Work out these multiplications. The first two have been started for you.

a 21 × 34
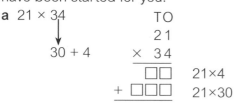

b 121 × 34
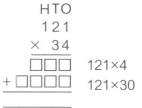

c 62 × 13   **d** 762 × 13   **e** 581 × 27

6 Work out these divisions. Some of them have been started for you.

a

b 3)246   **c** 4)148   **d** 3)141

e
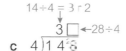
4)1 4²9 remainder 1   **f** 3)143

7 Work out these divisions. The first one has been started for you.

a

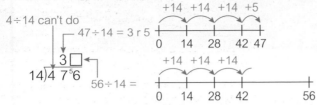

4 ÷ 14 can't do

47 ÷ 14 = 3 r 5

+14 +14 +14 +5

0 14 28 42 47

3☐

14)4 7⁵6

56 ÷ 14 =

+14 +14 +14

0 14 28 42   56

b 17)357   **c** 15)510   **d** 21)1092   **e** 21)1095

Mental work

1 Round 37 472 to the nearest

  a 10 000

├─┼─┼─┼─┼─┼─┼─┼─┼─┼─┤
30 000   40 000

  b 1000

├─┼─┼─┼─┼─┼─┼─┼─┼─┼─┤
37 000   38 000

2 Round 2 461 098 to the nearest

  a 1 000 000

├─┼─┼─┼─┼─┼─┼─┼─┼─┼─┤
2 000 000   3 000 000

  b 100 000

├─┼─┼─┼─┼─┼─┼─┼─┼─┼─┤
2 400 000   2 500 000

  c 10 000

├─┼─┼─┼─┼─┼─┼─┼─┼─┼─┤
2 460 000   2 470 000

3 Work out

a 7×4	**b** 7×40	**c** 7×400
d 6×3	**e** 6×30	**f** 600×3
g 5×2	**h** 5×20	**i** 5×200
j 8×6	**k** 8×60	**l** 80×6 **m** 80×60

> **Q3b hint** $7 \times 40 = 7 \times 4 \times 10$
> **Q3c hint** $7 \times 400 = 7 \times 4 \times 100$

4 Work out these multiplications.

  a 42×7

    $42 = 40 + \square$

    42×7

    40×7 $\square \times 7$

    $\square$ + $\square = \square$

  b 25×6

    $25 = \square + \square$

    25×6

    $\square \times 6$ $\square \times 6$

    $\square$ + $\square = \square$

  c 34×8 **d** 53×4 **e** 72×5

5 Work out these calculations. Use the priority of operations.

BIDMAS   BIDMAS   BIDMAS

  a $8 + 3 \times 2$ **b** $4 \times (5 + 3)$ **c** $12 \div 6 - 4$
  d $(10 - 4) \times 3$ **e** $(12 + 6) \div (12 - 9)$

> **Q5 hint** Use **BIDMAS** to remember the priority of operations.
> → **B**rackets
> → **I**ndices (powers)
> → **D**ivision and **M**ultiplication
> → **A**ddition and **S**ubtraction

6 These calculations *only* have × and ÷. Work from left to right.

   a $3 \times 6 \div 2$ **b** $10 \div 5 \times 3$

   c $28 \div 4 \times 6$ **d** $12 \times 2 \div 3$

7 These calculations *only* have + and −. Work from left to right.

   a $5 + 7 - 3$ **b** $17 - 8 + 2$ **c** $21 + 2 - 5$ **d** $13 + 47 - 25$

8 Use the thermometer to find the new temperature after these changes.

   a $3\,°C$ rises by $2\,°C$ **b** $0\,°C$ rises by $5\,°C$

   c $-3\,°C$ rises by $4\,°C$ **d** $-6\,°C$ rises by $11\,°C$

9 Use the thermometer in Q8 to calculate the new temperature after these changes.

   a $7\,°C$ falls by $2\,°C$ **b** $5\,°C$ falls by $4\,°C$

   c $0\,°C$ falls by $3\,°C$ **d** $7\,°C$ falls by $9\,°C$

10 Write the correct symbol, < or >, between each pair of numbers.

   a 8 … 5   **b** −6 … −2

   c −2 … −5   **d** 9 … −3

11 Write these numbers in order, smallest first.

   a 5, −3, 8, 9, −4, 7, 0

   b −4, −7, −8, 2, −3, 6

   c −5, 5, −3, 3, 2, −4, 1

12 This diagram shows the calculation −1 + 1.

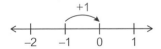

Copy and complete

$-1 + 1 = \square$

13 a These counters show the calculation −2 + 2.
Work out −2 + 2.

   b These counters show the calculation −4 + 3.
Work out −4 + 3.

   c Write the calculation shown by each set of counters.
Work out the answer to each calculation.

    i

    ii

    iii

    iv

Thermometer scale:
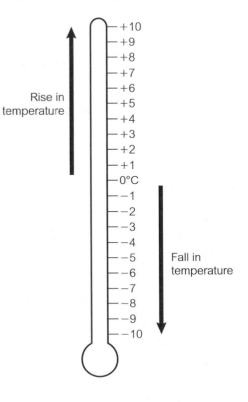
Rise in temperature
+10, +9, +8, +7, +6, +5, +4, +3, +2, +1, 0°C, −1, −2, −3, −4, −5, −6, −7, −8, −9, −10
Fall in temperature

14 Look at Tom's answer to this question.

Q: Is 4 a square number?

A: Yes, because I can make a square with 4 dots.

• •

• •

$2 \times 2 = 4$

Use Tom's method to decide whether each of these is a square number.

a 3　　　　　**b** 9　　　　　**c** 16　　　　　**d** 20

15 Copy and complete. The first one has been answered for you.

a $2^2 = 4$　　$\sqrt{4} = 2$　　　**b** $3^2 = \square$　　$\sqrt{\square} = 3$

c $4^2 = \square$　　$\sqrt{\square} = 4$　　**d** $5^2 = \square$　　$\sqrt{\square} = 5$

e $6^2 = \square$　　$\sqrt{\square} = 6$　　**f** $10^2 = \square$　　$\sqrt{\square} = 10$

16 Work out

> BIDMAS

a $3^2 + 2$　　　**b** $5^2 - 4$　　　**c** $2^2 \times 4$　　　**d** $12 \div 2^2$

17 Work out

> BIDMAS

a $(3 + 1)^2$　　**b** $(3 - 1)^2$　　**c** $(5 + 4)^2$　　**d** $(7 - 4)^2$

18 a Write the factor pairs of

　　i 8　　　　　**ii** 20

b Copy and complete these lists of factors

　　Factors of 8: 1, 2, ..., ...

　　Factors of 20: 1, 2, 4, ..., ..., ...

　　Common factors: 1, ..., ...

c What is the **highest** common factor of 8 and 20?

d Find the highest common factor of 15 and 30.

19 a Copy and complete these lists of multiples.

　　Multiples of 5: 5, 10, □, □, □, □, □, □, □, 50

　　Multiples of 7: 7, 14, □, □, □, □, □, □, □, 70

b What is the lowest common multiple of 5 and 7?

c Find the lowest common multiple of 6 and 7.

Problem-solving

1 John has £2.70, Richard has £12.84 and Lucy has £8.50.

a Round each amount to the nearest whole pound (£).

b Use your answers to part **a** to estimate how much John, Richard and Lucy have altogether.

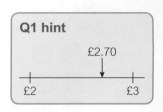

Q1 hint

£2.70

£2　　　　£3

2 Aled follows 246 people on Twitter.

Yasmin follows four times as many people.

Use the bar model to work out the number of people that Yasmin follows.

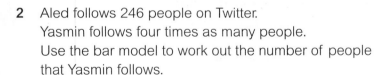

Yasmin

246

Aled

3 Ali earns £6.40 per hour.
One morning she earns £20.80.
How many hours did she work?

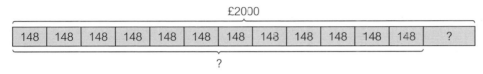

Q3 hint ☐.25 hours
= ☐ $\frac{1}{4}$ hours
= ☐ hours ☐ minutes

4 Owen takes out a business loan of £2000.
He makes 12 equal repayments of £148.
Use the bar model to work out how much he has left to repay.

£2000

| 148 | 148 | 148 | 148 | 148 | 148 | 148 | 148 | 148 | 148 | 148 | 148 | ? |

?

5 Motorbike insurance costs £1196.
Rena makes an initial payment of £260. She pays the remainder in 12 equal amounts.
How much is each payment?

2 Extend

1 a Work out
  i $(4 + 4) \div (4 + 4)$ **ii** $(4 \times 4) \div (4 + 4)$
  iii $(4 + 4 + 4) \div 4$ **iv** $4 \times (4 - 4) + 4$
  v $(4 \times 4 + 4) \div 4$ **vi** $4 + (4 + 4) \div 4$

  b What do you notice about your answers to part **a**?

  c **Reasoning** How could you use four 4s to make 7? ... to make 8?

2 a Which of these numbers are square numbers?
  123, 169, 101, 144, 230

  b **Reasoning** Gwynfor says, 'There is no square number between 122 and 140.'
  Is he correct?

3 Problem-solving / Reasoning
1 September 2010 was a Wednesday.
1 September 2011 was a Thursday.
When will 1 September next fall on a Wednesday?

> **Q3 hint** You need to take leap years into account.

4 Problem-solving / Reasoning Seating arrangements need to be made for 42 girls and 36 boys attending a school prom. All tables need to have the same number of girls. All tables need to have the same number of boys. All tables need to have at least one girl and one boy.

  a How many girls and boys could you have at one table?

  b Is there more than one answer to part **a**?

  c What is the maximum number of tables that can be used?

5 Reasoning Two of these calculations are wrong.
Use inverse operations to find out which two.

  A $12 \times 28 = 363$

  B $8896 \div 64 = 139$

  C $7881 \div 213 = 73$

> **Q5 hint**
> The inverse operation would be $363 \div 12$ or $363 \div 28$.

6 Problem-solving A laboratory experiment is carried out by 8 technicians.
The experiment takes 38 hours to complete.
Each technician works on their own.
They all work for the same number of hours.
How long does each technician work? Give your answer in hours and minutes.

7 Problem-solving 30 000 children need an MMR vaccination.
Surgery A receives 11 726 MMR vaccines.
Surgery B receives 6375 more vaccines than Surgery A.
Can all the children be vaccinated? Show working to explain your answer.

8 Problem-solving A pottery production line works continuously for 24 hours.
It produces 108 pieces of pottery each hour.
A total of 53 pieces are damaged.
How many undamaged pieces of pottery are produced in total in 24 hours?

9 Work out

a $(2 + 3)^2 \div (14 - 9)^2$

b $\dfrac{(2 + 3)^2}{(14 - 9)^2}$

c $(5^2 - 7) \div (2^2 - 1)$

Q9a hint
Use the bracket keys on your calculator.

d $\dfrac{(5^2 - 7)}{(2^2 - 1)}$

e $\dfrac{(4 + 8)^2}{(4^2 - 8)}$

f $(4 + 8)^2 \div (4^2 - 8)$

10 Work out

a $3947 - 907 - 81$

b $731 - 39 + 256$

c $48 + 2609 - 146 + 397$

11 Problem-solving A supermarket sells the same brand of chocolate in three different boxes.

Box 1: £1.86 for 12 chocolates

Box 2: £2.10 for 15 chocolates

Box 3: £3.59 for 24 chocolates

a Which box is the best value for money?

b Lulu has £25. What is the largest number of chocolates that she can buy?

12 Problem-solving Rufus is paid £150.40 for 16 hours' work.

Mamadou is paid £8.60 per hour for the first 22 hours.

He is then paid £9.90 per hour.

Who is paid more for working 32 hours? How much more?

13 Problem-solving A farmer has 513 eggs to pack.

She uses 12×12 and 15×15 trays.

How many trays should she use of each type, so that every tray is completely full?

14 Which two numbers does each of these square roots lie between? The first one has been done to help you.

a $\sqrt{15}$

$3^2 = 9$ and $4^2 = 16$

So the square root of 15 lies between 3 and 4.

b $\sqrt{70}$

c $\sqrt{28}$

d $\sqrt{91}$

e $\sqrt{39}$

f $\sqrt{60}$

15 Find the lowest common multiple of each of these sets of numbers.

a 4, 6 and 9

b 3, 4 and 5

16 Find the highest common factor of each of these sets of numbers.

a 12, 18 and 36

b 24, 32 and 40

Reflect   In this Extend lesson, do you think you did

A well   **B** OK   **C** not very well?

List three things that made a difference to how well you did.

They can be things that made you do better or worse.

Here are some things other students said, to give you some ideas.

'I didn't read some questions properly.'

'It helped that I can do long multiplication.'

'I chatted to John too much.'

How do you think you could have done even better in these lessons?

2 Unit test

1. Round 2 486 005 to the nearest 100 000.

2. Work out 1063 − 297

3. Work out
   a 82 × 6
   b 371 × 62

4. Work out
   a 168 ÷ 8
   b 8473 ÷ 13

5. A weather chart shows these temperatures.

   | 5°C, −1 °C, 3°C, −4 °C, 0 °C, −5 °C |

   a Which is the warmest temperature?
   b Is 0°C colder or warmer than −1°C?
   c Write down a temperature from the list that is colder than −1°C.

6. A computer costs £1420.
   Nathan pays a deposit of £900.
   He then pays the rest in four equal amounts.
   How much is each payment?

7. Karen uses a calculator to divide £81 by 18.
   Look at the calculator display.

   [4.5]

   Write this answer in pounds (£).

8. In Cardiff the temperature was −2 °C.
   In Edinburgh the temperature was 4 degrees colder.
   What was the temperature in Edinburgh?

9. Work out
   a 9 × 40
   b 300 × 7
   c 15 × 4000

10. Find the lowest common multiple of 20 and 12.

11 28 boxes of cookery books are delivered to a warehouse.
Each box contains 42 books.
How many books are delivered?

12 A florist sells roses in bunches of 12.
The florist takes delivery of 276 roses.
How many bunches of roses can the florist make?

13 Work out
  a $-10 + 6$
  b $-10 - 6$
  c 2×-6

14 Work out
  a $20 - 4 \times 5$
  b $20 \div 4 \times 5$
  c $7 \times (3 + 2)$
  d $7^2 + 6$
  e $(7 - 5)^2 + 4$

15 Shop 1 sells 2843 charity wristbands.
Shop 2 sells 479 fewer wristbands than Shop 1.
Shop 3 sells 99 more wristbands than Shop 2.
How many wristbands do they sell altogether?

16 Steven buys 36 blank CDs.
The CDs are sold in packs of four. Each pack costs £1.60.
Lucas also buys 36 blank CDs.
His CDs are sold in packs of six. Each pack costs £1.92.
Who pays more for the CDs? How much more?

Challenge

a Work out the square numbers from 1×1 up to and including 20×20.

b What is the pattern in the last digits of the square numbers?

c Which numbers are never the last digit of a square number?

d Test your answer to part **c** by finding three more square numbers greater than 400.

Reflect Look back at the work you have done in this unit.
When you answered questions, how did you decide which operation to use: addition, subtraction, multiplication or division?

3 Expressions, functions and formulae

Master   Check up p77   Strengthen p79   Extend p84   Unit test p86

3.1 Functions

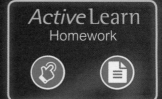

- Find outputs of simple functions written in words and using symbols
- Describe simple functions in words

Warm up

1 **Fluency** Work out
   a 3 add 19   **b** 30 divided by 5   **c** 9 subtract 4

2 Work out
   a 2 add 3 then multiply by 7   **b** 4 divided by 2 then subtract 2
   c 12 subtract 3 then add 11

3 Which operation is missing from each of these calculations: +, −, × or ÷?
   a 2 ☐ 3 = 6   **b** 5 ☐ 3 = 2   **c** 6 ☐ 3 = 2

Key point

A **function** is a relationship between two sets of numbers. The numbers that go into a **function machine** are called the **inputs**. The numbers that come out are called the **outputs**.

Worked example

Work out the outputs of this function machine.

input   output
3 →   → 7 ──── 3 + 4
5 → add 4 →   9 ───── 5 + 4
10 →   → 14 ──── 10 + 4

4 Work out the **outputs** of each **function machine**.

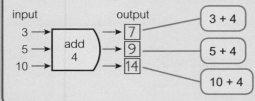

   a input   output
   7 →
   10 → subtract 3 →
   18 →

   b input   output
   6 →
   12 → divide by 2 →
   19 →

   c input   output
   4 →
   7 → +6 →
   11 →

   d input   output
   1 →
   2 → ×3 →
   3 →

5 Copy and complete the table to show the inputs and outputs of this function machine.

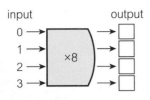

Input	0	1	2	3
Output				

6 Write down the **function** for each machine.

a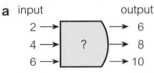

b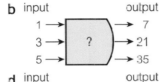

Q6a hint Is the function '×3', or '+4' or something else?

c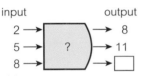

d input
12 → ? → 4
15 → → 5
21 → → 7

7 Problem-solving Work out the missing output of this function machine.

input
2 → ? → 8
5 → → 11
8 → → ☐

8 A café owner pays her staff £7 per hour.
Copy and complete this table to show how much she must pay them for shifts of

a 4 hours   **b** 6 hours   **c** 8 hours

Input (hours)	4	6	8
Output (£)			

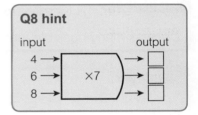
Q8 hint
input
4 →
6 → ×7
8 →
output

9 Texts cost 12p.
  a Copy and complete this function machine to show the number of texts for £4.80.

input   output
£4.80 → ? → ☐

  b Work out the number of texts for
    i £1.92   **ii** £4.08

10 Work out the outputs of each two-step function machine.

a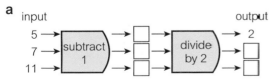

input
5 →
7 → subtract 1 → → divide by 2 → 2
11 →

b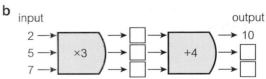

input
2 →
5 → ×3 → → +4 → 10
7 →

11 Work out the outputs of each two-step function machine.

a input

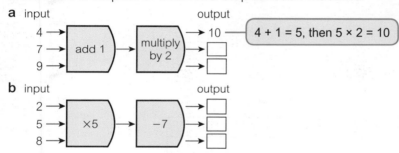

b input

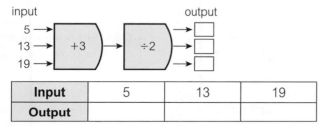

12 a Copy and complete the table for the inputs and outputs of this function machine.

input

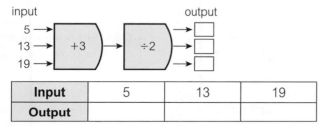

Input	5	13	19
Output			

b Reasoning Will the output numbers always be whole numbers, for *any* input numbers?

13 Problem-solving Work out the missing output of this function machine.

input   output

3 → [×3] → [?] → 8

7 →   20

10 →   □

Challenge   A two-step function machine has input 6. The output is 20.

Draw as many possible function machines as you can with this input and output.

Reflect

a Bella says, 'A function always has inputs and outputs.'
Is Bella correct?

b Kim says, 'Each input gives only one output.'
Is Kim correct?

c What else can you say about functions? Discuss with a classmate.

3.2 Simplifying expressions 1

*Active*Learn
Homework

- Use letters to represent unknowns in algebraic expressions
- Simplify linear algebraic expressions by collecting like terms

1 **Fluency** Work out

   a $4 - 7$ **b** $-3 + 5$ **c** $6 - 9$ **d** $-2 - 5$

2 Copy and complete: $4 + 4 + 4 = \square \times 4 = \square$

3 Work out

   a $4 + 5 - 3$ **b** $12 - 6 - 4$ **c** $8 - 6 + 2$ **d** $5 - 8 + 1$

Key point   In maths, if you do not know a value, you can use a letter to represent it.

Worked example

Simplify

a $p + p$

   $p + p = 2p$

b $2p + 3p$

   $2p + 3p = 5p$

4 Josie has rods of three different lengths.

The yellow rods are x cm long.

The blue rods are y cm long.

The grey rods are z cm long.

Simplify

   a $x + x + x + x$

   b $y + y$

   c $z + z + z$

   d $2x + x$

   e $3y + 2y$

   f $4z - 2z$

5 Match each calculation on the top row with its equivalent on the bottom row.

   $6 + 6 + 6$ $z + z + z + z + z$ $x + x$ $y + y + y + y$

     $2 \times x$ $4 \times y$ 3×6 $5 \times z$

An **expression** contains **terms** using numbers and letters.

Terms: $2m$ $13x$ y 7

Expressions: $2m + 7$ $3x + 2m$

6 Simplify

   a $4m + 3m$ **b** $7n + 2n$ **c** $9q + 3q$ **d** $8x + 2x + x$

$1y$ is written as y.

7 Simplify

   a $5y - 3y$ **b** $7y - 3y$ **c** $12b - 4b$ **d** $2r - 4r$

Worked example

Simplify $4b + 2b - b$

$4b + 2b - b = 6b - b$

             $= 5b$

> Work from left to right:
> First work out $4b + 2b$,
> then subtract b.

8 Simplify

   a $6m + 4m - 3m$ **b** $8x - 3x + 3x$ **c** $2x + 2x - x$ **d** $9p + p - 2p$

   e $2z + 3z - 10z$ **f** $3y - 5y + 6y$ **g** $8n - 3n - n$ **h** $3t + t - 5t$

9 **Problem-solving** Write three calculations that give an answer of $8x$.

Like terms contain the same letter (or contain no letter). For example, $5x$ and $7x$ are like terms, but $4x$ and $3y$ are not like terms. You **simplify** an expression by collecting like terms.

Worked example

Simplify $2b + 3r + 5b$

$2b + 3r + 5b = 2b + 5b + 3r$

               $= 7b + 3r$

> Think of some blue and red tiles.
> The question is:
> [B][B] + [R][R][R] + [B][B][B][B][B]
>
> You can collect the blue tiles together, but you can't collect the blue tiles with the red tiles. The answer is:
> [B][B][B][B][B][B][B] + [R][R][R]

10 Simplify by collecting **like terms**.

   a $4r + 5b + 6r$ **b** $8a + 3c + 5a$ **c** $3t + 9 + 7t + 2$ **d** $7x + 3y + 2x + y$

To find a **sum** you need to add. The sum of 2 and 7 is $2 + 7 = 9$.

11 Copy and complete this addition pyramid.

Each brick is the sum of the two bricks below it.

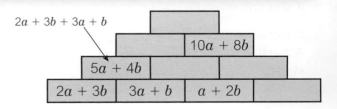

$2a + 3b + 3a + b$

$10a + 8b$

$5a + 4b$

$2a + 3b$ $3a + b$ $a + 2b$

12 Simplify by collecting like terms.

    a $8g + 5h - 3g = 8g - 3g + 5h = \square g + 5h$ **b** $6x - 2y - 3x = 6x - 3x - 2y = \square x - 2y$

    c $9d + 4k - 5d - 2k$ **d** $5x - 3y - 2x$

    e $4a + 5b - 8a$ **f** $7n + 5p - 5n - 8p$

    g $7w + 3u - 6u - 6w$ **h** $3 + 8 - 5 + b + b$

    i $17b + 14 - 6b + 12 - 8 - 3b$

13 Copy and complete these addition pyramids.

    a 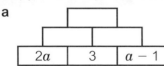   **b** 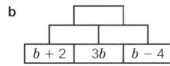   **c**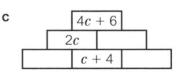

14 **Problem-solving** Which two of these expressions simplify to give the same answer?

    A $6x + 4y - 10x - 7y + 5x$ **B** $2x + 6x + y - 7x - 5y$ **C** $2x + 7y - x - 8y - 2y$

15 **Problem-solving** Show that the perimeter of this
    triangle is $16w + 15z$.

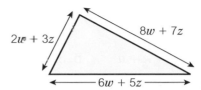

16 Simplify

    a $a + a + a + a - b - b$ **b** $5a + 6 - 3 + 2b$

    c $2a + 5b + 6a - 7b - 8$ **d** $17a + 9 - 5a + 4b$

Challenge   This square is divided into 16 small squares
The expression $16s$ describes the square.

 $= 16s$ $= 2\square$

Divide the square in some more ways. Label identical parts with the same letter.

Write expressions to describe the square.

Reflect   In algebra, letters are used to represent values you do not know.
This lesson might be the first time you have done algebra.

Choose A, B or C to complete each statement.

In this lesson, I did …   **A** well   **B** OK   **C** not very well

So far, I think algebra is …   **A** easy   **B** OK   **C** difficult

When I think about the next lesson, I feel …   **A** confident   **B** OK   **C** unsure

If you answered mostly As and Bs, did your experience surprise you? Why?

If you answered mostly Cs, look back at the questions you found most tricky.
Ask a friend or your teacher to explain them to you. Then complete the
statements above again.

3.3 Simplifying expressions 2

- Multiply and divide algebraic terms
- Use brackets with numbers and letters

*Active*Learn
Homework

Warm up

1 Fluency What is the formula for the area of a rectangle?

2 Write these using index form.
   a 5×5 **b** 11×11

3 Work out
   a $\frac{12}{2}$ **b** $\frac{12}{3}$ **c** $\frac{12}{4}$

4 Simplify
   a $n + n + n + n$ **b** $3x + 5x$
   c $4y + 2y - 5y$ **d** $7w + w - 10w$

Worked example

Simplify

a $4 \times y$
   $= 4y$ →   Write the number before the letter when multiplying.

b $y \times y$
   $= y^2$

c $4y \times 3$
   $= 12y$ $3 \times 4y = 4y + 4y + 4y = 12y$

d $y \times x$
   $= xy$ $y \times x = x \times y = xy$
   Always write the letters in alphabetical order.

e $\frac{20n}{10}$
   $= 2n$ $\dfrac{^2\cancel{20} \times n}{_1\cancel{10}} = 2n$

5 Simplify
   a $3 \times m$ **b** $n \times 7$ **c** $p \times 5$ **d** $e \times d$
   e $e \times g$ **f** $k \times e$ **g** $k \times k$ **h** $k \times a$

6 Match the equivalent expressions.

| $4a$ | $a + 4$ | $5b$ | $3a$ | $a \times b$ | $a \times 3$ | $b \times 3$ |

| $4 \times a$ | ab | $3 \times a$ | $4 + a$ | $3 \times b$ | $5 \times b$ |

7 Simplify these expressions.

a $5 \times 2a$ **b** $4 \times 8b$ **c** $3 \times 2y$ **d** $\frac{16y}{4}$

e $\frac{14a}{7}$ **f** $9c \times 5$ **g** $12b \div 2$ **h** $5z \times 2$

8 **Problem-solving** Write four multiplications or divisions that give each of these answers.

a $6a$ **b** $2c$ **c** $4t$ **d** $10s$

> **Key point** To find a **product** you need to multiply. The product of 7 and 2 is 14.

9 Copy and complete these multiplication pyramids.
Each brick is the product of the two bricks below.

a 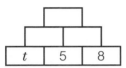 **b**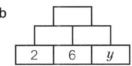

> **Key point** When you multiply out a bracket, multiply every number inside the bracket by the number outside the bracket.
> $5 \times (3 + 12) = 5 \times 3 + 5 \times 12$
> This is called the **distributive law**.

10 Work these out. The first one has been started for you.

a $3 \times (20 + 4) = 3 \times 20 + 3 \times 4 = \square$

b $5 \times (30 + 2)$ **c** $2 \times (10 + 9)$

> **Key point** $3(10 + 7)$ means $3 \times (10 + 7)$. You don't need to write the multiplication sign.

> **Worked example**
>
> Work out $4(10 + 6)$
>
> $4(10 + 6) = 4 \times 10 + 4 \times 6$
> $\qquad = 40 + 24$
> $\qquad = 64$
>
>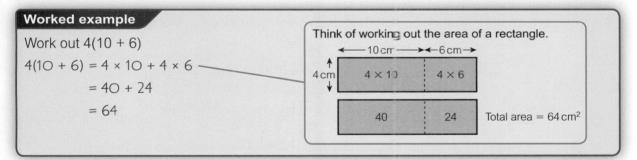
>
> Think of working out the area of a rectangle.
> Total area = 64 cm²

11 Sort these calculations into equivalent pairs.

| 7×15 | 4×13 | 4×17 | 3×39 | 5×17 |

| $5(7 + 10)$ | $7(5 + 10)$ | $4(3 + 10)$ | $4(20 - 3)$ | $3(40 - 1)$ |

12 Use brackets and the **distributive law** to work out these multiplications.
The first one has been done for you.

a $4 \times 59 = 4 \times (60 - 1)$
$\qquad = 4 \times 60 \quad 4 \times -1$

$\qquad \quad 240 \quad - \quad 4 \quad = 236$

b 3×29 **c** 7×28 **d** 9×49

13 This large rectangle has been split into two smaller rectangles.

    a Write an expression for the width of the large rectangle.

    b Use your answer to part **a** to complete this sentence: area = ☐ (☐ + ☐)

    c Write expressions for the areas of the two smaller rectangles.

    d Copy and complete: $4(p + 3) = \Box\, p + \Box$

p	3
		4

Key point   To multiply out or **expand** expressions with brackets, multiply everything inside the bracket by the number outside.

Worked example

Expand $3(x + 2)$

$3(x + 2) = 3 \times (x + 2)$

    $= 3 \times x + 3 \times 2$

    $= 3x + 6$

> Multiply everything inside the brackets by 3.
> $3 \times x = 3x$ and $3 \times -4 = -12$

Expand $3(x - 4)$

$3(x - 4) = 3 \times (x - 4)$

    $= 3 \times x + 3 \times -4$

    $= 3x - 12$

14 Multiply out the brackets.

    a $2(x + 7)$ **b** $7(c - 2)$ **c** $7(c + 2)$ **d** $2(x - 7)$

    e $6(2x - 1)$ **f** $2(6x - 1)$ **g** $5(3x + 1)$ **h** $3(5x + 1)$

15 Expand and simplify. Part **a** has been done for you.

    a $2(x + 1) + 4x = 2x + 2 + 4x = 6x + 2$ **b** $6(x + 4) + 5x$

    c $2(x + 8) + 4(x + 3)$ **d** $5(2x + 3) + 3(x - 2)$

Challenge

a Choose a number from cloud 1 and an expression from cloud 2. Multiply them together.

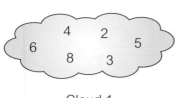

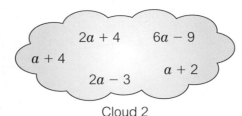

    Cloud 1   Cloud 2

b Repeat this five times.

c Find two calculations that give the same answer.

Reflect   Lilina says, 'Simplifying algebra means make the expression so that it can't be added, multiplied, divided or subtracted any more.'

Howard says, 'Simplifying algebra means getting as few terms as possible.'

Do you agree with Lilina?

Do you agree with Howard?

Write a definition for yourself that begins 'In algebra, 'simplify' means ...'.

3.4 Writing expressions

- Write expressions from word descriptions using addition, subtraction, multiplication and division
- Write expressions to represent function machines

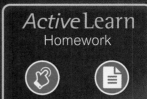

Active Learn
Homework

Warm up

1 Fluency Work out
  a 12 add 18
  b the difference between 25 and 7
  c the total of 4, 6 and 12
  d 14 less than 20
  e half of 16
  f the sum of 5 and 9
  g the product of 5 and 9
  h double 6

2 Simplify
  a $x + 2x$
  b $7y - 5y$
  c $3t - 5t$
  d $5p - 3 + 6p + 7$
  e $2a + 3b + 4a - b$

Key point

You write an algebraic expression by using letters to stand for numbers. The letter is called a **variable** because its value can change or vary.

Worked example

Alice is x years old.

Write expressions for Ben's and Carl's ages.

a   Ben is 5 years older than Alice.
   Ben's age is $x + 5$

b   Carl is 3 years younger than Alice.
   Carl's age is $x - 3$

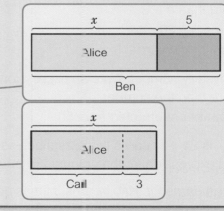

3 Danielle is d years old. Amber is 7 years older than Danielle. Jenny is 9 years younger than Danielle.
Write an expression for
  a Amber's age
  b Jenny's age

> **Q3 hint** Draw a bar model, like the one in the Worked example to help you.

4 The top stick is 3 cm longer than the bottom stick.
Write an expression for the length of the bottom stick.

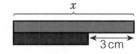

5   Ava has y books. Write an expression for the number of books each person has.
    a   Tim has 2 more books than Ava.
    b   Erin has 6 more books than Ava.
    c   Danel has 3 fewer books than Ava.
    d   Gilen has 9 fewer books than Ava.

Q5a hint
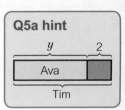

6   A red envelope contains r counters.
    A yellow envelope contains y counters.
    A white envelope contains w counters.

    Write an expression for the number of counters in
    a   the white envelope and 3 more counters
    b   the red envelope and 10 more counters
    c   the yellow envelope with 7 taken out
    d   the white envelope with 8 added
    e   the white envelope with 3 taken out.

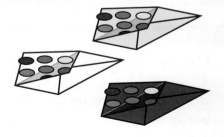

Worked example

Draw a function machine, then write an expression for half of y.

input   output

$y \rightarrow \boxed{\div 2} \rightarrow \dfrac{y}{2}$ —————— To calculate half of a number divide by 2.

7   Draw a function machine, then write an expression for
    a   4 more than a b   3 less than b
    c x take away 20   d y with 7 added on
    e m with 4 removed   f l divided by 2

8   n is a mystery number. Write an expression for
    a   3 more than the number   b   21 less than the number
    c   the number subtracted from 50 d   the number with 8 added

9   Katy earns £x. Lily earns twice as much as Katy.
    Write an expression for the amount Lily earns.

10 a   George earns £y. Kabir earns 3 times as much as George.
        Write an expression for the amount Kabir earns.
    b   Zac earns half as much as George.
        Write an expression for the amount Zac earns.

11 **Problem-solving** In a Year 7 class, there are twice as many Chelsea supporters as Spurs
    supporters. There are 3 times as many Arsenal supporters as Chelsea supporters.
    There are x Spurs supporters.
    Write an expression for the number of Arsenal supporters in terms of x.

12 **Problem-solving** Zoe charges £9 per student to attend a dance class.
    There are y students in the class.
    a   Write an expression for the total amount she charges for a class.
    b   How much does she charge for 5 students?

13 Sam writes down two numbers, x and n.

Write an expression for

a double x **b** 5 times x

c the sum of the two numbers   **d** the product of the two numbers

e half of n **f** one third of n

14 Match each yellow description card to its correct blue expression card.

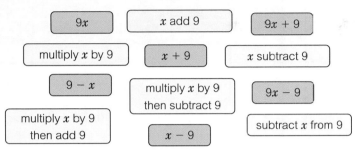

$9x$	x add 9	$9x + 9$
multiply x by 9	$x + 9$	x subtract 9
$9 - x$	multiply x by 9 then subtract 9	$9x - 9$
multiply x by 9 then add 9	$x - 9$	subtract x from 9

15 A packet of Biscos biscuits contains b biscuits.

A packet of Yum biscuits contains y biscuits.

A packet of Giants biscuits contains g biscuits.

Write an expression for the number of biscuits in

a three packets of Giants

b five packets of Biscos

c seven packets of Biscos

d a packet of Biscos and a packet of Giants

e the total of a packet of Yum, a packet of Giants and three more biscuits

Challenge   Greg travels to work each day from Monday to Friday. His normal morning journey time is x minutes. His normal evening journey time is 10 minutes longer.

a Write an expression for the total amount of time Greg spends travelling in a week, using his normal morning and evening times.

This is Greg's journey time information for one week in May.

	Morning journey time	Evening journey time
Monday	twice as long as normal	normal
Tuesday	5 minutes longer than normal	normal
Wednesday	normal	25 minutes less than normal
Thursday	10 minutes longer than normal	twice as long as normal
Friday	day off – no journey	day off – no journey

Greg says, 'Even though I didn't go to work on Friday, I still spent the same amount of time travelling this week as I do in a normal week.

b Is Greg correct? Show all your working and explain your answer.

Reflect   This lesson suggested using bar models and function machines to help you with writing expressions. Did they help you? How?

Did you use any other methods? Explain the method(s) you used.

3.5 Substituting into formulae

- Substitute positive whole numbers into simple formulae written in words
- Substitute positive whole numbers into formulae written with letters

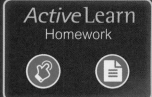

Active Learn
Homework

Warm up

1 **Fluency** Work out
   a 3×6 **b** 4×7 **c** $30 \div 5$ **d** $24 \div 8$

2 Work out
   a 60×3 **b** 70×4 **c** 65×3 **d** $\frac{20}{10}$ **e** $\frac{60}{15}$

3 Simplify
   a $2 \times t$ **b** $x \times y$

Key point A **formula** shows the relationship between different variables, written as words or letters. You can use a formula to work out an **unknown** value by **substituting** the values that you do know into the formula.

Worked example

The **formula** to work out the pressure when a force is applied to an area of $5\,\text{cm}^2$ is

pressure = force ÷ 5

Work out the pressure when the force is 30.

   pressure = force ÷ 5 ———— [Write the formula first.]

   $\quad\quad\quad = 30 \div 5$ ———— [Substitute the value for the force into the formula.]

   $\quad\quad\quad = 6$

4 The formula to work out the pressure when a force is applied to an area of $8\,\text{cm}^2$ is

   pressure = force ÷ 8

   Work out the pressure when the force is
   a 16   **b** 24   **c** 40

5 The formula to work out the amount of painkiller to give a child is

   amount of painkiller (mg) = $\frac{m}{10}$

   where m = mass of child in kg.

   Work out the amount of painkiller to give a child of mass
   a 30 kg   **b** 40 kg

6 The formula for the total resistance in a circuit is

total resistance = $R + 2$

where R is the resistance.

Use this formula to work out the total resistance in a circuit when

a $R = 3$ **b** $R = 6$

7 Naomi uses this formula to work out her pay (£P)

$P = 8.25h$

where h is the number of hours worked.

Work out her pay

a for a 7-hour shift   **b** for a 40-hour week

8 The formula to work out the voltage (V) needed for a lamp is

$V = 2 \times$ resistance

Work out the voltage needed when the resistance is

a 3   **b** 4   **c** 5

9 The formula to work out the distance a car travels when you know the speed of the car and the time taken is

$d = s \times t$

a **Reasoning** What do you think the letters d, s and t stand for?

b Copy and complete the workings to find the distance when

  i $s = 50$ km and $t = 2$ hours

$d = s \times t$
$= 50 \times 2$
$= \square$ km

  ii $s = 60$ km and $t = 3$ hours

$d = s \times t$
$= \square \times \square$
$= \square$ km

10 Use the formula $d = st$ to work out the distance a car travels when it moves at

a a speed of 50 miles per hour for 2 hours

b a speed of 70 miles per hour for 4 hours

c a speed of 65 miles per hour for 3 hours

11 Use the formula $s = \dfrac{d}{t}$

to work out the speed in m/s of an animal that runs

a 20 m in 10 seconds

b 36 m in 12 seconds

c 100 m in 25 seconds

12 A formula used in science to work out the force (F) on an object is

$F = ma$

where m is the mass and a is the acceleration of the object.

Work out the value of the force when

a $m = 20$ and $a = 4$

b $m = 50$ and $a = 2$

c $m = 12$ and $a = 6$

13 Sophie uses this formula to work out the pressure when a force F is applied to an area A:

$$P = \frac{F}{A}$$

Work out the pressure when

a $F = 30$ and $A = 15$ **b** $F = 36$ and $A = 18$ **c** $F = 240$ and $A = 30$

14 The formula for mobile phone charges, in pence, is

$$C = 3m + 500$$

Work out the charge, C, for

a 100 messages   **b** 220 messages

15 A formula used in science to work out the final velocity, v, of an object that starts from rest is

$$v = at$$

where a is the acceleration and t is the time.

Work out the final velocity (v) when

a $a = 5$ and $t = 4$ **b** $a = 8$ and $t = 6$

c $a = 2$ and $t = 12$ **d** $a = 3.5$ and $t = 10$

16 Amy organises a charity concert. She uses this formula to work out the total amount of money ($£T$) she takes from ticket sales:

$$T = 4C + 8A$$

where C is the number of child tickets she sells and A is the number of adult tickets she sells.

Amy sells 80 child tickets and 100 adult tickets.

What is the total amount of money that she takes?

17 Formulae for predicting the adult heights of girls and boys are

Girl: (father's height − 13 + mother's height) ÷ 2
Boy: (father's height + 13 + mother's height) ÷ 2

Mr Singh is 180 cm tall and Mrs Singh is 168 cm tall.
Work out the predicted adult heights of their daughter and their son.

Challenge   In a science experiment, four students measured their hand span and their height. The table shows the results.

Hand span (cm)	15	13	16	14
Height (cm)	145	128	155	138

Sadie thinks that the best formula you could use to work out the height (H) of a person when you know their hand span (S) is $H = 9S$.

Zosha thinks that the best formula is $H = 10S$.

Who is correct? Explain how you worked out your answer.

Reflect   Look back at the formula in Q12.

a Would it matter if this formula used the letters x and y instead of m and a?

b Do the letters help you to understand a formula? Explain.

3.6 Writing formulae

- Write simple formulae in words
- Write simple formulae using letter symbols
- Identify formulae and functions
- Identify the unknowns in a formula and a function

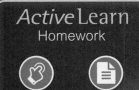

Active Learn
Homework

Warm up

1 **Fluency** Simplify
   a $3 \times k$ **b** $4 \times m$ **c** $n \times 2$ **d** $y \times 5$

2 Lin is y years old. Write an expression for the age of each of these people.
   a Alice is 5 years older than Lin.
   b Joe is 2 years younger than Lin.
   c Kai is twice as old as Lin.

3   **a** $x = y + 2$ Work out x when $y = 6$.
   b $m = 2n$ Work out m when n is 5.
   c $F = 5p$ Work out F when $p = 4$.

Key point   You can write a formula in words to work out an amount, then use letters to represent the variables.

4 A swimming coach always brings 2 more floats than the number of students in the class.
   a How many floats does she bring when the number of students is
     i 1   **ii** 2   **iii** 3?
   b What do you do to the number of students to find the number of floats?
   c Copy and complete this word formula for the number of floats.
     number of floats = _____ + _____
   d Copy and complete this formula that connects the number of students, s, and the number of floats she brings, f.
     $f = \square + \square$

5 All items in a sale are reduced by £10.
   a Work out the sale price of items when the original price is
     i £15   **ii** £20   **iii** £25
   b Copy and complete this word formula for the sale price of an item.
     sale price = _____ − _____
   c Copy and complete this formula that connects the sale price of an item, x, and the original price of an item, y.
     $x = \square - \square$

Worked example

Aisha is making a clock. The length of wood she needs is 10 cm more than the width of the clock.

Write a formula that connects the length of wood, l, to the width of the clock, w.

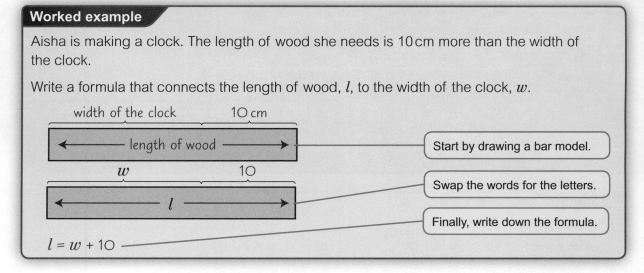

Start by drawing a bar model.

Swap the words for the letters.

Finally, write down the formula.

$l = w + 10$

6 Every day, a baker makes 12 more bread rolls than customers have ordered. Write a formula that connects the number of rolls made, M, to the number of rolls ordered, R.

7 **Reasoning** Ali and Shiraz have decided to put together what's left of their pocket money to buy a present. Write a formula that connects the total amount of money, M, to the amount of money Ali has, y, and the amount of money Shiraz has, z.

Worked example

Write a formula to work out the total number of players at a 5-a-side football tournament, when you know the number of teams.

1 team = 5 players,
2 teams = 2 × 5 = 10 players,
3 teams = 3 × 5 = 15 players, ...

Start by trying different numbers to see the pattern.

number of teams × 5 = number of players

Write the rule for the pattern in words or as a function machine.

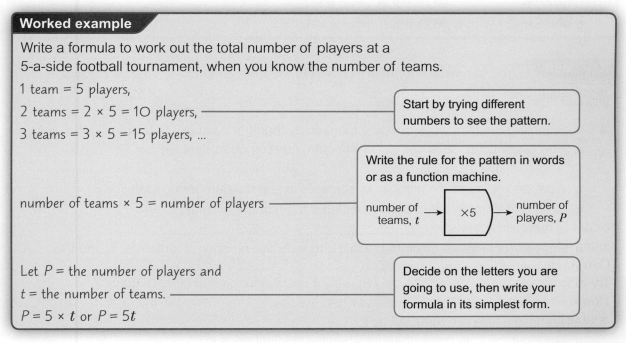

Let P = the number of players and
t = the number of teams.
$P = 5 \times t$ or $P = 5t$

Decide on the letters you are going to use, then write your formula in its simplest form.

8 To make cheese scones you need 4 grams of cheese per scone.
   a Copy and complete this function machine.

   number of scones → × ☐ → total amount of cheese

   b Write a formula to work out the amount of cheese, C, you use when you know the number of scones, s.

9 To make fruit buns you need 12 grams of fruit per bun.
   Write a formula that connects the amount of fruit, F, to the number of buns, b.

10 Simon is paid £8 per hour. Write a formula that connects the total he is paid, P, with the number of hours he works, h.

11 To work out the amount of food in kilograms a horse needs per day, you divide the mass of the horse in kilograms by 40.

a Write a formula that connects the amount of food, F, to the mass of a horse, M.

The table shows the masses of six horses at a riding stables.

Name	Mass (kg)
Aurora	520
Bluegrass	500
Flanagan	480
Phantom	600
Summer	360
Tonto	460

b Use your formula to work out the amount of food that each horse needs.

12 Problem-solving   **a** Write an expression for the area of this rectangle, using p.
Multiply out your answer.

b The value of p is 5 cm. What is the area of the rectangle?

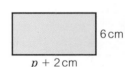

6 cm

$p + 2$ cm

Challenge   When one cube is placed on a table, you can see five of its six faces.

When two cubes are placed side-by-side on a table, you can see eight of their faces.

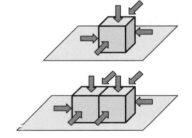

Count how many faces you can see when there are three, four and five cubes.
Can you see the pattern?
Try to write a formula connecting the number of faces and the number of cubes.
How many faces can you see when there are 50 cubes placed side-by-side in a row on a table?

Reflect   In lesson 3.5, you were given formulae to work with. In this lesson you wrote your own formulae.

a Which did you find more difficult?

b What made it more difficult?

c Are there particular kinds of questions you need more practice on? If so, what kinds?

3 Check up

Functions

1 Work out the outputs of each function machine.

a input   output
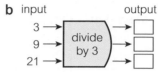

b input   output

2 Write down the function for each machine.

a input   output
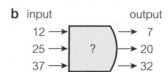

b input   output

3 Work out the outputs of each two-step function machine.

a input   output
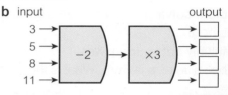

b input   output

Expressions

4 Simplify these expressions.

  a $6k + 5k$ **b** $18h - 6h$ **c** $7b - 2b + b$ **d** $8y - y$

  e $2 \times x$ **f** $z \times 9$ **g** $g \times f$ **h** $p \times q$

  i $8 \times 3c$ **j** $9t \times 3$ **k** $\dfrac{21y}{7}$ **l** $m \times m$

5 Gill writes down two numbers, a and b.
Write an expression for

  a 4 multiplied by a

  b b multiplied by 9

  c the sum of the two numbers

  d the product of the two numbers

6 Work out

  a $4 \times (10 + 4)$ **b** $3(40 - 2)$

7 Simplify these expressions.

  a $c + 9 + 5c + 3$ **b** $14d + 11 - 4d + 2 - 6 - 2d$

8 Multiply out the brackets.

  a $4(x + 4)$ **b** $6(x - 2)$ **c** $8(2x + 5)$

9 Expand and simplify $3(5x + 2) + 2(x - 1)$.

Formulae

10 The formula for working out the amount of pasta, in grams, for a meal is

    amount of pasta = 125 × number of people

  Work out the amount of pasta needed for

  a 2 people   **b** 10 people

11 A formula used in science is $m = dV$.
  Use the formula to work out m when
  a $d = 15$ and $V = 3$
  b $d = 25$ and $V = 6$

12 This bar model shows the difference in price (in pounds) between weekend tickets (w) and day tickets (d) for a festival.

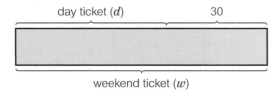

  Write a formula to work out the price of a weekend ticket when you know the price of a day ticket.

13 James works out how much money he has left at the end of every month, by working out the difference between the amount he earns and the amount he spends. All amounts are in pounds (£).
  Write down a formula that connects the amount of money he has left, M, to the amount of money he earns, E, and the amount of money he spends, S.

Challenge

a Shelby says that this function machine only gives outputs ending in 0 or 5.
Is she correct? Explain your answer.

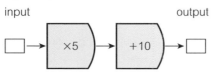

b In this two-step function machine, the input is 6 and the output is 20.

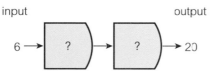

Write down three different two-step rules that this function machine could have.

Reflect

How sure are you of your answers? Were you mostly

😞 **Just guessing**   😐 **Feeling doubtful**   🙂 **Confident**

What next? Use your results to decide whether to strengthen or extend your learning.

3 Strengthen

Functions

1 Work out the output of each function machine.

a input   output

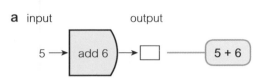

b input   output

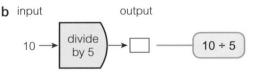

2 Work out the outputs of each function machine.

a input   output

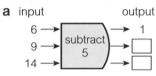

b input   output

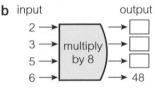

3 Follow these steps to find the function of this machine.

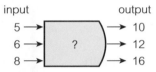

input   output
5 →   → 10
6 →   ?   → 12
8 →   → 16

a Copy and complete to find two possible functions for the first input and output.

$5 \times \square = 10$ and $5 + \square = 10$

b Try both possible functions for 6 → ? → 12

Which one works?

c Check the function works for 8 → ? → 16

d Write the function that works for all the inputs and outputs.

4 Write the function for each function machine.

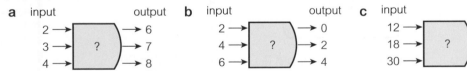

a input   output   **b** input   output   **c** input   output
2 →   → 6   2 →   → 0   12 →   → 2
3 → ? → 7   4 → ? → 2   18 → ? → 3
4 →   → 8   6 →   → 4   30 →   → 5

5 Work out the outputs of each two-step function machine.

a

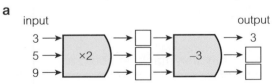

input   output
3 →   → 3
5 → ×2 → → −3 →
9 →   →

b

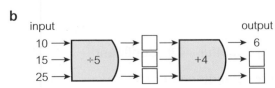

input   output
10 →   → 6
15 → ÷5 → → +4 →
25 →   →

Expressions

1 Work out

a $4 \times (6 + 3)$

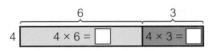

b $3 \times (9 + 2)$

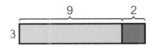

c $8 \times (11 + 3)$

d $9 \times (5 + 2)$

2 Copy and complete

a $2 \times (20 - 1) = \underline{2 \times 20} - \underline{2 \times 1}$
  $= \square - \square = \square$

b $5 \times (30 - 1) = \underline{\square \times 30} - \underline{5 \times \square}$
  $= \square - \square = \square$

c $3 \times 29 = 3 \times (30 - 1) = 3 \times \square - 3 \times \square$
  $= \square - \square = \square$

3 Simplify

a $b + b$ $\boxed{b} + \boxed{b}$

b $b + b + b$ $\boxed{b} + \boxed{b} + \boxed{b}$

c $4b + 3b$ $\boxed{b}\boxed{b}\boxed{b}\boxed{b} + \boxed{b}\boxed{b}\boxed{b}$

d $5b + b$ $\boxed{b}\boxed{b}\boxed{b}\boxed{b}\boxed{b} + \boxed{b}$

e $3b + b + 2b$ $\boxed{b}\boxed{b}\boxed{b} + \boxed{b} + \boxed{b}\boxed{b}$

4 Simplify

a $b + b + b + b$ $\boxed{b} + \boxed{b} + \boxed{b} + \boxed{b}$

b $3b + 2b$ $\boxed{b}\boxed{b}\boxed{b} + \boxed{b}\boxed{b}$

c $5b - 2b$ $\boxed{b}\boxed{b}\boxed{b}\boxtimes\boxtimes$

d $4b - 3b$ $\boxed{b}\boxtimes\boxtimes\boxtimes$

e $7b - b$ $\boxed{b}\boxed{b}\boxed{b}\boxed{b}\boxed{b}\boxed{b}\boxtimes$

f $6b - b - 4b$ $\boxed{b}\boxtimes\boxtimes\boxtimes\boxtimes\boxtimes$

5 Ellen uses this method to simplify the expression $8x + 5y - 12x$

$\widehat{8x} + \widehat{5y} \; \widehat{-12x} = 8x - 12x + 5y$
  $= -4x + 5y$
  $= 5y - 4x$

Simplify these expressions.

a $6x + 3y - 9x$

b $5a + 4b - 8a$

c $6z + 8w - 12w$

d $9x + 2y - 3x - 7y$

e $7a + 5b + a - 9b$

f $8p + 5q - 20p - 3q$

g $13 + 11z - 5z - 3$

h $3d + 5 + e - d + 4e$

6 Simplify

a $x \times 4$

b $z \times 2$

c $n \times 6$

d $p \times 4$

e $3 \times y$

f $11 \times w$

g $b \times a$

h $a \times c$

i $t \times a$

> **Q6 hint** Write the number before the letter. Write letters in alphabetical order.

7 Simplify

   a $4 \times 3x$ **b** $8 \times 2y$ **c** $6b \times 5$

   d $2c \times 2.5$ **e** $\dfrac{12y}{3}$ **f** $\dfrac{16f}{8}$

   g $\dfrac{20z}{10}$ **h** $\dfrac{4g}{4}$

Q7a hint

Q7e hint

8 Match each story to a calculation.

   $\boxed{x - 2}$ $\boxed{x + 3}$ $\boxed{3 - 1}$ $\boxed{3 + 2}$

   a Joe has 3 sweets. He gets 2 more.
   b Penny has 3 texts. She deletes 1.
   c Suha has x sweets. She gets 3 more.
   d Kunal has x texts. He deletes 2.

9 **a** Ella has a bag containing 5 apples. She puts in 4 more.
      Write a calculation to show this.
   b Will has a bag containing x apples. He puts in 4 more.
      Write an expression to show this.

10 **a** Jamie has a box containing 9 pens. He takes out 5.
      Write a calculation to show this.
   b Katya has a box containing y pens. She takes out 3.
      Write an expression to show this.

11 Write an expression for
   a 6 more than x **b** 5 less than x **c** x with 4 added on   **d** x less than 10

12 Match each story to a calculation.

   $\boxed{x \times 2}$ $\boxed{x \times 5}$ $\boxed{8 \times 2}$ $\boxed{4 \times 5}$

   a Joe has £8. Alice has twice as much.
   b Paula has 4 books. Hans has 5 times as many.
   c Sham has x sweets. Sita has twice as many.
   d Erin has x books. Tao has 5 times as many.

13 **a** It takes Emily 6 minutes to complete a puzzle. It takes Aaron 4 times as long.
      How long does it take Aaron to complete the puzzle?
   b It takes Emily r minutes to complete a crossword. It takes Aaron 4 times as long.
      Write an expression for the time it takes Aaron to complete the crossword.

14 **a** Harsha earns £20. Pavel earns twice as much.
      How much does Pavel earn?
   b Rhian earns £h. Louis earns 3 times as much.
      Write an expression for the amount Louis earns.

15 Write an expression for
   a double y **b** 6 multiplied by y **c** the sum of x and y **d** the product of x and y

16 Multiply out the brackets.

   a $3(x + 2)$ **b** $4(5 + y)$ **c** $2(3c + 4)$ **d** $4(4 + 3p)$

     $= \underline{3 \times x} + \underline{3 \times 2}$
        $\square x + \square$

17 Expand

   a $2(w - 5)$ **b** $4(2u - 3)$ **c** $3(5 - a)$ **d** $4(3 - 5b)$

     $= \underline{2 \times w} - \underline{2 \times 5}$
        $\square w - \square$

Formulae

1 The formula to work out the number of nails you need to shoe a horse is

     number of nails = 7 × number of horse shoes

   Copy and complete the workings to find the number of nails needed for

   a 4 horse shoes   **b** 8 horse shoes
     number of nails = 7 × 4 = $\square$ number of nails = 7 × 8 = $\square$

   c 12 horse shoes
     number of nails = 7 × $\square$ = $\square$

2 The formula to work out the number of shoes, s, you need for h horses is

     $s = 4 \times h$

   Copy and complete the workings to find the number of shoes needed for

   a 3 horses   **b** 5 horses
     $s = 4 \times h = 4 \times 3 = \square$ $s = 4 \times h = 4 \times \square = \square$

   c 9 horses
     $s = 4 \times h = 4 \times \square = \square$

3 Copy and complete the workings using the formula $W = V + 7$ to find the value of W when

   a $V = 3$ **b** $V = 9$
     $W = V + 7 = 3 + 7 = \square$ $W = V + 7 = 9 + 7 = \square$

   c $V = 23$
     $W = V + 7 = 23 + \square = \square$

4 Copy and complete the workings using the formula $T = P + S$ to find the value of T when

   a $P = 8$ and $S = 4$ **b** $P = 12$ and $S = 15$
     $T = P + S = 8 + 4 = \square$ $T = P + S = 12 + \square = \square$

   c $P = 9$ and $S = 0$
     $T = P + S = \square + \square = \square$

5 Copy and complete the workings using the formula $X = 6Y$
   to find the value of X when

   a $Y = 3$ **b** $Y = 5$
     $X = 6Y = 6 \times 3 = \square$ $X = 6Y = 6 \times \square = \square$

   c $Y = 10$
     $X = 6Y = 6 \times \square = \square$

> **Q5 hint**
> $6Y$ means $6 \times Y$

6 A formula for a rough conversion from weight (W) to mass (M) is

$$M = \frac{W}{10}$$

Work out the value of M when

    a $W = 60$ **b** $W = 500$ **c** $W = 90$

> **Q6a hint**
> $M = \frac{W}{10} = \frac{60}{10} =$

7 In a fun run there are 5 more medals than the number of runners.
    a Copy and complete the correct function machine to find the number of medals.

number of runners $\rightarrow$ [   ] $\rightarrow$ number of medals

number of medals $\rightarrow$ [   ] $\rightarrow$ number of runners

    b How many medals will there be for 20 runners?
    c Copy and complete the formula.
       Number of _____ = number of _____ + ☐

8 Shem spends half of his pocket money, P, on music downloads.
    a Write an expression for half of his pocket money.
    M is the amount he spends on music downloads.
    b Copy and complete the function machine.

☐ $\rightarrow$ [$\div$ ☐] $\rightarrow M$

    c Write the formula: $M =$ _____

9 At a party, there are 6 more cans of drink than the number of children coming to the party. Write a formula to work out the number of cans of drink, d, when you know the number of children, c.

Challenge   This is what Callum writes in his exercise book.

> When I expand $3(2x + 4)$ I get the same expression as when I expand $2(\blacksquare x + \bullet)$.
>
> When I expand $4(9 - 3y)$ I get the same expression as when I expand $6(\blacksquare - \blacksquare y)$.

Callum has spilt tea on his homework.
Work out what numbers are under the tea stains.

Reflect   Look back at the questions you answered in these lessons.
a Which hints were most useful? What made them more useful?
b Which hints were least useful to you? What made them less useful?
c What do your answers tell you about how you best learn maths?

3 Extend

1 **Problem-solving** Sanjay has spilt tomato sauce on his homework.

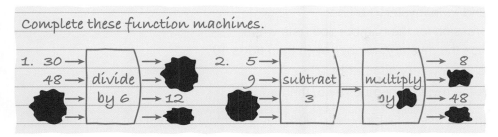

a Copy and complete the function machines.
   Work out the missing numbers underneath the tomato sauce.
b Is it possible to work out all the missing numbers?
   If not, suggest numbers that the missing numbers could be.

2 **Reasoning** The diagram shows four cards. a is a positive number.

| $5a$ | $3a - a$ | $6a - 2a$ | $10a$ |

Tom chooses these two cards

| $6a - 2a$ | $10a$ |

and adds the **expressions**.
$6a - 2a + 10a = 4a + 10a$
$\qquad\qquad\qquad = 14a$

a Choose two other cards and add the expressions.
b Choose two different cards and add the expressions.
c What is the greatest total you can get by adding the expressions from two cards? Show
   how you worked out your answer.

3 For a science project, some students work out how much they have grown in Year 7.

Girls	4.7 cm	5.6 cm	4.9 cm	6.1 cm	5.2 cm
Boys	5.2 cm	4.9 cm	5.8 cm	6.2 cm	5.7 cm

a Work out the range for
   i the girls   ii the boys
b Write a formula in words for working out the range.

4 **Problem-solving** Sort these cards into groups that simplify to the same expression.

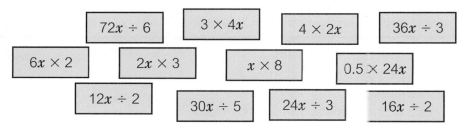

5 Sam thinks of a number, n.
   a Write an expression for the number Sam gets when he multiplies the number by 2.
   b Copy and complete this expression for the number Sam gets when he multiplies the number by 2 and then adds 5.

   $\Box n + \Box$

6 Use m to stand for a mystery number.
   Write an expression for the number Joe gets when
   a he multiplies the mystery number by 3, then adds 10
   b he divides the mystery number by 2
   c he divides the mystery number by 5, then adds 3

7 A personal trainer uses this formula to work out an athlete's maximum heart rate, H_{max}

   $H_{max} = 220 - A$

   where A is the athlete's age in years.
   Work out the maximum heart rate for athletes with these ages.
   a 18 years   **b** 35 years   **c** 50 years

8 To work out an approximate time (in hours) it will take to walk a distance (in kilometres), you divide the distance by 4.
   a Write a formula to work out the time, t, when you know the distance, d.

   Bill plans a walking holiday. The table shows the distances he plans to walk each day.
   b Use your formula to work out the time it will take Bill to do his walk each day.

Day	Distance (km)
Monday	20
Tuesday	24
Wednesday	16
Thursday	18
Friday	26

Challenge   Look at the number grid on the right.
There are green and yellow L shapes on the grid.
You can work out the L-value of a shape using this rule:

   L-value = total of bottom numbers – top number

1	2	3	4	5
6	7	8	9	10
11	12	13	14	15
16	17	18	19	20
21	22	23	24	25

So the yellow L-value is $12 + 13 - 7 = 18$.
a Work out the green L-value.
b Choose two more L shapes on the grid and work out their L-values.
c **i** What do you notice about each L-value and the number at the top of the L shape?
   ii Copy and complete this rule: L-value = top number + $\Box$
d **i** Copy and complete the expressions in this L shape.
   ii Write an expression for the L-value of this L shape. Simplify your expression.
e What do you notice about your answers to parts c and d?
f What will be the L-value of the L shape that has 42 as its top number?

L shape: top cell n; bottom row $n + 5$ | $n + \Box$

Reflect   What kinds of jobs involve using formulae?
What careers are you interested in?
Do you think you will need to use formulae in your job? How?
What professionals are you likely to meet who might use formulae in their work?

3 Unit test

1 Work out the outputs of each function machine.

a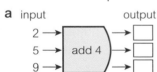

b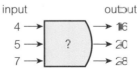

2 Write the function for this machine.

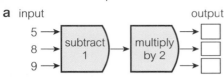

3 Work out the outputs of each two-step function machine.

a input

b input

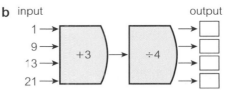

4 Simplify

   a $x + x + x$ **b** $4m + 3m$ **c** $2q - q$ **d** $5a - 8a + a$

5 Work out $4 \times (30 + 2)$.

6 The cost of using broadband in a hotel is £3 per hour.
   The formula to work out the total cost is
      cost in pounds = 3 × number of hours
   Work out the total cost of using broadband for
      a 2 hours   **b** 5 hours

7 Alex thinks of three different ways of using function machines to change 5 into 20.
   Work out the missing numbers.

8 Simplify

   a $4 \times y$ **b** $z \times 8$ **c** $m \times n$ **d** $c \times b$ **e** $x \times x$

9 Simplify

   a $8a + 12 - 8 + 7b$ **b** $9a + 8b + 5a - 2b - 4$

10 Bryn is b years old. Jez is 2 years older than Bryn. Helen is 5 years younger than Bryn.
   Write an expression for
      a Jez's age   **b** Helen's age

11 Ben earns £d per week.
   Amy earns twice as much as Ben.
   Write an expression for the amount Amy earns per week.

12 $P = T + R$
   Work out the value of P when $T = 18$ and $R = 6$.

13 Every day, a butcher makes 30 more sausages than customers have ordered.
   Write a formula to work out the number of sausages made, S, when customers have
   ordered R sausages.

14 A tennis coach always brings 4 more racquets to the lesson than the number of students.
  a Write a formula to work out the number of racquets, r, the coach brings when there are s students.
  b Use your formula to work out the number of racquets the coach brings to the lesson when there are 12 students.

15 A tutor is paid £20 per hour.
  Write a formula to work out the total she is paid, T, when you know the number of hours she works, h.

16 Use the formula $M = fd$ to work out the value of M when $f = 5$ and $d = 12$.

17 Simplify these expressions.
  a $6 \times 3a$ b $4b \times 9$ c $\dfrac{24c}{6}$

18 Expand the brackets.
  a $4(x + 2)$ b $5(3y + 4)$ c $2(z - 6)$

19 Write an expression for
  a 5 times m
  b the sum of m and n
  c the product of m and n

20 Amy uses blue and yellow thread to make friendship bracelets.
  She uses b cm of blue thread and y cm of yellow thread.
  Write a formula to work out the total length of thread, T, in centimetres.

Challenge   Here are seven formula cards.

$$F = 2G + 4 \qquad B = 4G \qquad D = A + 3$$

$$C = \frac{E}{10} \qquad E = AD \qquad G = D - A \qquad H = 5C + F$$

a Work out the value of H when $A = 5$.
  Use a table like this one to record the value of each letter as you find it.

A	B	C	D	E	F	G	H
5							

b Write down the order in which you used the formulae in part **a**.
  Could you have used the formulae in a different order?
c Which formula don't you need to use to work out H?
d Write a formula that connects A to H.

Reflect   Look back at the work you have done in this unit.
Find a question that you could not answer straight away, but that you really tried at, and then answered correctly.
How do you feel when you struggle to answer a maths question?
Write down the strategies you use to overcome your difficulty.
How do you feel when you eventually understand and get the correct answer?

4 Decimals and measures

Master

Check up p111

Strengthen p113

Extend p118

Unit test p120

4.1 Decimals and rounding

- Measure and draw lines to the nearest millimetre
- Write decimals in order of size
- Round decimals to the nearest whole number and to 1 decimal place
- Round decimals to make estimates and approximations of calculations

Active Learn
Homework

Warm up

1 Fluency How many mm in 1 cm?

2 a Round each amount to the nearest £.
  i £47.53   **ii** £265.10   **iii** £496.97

  b Round these calculator money answers (in £) to the nearest penny.
  i
  ```
  4.176
  ```
  ii
  ```
  31.669
  ```
  iii
  ```
  47.103
  ```

3 Which of the numbers in this list are square numbers?
12, 25, 55, 36, 64, 18, 99, 81, 9, 48

4 Write the measurement shown by each arrow. Give your answer
  i in centimetres   **ii** in millimetres.
  The first one has been done for you.

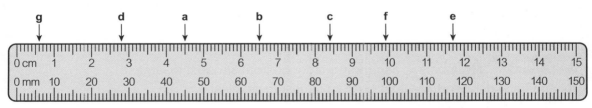

  a i 4.5 cm   **ii** 45 mm

5 Measure the length of each line. Give your answer
  i in centimetres   **ii** in millimetres.

  a _____

  b _____   **c** _____

Q5 hint Line up 0 with the start of the line.

6 Draw lines measuring
  a 5.4 cm   **b** 7.5 cm   **c** 2.6 cm
  d 9.1 cm   **e** 42 mm   **f** 18 mm

Q6 hint Make sure you start your line at 0 on your ruler.

H	T	O	.	$\frac{1}{10}$	$\frac{1}{100}$	$\frac{1}{1000}$
		0	.	1		
		0	.	0	1	
		0	.	0	0	1

$0.1 = \frac{1}{10}$ (one tenth)
$0.01 = \frac{1}{100}$ (one hundredth)
$0.001 = \frac{1}{1000}$ (one thousandth)

7   Write in words the value of the 1 in each of these decimal numbers.
    The first one has been done for you.
    a 72.16   *one tenth*
    b 13.62   **c** 0.41   **d** 24.441   **e** 300.01

Worked example

Write > or < between each pair of numbers.
5.4 ... 5.38
5.4 > 5.38

The symbol '>' means 'is greater than'.
The symbol '<' means 'is less than'.

There are more tenths in 5.4 than in 5.38:
$\frac{4}{10} > \frac{3}{10}$

8   Write > or < between each pair of numbers.
    a 6.7 ☐ 6.9   **b** 6.6 ☐ 6.2   **c** 1.7 ☐ 1.4   **d** 3.7 ☐ 3.86
    e 10.09 ☐ 10.9   **f** 3.9 ☐ 3.107   **g** 21.299 ☐ 21.92   **h** 0.400 ☐ 0.6

9   **Reasoning**   Noah says, '9.35 m is greater than 9.5 m, because 35 is greater than 5.'
    Is Noah right? Explain.

10 Write > or < between each pair of numbers.
    a 6.08 ☐ 6.03   **b** 6.01 ☐ 6.06
    c 12.371 ☐ 12.38   **d** 0.42 ☐ 0.419
    e 7.624 ☐ 7.621   **f** 9.909 ☐ 9.099

Q10 hint
1 First compare the whole number parts.
2 If these are the same, compare the tenths.
3 If these are also the same, compare the hundredths and so on.

11 Write these decimals in **ascending** order.
    a 5.25   5.28   5.23   5.21   **b** 4.3   4.67   4.7   4.19

12 **Problem-solving**   Order these athletics scores to show
    who came first, second, third and fourth.

Q12 hint The fastest run and the longest jump win.

Name	C Black	R Brown	B Green	D White
100 m sprint time (in seconds)	16.4	16.39	16.28	16.3
Triple jump length (in metres)	14.1	14.08	14.23	14.2

13 Write these decimals in **descending** order.
    a 0.01   0.10   0.11   0.111   **b** 9.89   9.98   9.9   9.809

Key point For rounding to the nearest whole number, look at the tenths.
- 5 tenths and above round up.
- 4 tenths and below round down.

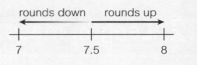

rounds down   rounds up

7   7.5   8

14 Round each number to the nearest whole number. The first one has been done for you.
- **a** 7.8 → 8
- **b** 9.2
- **c** 5.3
- **d** 7.36
- **e** 14.82
- **f** 18.05
- **g** 3.02
- **h** 7.999

Key point A number rounded to **1 decimal place** (1 d.p.) has only one digit after the decimal point.

15 Which of these numbers are given to 1 decimal place?

3.13   13.9   1.8   14.32   6   2.0   91.49   5.1

Key point For rounding to 1 decimal place, look at the hundredths.
- 5 hundredths and above round up.
- 4 hundredths and below round down.

rounds down   rounds up

3.1   3.15   3.2

16 Round each number to 1 decimal place. The first one has been done for you.
- **a** 3.17 → 3.2 (1 d.p.)
- **b** 8.23
- **c** 4.78
- **d** 3.143
- **e** 15.621
- **f** 0.557
- **g** 3.02
- **h** 7.999

> **Q16h hint**
>
> 7.9   8.0

17 Work out an estimate for the answer to each calculation by rounding each number.
- **a** 7.8 × 5.3
- **b** 4.7 × 6.2
- **c** 7.36 × 3.02
- **d** 8.08 × 3.95
- **e** $\sqrt{10.04}$
- **f** $\sqrt{23.8} \times 4.05$
- **g** 12.55 ÷ 5.92
- **h** 14.83 ÷ 4.47

> **Q17e hint** Round 10.04 to 9 as you can easily work out $\sqrt{9}$.

> **Q17g hint** Round 12.55 to 12. Round 5.92 to 6. This gives a calculation that is easy to work out.

18 Problem-solving Sally tries working out an estimate of 3.45 ÷ 3.06 by writing 8 ÷ 3. What is an easier way of rounding these numbers to work out an estimate of 8.45 ÷ 3?

Challenge

a What are the different values of 8 in the number 800.080?

b Why does 8.8 have a different value to 8.08?

c Why does 8.8 have the same value as 8.80?

d Is this correct?

8.88 < 8.9

Explain why.

Reflect In this lesson you were rounding and estimating
Are rounding and estimating the same thing or different? Explain.
Could rounding and estimating be useful in other subjects?

4.2 Length, mass and capacity

- Multiply and divide by 10, 100 and 1000
- Convert measurements into the same units to compare them
- Solve simple problems involving units of measurement in the context of length, mass and capacity
- Convert between metric units of length, mass and capacity

*Active*Learn
Homework

Warm up

1 Fluency What is
   a 7 × 10   **b** 7 × 100   **c** 80 ÷ 10   **d** 800 ÷ 10

2 Work out
   a 230 × 10   **b** 23 × 100   **c** 230 × 100   **d** 18 × 1000
   e 64 000 ÷ 1000

3 Write these units of length in order, smallest first.
   metre   centimetre   kilometre   millimetre

4 Work out these calculations. Write the answers in a place value table like this.

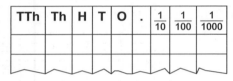

TTh	Th	H	T	O	.	$\frac{1}{10}$	$\frac{1}{100}$	$\frac{1}{1000}$

   a 4.521 × 10   **b** 23.67 × 10
   c 4.521 × 100   **d** 23.67 × 100
   e 4.521 × 1000   **f** 23.67 × 1000
   g 1308 ÷ 10   **h** 87 ÷ 10
   i 1308 ÷ 100   **j** 870 ÷ 100
   k 1308 ÷ 1000   **l** 870 ÷ 1000

> **Q4l hint** Write a 0 in the ones column.

5 Copy and complete to describe how the digits move.
   a × 10   1 place to the _____
   b × 100   ☐ places to the left
   c × 1000   ☐ places to the _____
   d ÷ 10   ☐ place to the _____
   e ÷ 100   ☐ places to the _____
   f ÷ 1000   ☐ places to the _____

> **Q5 hint** Use your answers to Q4.
> For example:
>
> | TTh | Th | H | T | O | . | $\frac{1}{10}$ | $\frac{1}{100}$ | $\frac{1}{1000}$ |
> |-----|----|----|----|----|----|----|----|----|
> | | | | | 4 | . | 5 | 2 | 1 |
> | | | | 4 | 5 | . | 2 | 1 | |

6 Work out
   a 5.6 × 10   **b** 4.55 × 100   **c** 12.02 × 1000   **d** 0.4 × 100
   e 543 ÷ 100   **f** 2750 ÷ 1000   **g** 20.4 ÷ 10   **h** 360 ÷ 1000

7 Problem-solving Sarah multiplies a decimal by 100.
Her answer is 31.
Write down her calculation.

8 Problem-solving Philip divides a number by 1000.
His answer is 4.2.
Write down his calculation.

9 Copy and complete.
  a $6\,cm = 6 \times \square = \square\,mm$ b $8\,m = 8 \times \square = \square\,cm$
  c $9\,km = 9 \times \square = \square\,m$ d $400\,cm = 400 \div \square = \square\,m$
  e $80\,mm = 80 \div \square = \square\,cm$ f $25\,000\,m = 25\,000 \div \square = \square\,km$

Q9a and e hint
$\times\,10$
$1\,cm = 10\,mm$
$1\,cm = 10\,mm$
$\div\,10$

10 Reasoning Do you multiply or divide when converting
  a from cm to mm   b from m to cm
  c from km to m   d from mm to cm
  e from cm to m   f from m to km?

11 For each pair of lengths, work out which is shorter.
  a 5m or 400cm   b 300mm or 15cm   c 6km or 8000m

12 Copy and complete.
  a $2.5\,m = 2.5 \times 100 = \square\,cm$
  b $12.5\,km = 12.5 \times \square = \square\,m$
  c $88\,mm = 88 \div \square = \square\,cm$
  d $160\,cm = 160 \div \square = \square\,m$
  e $7.3\,m = \square\,cm$
  f $67\,mm = \square\,cm$

Q12a hint Use a place value table to help you.
As you are *multiplying* by 100, move every digit 2 places to the *left*.

H	T	O	·	$\frac{1}{10}$
		2	·	5
2	5	0	·	

Write a 0 in the ones column.

13 Write these lengths in order, shortest first.
  a 13cm   103mm   301mm   31cm
  b 400cm   3.5m   345cm   4.2m
  c 678cm   6.7m   6.87cm   675cm
  d 2.12m   3m   234cm   303cm

14 Problem-solving A shed is 1200mm wide.
Will a bike of length 135cm fit along one side?

Key point Metric units of **mass** include the **gram (g)** and **kilogram (kg)**.
$1000\,g = 1\,kg$
Metric units of **capacity** include the **millilitre (ml)** and **litre**.
$1000\,ml = 1\,litre$

15 Copy and complete.
  a $5\,kg = 5 \times \square = \square\,g$
  b $7\,litres = 7 \times \square = \square\,ml$
  c $15\,000\,ml = 15\,000 \div \square = \square\,litres$
  d $6000\,g = 6000 \div \square = \square\,kg$

Q15 hint
$\times\,1000$
$1\,kg = 1000\,g$
$1\,kg = 1000\,g$
$\div\,1000$

$\times\,1000$
$1\,litre = 1000\,ml$
$1\,litre = 1000\,ml$
$\div\,1000$

16 Copy and complete.

    a $4.2\,kg = 4.2 \times \square = \square\,g$

    b $0.75\,litres = 0.75 \times \square = \square\,ml$

    c $4250\,ml = 4250 \div \square = \square\,litres$

    d $875\,g = 875 \div \square = \square\,kg$

    e $9.5\,kg = \square\,g$

    f $1260\,ml = \square\,litres$

17 Write these masses in order, lightest first.

    a 475 g   1.45 kg   1475 g   0.54 kg

    b 0.03 kg   300 g   30.3 g   0.303 kg

18 Problem-solving   A nutmeg has a mass of 10 g.
A nutmeg tree produces 8000 nutmegs a year.
What is the total mass of nutmegs produced by this tree each year?
Give your answer

    a in grams   **b** in kilograms.

19 Problem-solving   A jug holds 2.5 litres.
Will the jug hold 2050 ml of orange juice?

Challenge   **Problem-solving**   The diagram shows three jugs full of water.

Jug 1   Jug 2   Jug 3

3 litres   750 ml   500 ml

Using only these jugs, how can you end up with

a   2500 ml in the largest jug

b   1750 ml in the largest jug?

Show your working and explain your method.

Reflect   Look back at Q11.

  • For part **a**, did you convert 5 m to cm or 400 cm to m?

  • For part **b**, did you convert 300 mm to cm or 15 cm to mm?

  • For part **c**, did you convert 6 km to m or 8000 m to km?

How did you decide which measure to convert for each question?
Which is easier – converting smaller measures to larger measures (like cm to m) or
converting larger measures to smaller measures (like cm to mm)? Explain.

4.3 Scales and measures

- Use scale diagrams
- Read scales
- Write decimal measures as two related units of measure
- Interpret metric measures displayed on a calculator

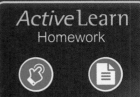

Active Learn
Homework

Warm up

1 Fluency How many
  a mm in 1 cm   **b** cm in 1 m   **c** m in 1 km   **d** g in 1 kg   **e** ml in 1 litre

2 Work out
  a 4.8 × 10   **b** 3.7 × 100   **c** 9.1 × 1000

Key point   A **scale drawing** shows a large real-life measure in a small space. It needs a **scale** to tell you what 1 cm on the drawing represents in real life.

3 This diagram is on a 1 cm squared grid. 1 cm on the grid represents 1 m in real life. Write down the length in metres of
  a line A   **b** line B

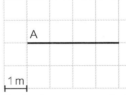

Scale: 1 cm represents 1 m

4 These rectangles are drawn on 1 cm squared grids.
Write down the length and width of each rectangle in metres. Read the scales carefully.

a   **b**   **c**

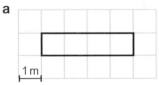 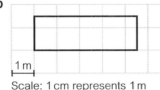 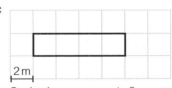

1 m   1 m   2 m
Scale: 1 cm represents 1 m   Scale: 1 cm represents 1 m   Scale: 1 cm represents 2 m

5 On a scale drawing, 1 cm represents 1 m. Use a ruler to draw
  a a line to represent 8 m   **b** a line to represent 6.5 m.

6 Copy and complete the number line.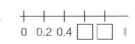
                           0 0.2 0.4 ☐ ☐

7 Write the value that the arrows on each scale point to.

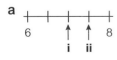

> **Q7b hint** Find the halfway value first.

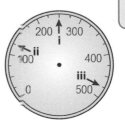

8 Write the measure shown on each scale.

a
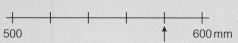
— 500 ml
— 400 ml
— 300 ml
— 200 ml
— 100 ml

b pints — 4, 3, 2, 1

c 10, 0, −10 °C

d
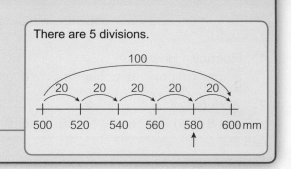
200 300 400
100   500
    g
  0   600

Worked example

Write the value the arrow is pointing to.

500 ————↑———— 600 mm

There are 5 divisions.

From 500 to 600, there is a jump of 600 − 500 = 100

100 ÷ 5 = 20 for each division.

The arrow points to 580 mm.

100
20 20 20 20 20
500 520 540 560 580 600 mm
                  ↑

---

9 Write the measure shown on each scale.

a

50
40 60
mph
30   70

b
— 500
←
— 250
— 0 ml

c 0 ———↑——— 1 m

°F   °C
40 — — 10
20 — — 0
      — −10
0 — — −20
−20 — — −30
−40 — — −40

10 An industrial freezer needs to be kept at a constant temperature of −22 °C.

The thermometer shows the temperature inside the freezer. Is this freezer at the correct temperature? If not, what temperature does it show?

Worked example

What value is the arrow pointing to?
Give your answer in grams.
1 kg = 1000 g
1000 ÷ 5 = 200 g — There are 5 sections.

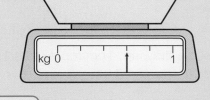
kg 0 —————↑———— 1

kg 0 200g 400g 600g 800g 1 — Label the points on the scale.
                ↑
So arrow points to 600 g

---

11 What value is each arrow pointing to?

a Give your answer in centimetres.

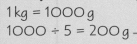
m 0 ——↑————— 1

b Give your answer in grams.

kg 2 —↑————— 3

c Give your answer in metres.

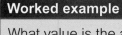
km 2 ——↑—— 3

d Give your answer in millimetres.

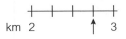
cm 5 ————↑— 6

12 What value is the arrow pointing to? Give your answer in millilitres.

13 Copy and complete.

   a 8.6 cm = 86 mm
           = 80 mm + 6 mm
           = □ cm 6 mm

   b 1.5 m = 150 cm
         = 100 cm + 50 cm
         = □ m 50 cm

   c 3.5 km = 3500 m
          = □ m + 500 m
          = □ km 500 m

   d 2.41 m = 241 cm
         = □ cm + 41 cm
         = □ m 41 cm

   e 4.7 litres = 4700 ml
         = □ ml + 700 ml
         = □ litres 700 ml

   f 9.3 cm = □ cm □ mm

   g 10.8 kg = □ kg □ g

   h 20.6 m = □ m □ cm

   i 7.3 litres = □ litres □ ml

> **Key point**   When you use a calculator to solve problems involving measures, make sure you understand the result on the calculator display.
>
> The result | 9.5 | can mean different things.
> - For money in pounds, it is £9.50.
> - For time in minutes, it is 9 minutes 30 seconds.
> - For length in metres, it is 9.5 m or 9 m 50 cm.

14 Reasoning Andy works out the answer, in metres, to a problem. The calculator display shows | 12.4 |

   Which of these statements are correct?

   A The answer is 12 m 4 cm.
   B The answer is 12.4 m.
   C The answer is 12 m 400 cm.
   D The answer is 12 m 40 cm.

15 Reasoning Use a calculator to work these out.

   Give your answer using the units of measure in brackets.

   a 7.5 cm × 11   (□ cm □ mm)
   b 3.04 m × 8   (□ m □ cm)
   c 1.7 kg × 3   (□ kg □ g)
   d 2.61 litres × 4   (□ litres □ ml)

16 Problem-solving A bottle contains 0.475 litres of olive oil.
   A restaurant kitchen has 12 full bottles.
   How much olive oil do they have? Give your answer in □ litres □ ml.

17 Problem-solving A roll of parcel tape has 4.5 m of tape.
   A pack includes 6 rolls plus 1 free.
   What length of tape do you get in 1 pack? Give your answer in □ m □ cm.

> **Challenge**   Draw a speedometer with arrows on it to show
>   **a** 40 mph   **b** 58 mph   **c** 35 mph

> **Reflect**   Look back at Q7, Q8, Q9 and Q10, about reading scales.
> Write the steps you took to read the scales. Make sure you include every step.
> You may begin with, 'Step 1: I looked at the values on the scale either side of the arrow.'
> What maths operations (addition, subtraction, multiplication, or division) did you use?

4.4 Working with decimals mentally

- Multiply decimals by multiples of 10, 100 and 1000
- Multiply decimals mentally
- Check a result by considering whether it is of the right order of magnitude
- Understand where to position the decimal point by considering equivalent calculations

*Active*Learn
Homework

Warm up

1 **Fluency** Match each calculation to the correct answer.
  a $2 \div 10$ **b** $22 \div 100$ **c** $2 \div 100$
  d $22 \div 1000$ **e** $2 \div 1000$

  | 0.22 | 0.02 | 0.002 | 0.022 | 0.2 |

Q1 hint

T	O	.	$\frac{1}{10}$	$\frac{1}{100}$	$\frac{1}{1000}$
	2				
2	2				

2 Copy and complete these multiplications.
  a $4 \times 20 = 4 \times 2 \times 10 = 8 \times 10 =$
  b $6 \times 300 = 6 \times 3 \times 100 =$
  c $40 \times 8 = 4 \times 10 \times 8 = 32 \times 10 =$

3 Copy and complete these multiplications using partitioning.
  a $3 \times 24 = 3 \times 20 + 3 \times 4 = \square + \square = \square$
  b $8 \times 26 = \square \times \square + \square \times \square = \square + \square = \square$
  c $42 \times 7 = 40 \times 7 + 2 \times 7 = \square + \square = \square$
  d $56 \times 3 = \square \times \square + \square \times \square = \square + \square = \square$

Key point

You can use multiplication facts to work out decimal multiplications.

Worked example

Work out 4×0.2

$4 \times 0.2 = 4 \times 2 \div 10$ ——— $0.2 = 2 \div 10$

$ = 8 \div 10 = 0.8$

4 Use a mental method to work out
  a 3×0.2 **b** 6×0.3 **c** 4×0.7
  d 8×0.4 **e** 9×0.6 **f** 0.8×2
  g 0.7×8 **h** 0.4×5 **i** 0.8×6

Q4f hint $0.8 \times 2 = 2 \times 0.8$

5 Use a mental method to work out these calculations. The first one has been started for you.
  a $7 \times 0.05 = 7 \times 5 \div 100$
  b 4×0.03 **c** 3×0.06 **d** 0.01×9

6 **Problem-solving** When Aaron goes on holiday to Australia, a meal costs him 7 Australian dollars (AUD).
1 Australian dollar (AUD) is worth 0.6 British pounds (GBP).
How much does the meal cost Aaron in GBP?

Q6 hint 1 AUD = 0.6 GBP
7 AUD = ☐ GBP

7 Use a mental method to work out these calculations. The first one has been started for you.
  a $0.2 \times 0.6 = 2 \div 10 \times 6 \div 10$
                $= 2 \times 6 \div 10 \div 10$
                $= 2 \times 6 \div 100$

  b 0.3×0.9 **c** 0.5×0.7 **d** 0.6×0.4
  e 0.9×0.8 **f** 0.11×0.4 **g** 0.2×0.12

8 **Reasoning** Sam says, '0.8×0.5 gives the answer 0.4'.
  a Is Sam correct? Show your working to explain.
  b Write a calculation that gives the answer 0.4

Worked example

Work out 60×1.2

$60 \times 1.2 = 6 \times 10 \times 1.2$ ← Rewrite 60 as ☐ × ☐
        $= 6 \times 1.2 \times 10$ ← $10 \times 1.2 = 1.2 \times 10$
        $= 6 \times 12$ ← Work out 1.2×10
        $= 72$

---

9 Work out
  a 30×1.1 **b** 20×3.2 **c** 40×2.1 **d** 70×1.2

Key point   You can also use **partitioning** to work out decimal multiplications. You can check your answer by using an **approximate** calculation.

10 Copy and complete these multiplications. Check each answer.
The first one has been done for you.
  a $22 \times 3.4 = 20 \times 3.4 + 2 \times 3.4 = 68 + 6.8 = 74.8$
    Check: $22 \approx 20$, $3.4 \approx 3.5$, $20 \times 3.5 = 70$ ✓
  b $32 \times 1.2 = ☐ \times 1.2 + ☐ \times 1.2 = ☐ + ☐ = ☐$
  c $21 \times 4.1 = ☐ \times 4.1 + ☐ \times 4.1 = ☐ + ☐ = ☐$
  d $43 \times 3.1 = ☐ \times 3.1 + ☐ \times 3.1 = ☐ + ☐ = ☐$

11 The exchange rate for pounds to US dollars is £1 to $1.60.
Use **partitioning** to work out how many US dollars you get for £42.
Show how you checked your answer.

Q11 hint
× 42 (£1 = $1.60) × 42
     £42 = $☐

12 In one tonne of ore there is 6.5 g of gold.
A lorry carries 64 tonnes of ore.
Use partitioning to work out how much gold is in the lorry.
Show how you checked your answer.

You can use an answer to a decimal multiplication to work out the answers to related decimal multiplications.

13 $0.93 \times 25 = 23.25$

Use this fact to work out these multiplications. Check your answers are approximately correct. The first two have been started for you.

a $9.3 \times 25 = \square$

$0.93 \times 25 = 23.25$

$\times 10\downarrow \qquad \times 10\downarrow$

so $9.3 \times 25 = \square$

Check $9.3 \times 25 \approx \square \times 2.5 = \square$

c 93×25 **d** 0.093×25

b $0.93 \times 2.5 = \square$

$0.93 \times 25 = 23.25$

$\div 10\downarrow \qquad \downarrow \div 10$

so $0.93 \times 2.5 = \square$

Check $0.93 \times 2.5 \approx \square \times 2.5 = \square$

e 0.93×0.25

14 Reasoning Larry says, '0.93×25 has the same answer as 9.3×2.5'.
Is Larry correct? Show your working to explain.

15 Problem-solving / Reasoning

a David says, '7.4×0.28 is approximately 21.'
Without working out the answer, explain why David is wrong.

b Donna knows that $8.4 \times 0.17 = 1.428$
She says, 'It's easy to work out $0.84 \times 1.7 = 1.428$'
Explain how Donna can easily work this out.

Challenge   Here is a number pattern.

$0.3 \times 1.1 = 0.33$

$3.3 \times 1.1 = 3.63$

$33.3 \times 1.1 = 36.63$

$333.3 \times 1.1 = 366.63$

a Use the number pattern to write down the answer to 3333.3×1.1.

b Write down the first four lines of a similar number pattern starting 0.3×0.11.

c What is the answer to 3.333×1.1? Explain how you worked out your answer.

Reflect   In this lesson you have learned two methods to help you multiply decimals mentally:

- multiplication facts
- partitioning

For each method, make up your own calculation to show how it works.
Work out the answer to each of your calculations. Show your working.
Explain how you chose the numbers for your calculations.

4.5 Working with decimals

*Active*Learn
Homework

- Add and subtract decimals
- Multiply and divide decimals by single-digit whole numbers
- Divide numbers that give decimal answers

Warm up

1 Fluency What is
a $4 + 9$
b $11 - 6$
c 8×8
d $35 \div 5$

2 Work out
a $67 + 58$
b $245 + 39$
c $392 - 186$
d $624 - 87$

3 Work out
a 48×6
b 8×234
c $65 \div 5$
d $144 \div 3$

4 Work out
a $4.6 + 2.7$
b $5.6 + 2.3$
c $5.6 + 2.4$
d $5.6 + 2.7$
e $5.6 + 2.9 + 1.2$
f $6.6 + 2.9 + 1.4$

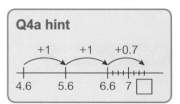
Q4a hint

5 Work out
a $4.5 - 3.2$
b $4.5 - 3.5$
c $4.5 - 3.6$
d $5.5 - 3.6$
e $5.5 - 3.7 - 1.4$
f $5.5 - 3.8 - 1.3$

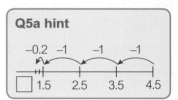
Q5a hint

6 Problem-solving What is the total length of two pipes measuring 1.6 m and 2.5 m?

7 Problem-solving Ahmed lives 6.4 km away from school and Jenna lives 3.7 km away. How much further away does Ahmed live?

Worked example

Use the column method to work out $39.82 - 8.54$.
Use an estimate to check your answer.

```
    3  9 . ⁷8̸ ¹2
 -     8 .  5  4
    3  1 .  2  8
```

Line up the decimal points first. Then line up 10s with 10s, ones with ones, tenths with tenths and so on.

Start with the ones column.

Check: $39.82 \approx 40$ and $8.54 \approx 9$, $40 - 9 = 31$, so 31.28 is close to 31.

8 Use the column method to work out these calculations. Use an estimate to check your answers.
a $44.78 + 19.53$
b $21.9 - 17.4$
c $34.51 - 23.66$
d $98.67 + 32.29$

9 **Problem-solving** Craig is a plumber. He has 7.45 m of tubing. He uses 3.75 m of the tubing. How much tubing does he have left over?

10 Use the column method to work out these calculations.
   Use an estimate to check your answers.
   a 4.52 + 8.6   b 19.8 + 2.45   c 45.82 − 1.5   d 8.9 − 6.76
   e 7.654 + 9.86   f 32.679 − 4.8   g 164.9 − 85   h 45 − 23.41

11 Use a number line or the column method to work out
   a 1 − 0.6   b 1 − 0.3   c 1 − 0.13   d 1 − 0.62

12 **Reasoning** How can you use number bonds to 10 and 100 to subtract a decimal from 1?

Worked example

Work out 5.6 × 4
Use an estimate to check your answer.

$$\begin{array}{r} 5\ 6 \\ \times\qquad 4 \\ \hline 2\ 2\ 4 \\ {}^2 \end{array}$$

56 × 4 = 224 ⟶ *Ignore the decimal point and work out 56 × 4.*

↓ ÷ 10   ↓ ÷ 10

5.6 × 4 = 22.4 ⟶ *56 ÷ 10 = 5.6, so work out 224 ÷ 10 to get the final answer.*

Check: 5.6 ≈ 6, 6 × 4 = 24

22.4 is close to 24

13 Use column multiplication to work out these multiplications.
   Use an estimate to check your answers.
   a 6.1 × 3   b 4.4 × 6   c 8 × 7.4
   d 2.9 × 9   e 7 × 2.8   f 3 × 2.31
   g 8 × 8.12   h 3 × 23.45   i 79.63 × 5

> **Q13c hint** 8 × 7.4 is the same as 7.4 × 8

Worked example

Work out 73.5 ÷ 3

$$\begin{array}{r} 2\ 4\ .\ 5 \\ 3\overline{)7\ {}^13\ .\ {}^15} \end{array}$$

First write the decimal point for the answer above the decimal point in the question. Then divide using short division as usual, starting on the left.

73.5 ÷ 3 = 24.5

14 Use short division to work out

  a $85.5 \div 5$ **b** $98.4 \div 3$ **c** $69.2 \div 4$ **d** $99.2 \div 8$

  e $38.5 \div 7$ **f** $75.24 \div 6$ **g** $601.86 \div 6$ **h** $283.76 \div 8$

15 Problem-solving LED party lights cost £3.84 per metre. What is the cost of 8 m of LED party lights?

> **Q15 hint** Think carefully! Do you need to multiply or divide to work out the answer?

16 Problem-solving A piece of Christmas tinsel 55.5 cm long is cut into 3 equal length pieces. How long is each piece?

17 Problem-solving In a science experiment, Katrina needs to know the mass of a seed pod. She measures the mass of 8 seed pods as 10.4 g. What is the mass of 1 seed pod?

18 Copy and complete the working for $23.8 \div 4$ and $23 \div 4$.

  a $\begin{array}{r} 5.9\square \\ 4\overline{\smash{)}2\,3.^3\!8\,^2\!0} \end{array}$ **b** $\begin{array}{r} 5.\square\square \\ 4\overline{\smash{)}2\,3.^3\!0\,\square} \end{array}$

> **Q18 hint** 23.8 and 23.80 have the same value. They both have no hundredths. 23, 23.0 and 23.00 have the same value. They all have no tenths or hundredths.

19 Work out the answers to these calculations. Round your answers to 1 decimal place.

  a $4\overline{\smash{)}24.6}$ **b** $4\overline{\smash{)}37}$ **c** $8\overline{\smash{)}258}$ **d** $12\overline{\smash{)}141}$

Challenge

1 Paula has these cards:

  | 2 | | 4 | | 8 | | • |

  a Use the cards to make six different numbers that have two decimal places. Use all the cards in each number.

  b Add the smallest decimal and the largest decimal.

  c Subtract the smallest decimal from the largest decimal.

  d Multiply the largest decimal by

    i 2   **ii** 4   **iii** 8

  e Divide the smallest decimal by

    i 2   **ii** 4   **iii** 8

2 Swap the card 8 for 0. Repeat parts **a** to **e**.

Reflect In this lesson, you added, subtracted, multiplied and divided with decimals. Which did you find most difficult? What made it most difficult? Do you need more practice on any kinds of question? If so, which kinds?

4.6 Perimeter

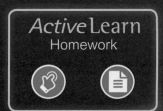

Active Learn
Homework

- Work out the perimeters of composite shapes and polygons
- Solve perimeter problems

Warm up

1 **Fluency** Name each of these **regular polygons** and state how many sides it has.

a   b   c   d   e

2 Work out the perimeter of each of these shapes.
The first one is drawn on centimetre squared paper.

a 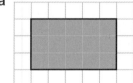   b 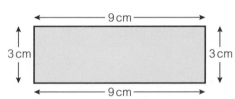   c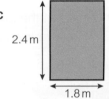

Key point   The **perimeter** is the total distance around the edge of a shape. To work out the perimeter of any shape, add up the lengths of all the sides.

3 These shapes are drawn on centimetre squared paper.
Work out the perimeter of each shape.

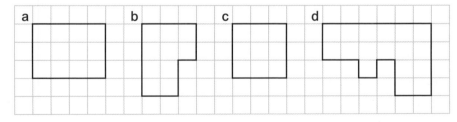

4 Work out the perimeter of this shape in centimetres.

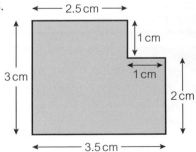

5 Calculate the **perimeter** of each regular polygon using the given side lengths.

Q5a hint perimeter = 3 × 2 cm = ☐ cm

a 2 cm

b 6 cm

c 5 cm

d 4 cm

6 Problem-solving A 50p coin has the shape of a heptagon. Each edge is 1.2 cm long. Find its perimeter.

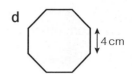

7 Find the missing lengths marked with letters.

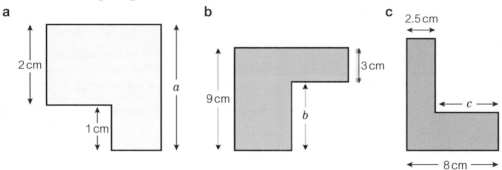

a 2 cm, a, 1 cm

b 9 cm, 3 cm, b

c 2.5 cm, c, 8 cm

8 These shapes are made from rectangles.

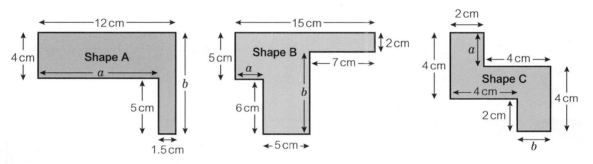

Shape A: 12 cm, 4 cm, a, b, 5 cm, 1.5 cm

Shape B: 15 cm, 5 cm, 2 cm, a, 7 cm, b, 6 cm, 5 cm

Shape C: 2 cm, 4 cm, a, 4 cm, 4 cm, Shape C, 2 cm, 4 cm, b

For each shape
a work out the missing lengths a and b **b** work out the perimeter of the shape

9 The diagram shows the dimensions of a bedroom that is going to be carpeted. Carpet gripper is put around the perimeter of the room before the carpet is laid.
a Work out the missing lengths marked x and y.
b How much carpet gripper is needed for this room?

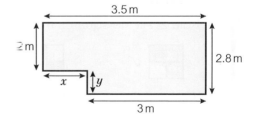

3.5 m, 2 m, 2.8 m, x, y, 3 m

10 Problem-solving / Reasoning A rectangle measuring 6 cm by 5 cm has a smaller rectangle measuring 3 cm by 2 cm cut out of one corner.
a Show that the remaining piece has the same perimeter as the original rectangle.
b Explain why this is.

Q10 hint Draw a diagram of the shape first.

11 Jenny calls one side of this rectangle x.

She knows the other side is 2 cm more than x.

   a Copy Jenny's rectangle and label all the sides.

   b Work out the perimeter, giving your answer as $\square x + \square$ cm.

12 **Problem-solving / Reasoning** Work out the perimeters of these shapes.

Simplify your expressions.

A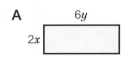

B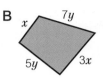

C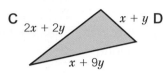

D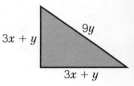

Which is the odd one out? Explain the method you used.

Worked example

A regular hexagon has perimeter 42 cm.
What is the side length of the hexagon?

$42 \div 6 = 7$ cm

All six sides of a regular hexagon are the same length, so divide 42 by 6 to work out the length of each side.

---

13 Each shape in a set of regular polygons has perimeter 24 cm.

Copy and complete the table.

Number of sides	3	4		8		24
Side length (cm)			4		2	

14 **a** A regular hexagon has perimeter 32.4 cm.

     What is the length of each side?

   b A regular octagon has perimeter 73.6 cm.

     What is the length of each side?

Challenge

a On a piece of centimetre squared paper, shade in
a shape made from eight squares.
Here are two examples.
Rule 1: Your shape must be made from whole squares.
Rule 2: All the squares must touch side to side, not corner to corner.

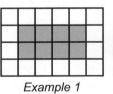

Example 1

Example 2

 ✓   ✗

b Work out the perimeter of your shape in centimetres.

c Try other shapes made from eight squares and work out the perimeter of each one.

d Draw a shape made from eight squares with the smallest possible perimeter.

e Draw a shape made from eight squares with the largest possible perimeter.

Reflect In this lesson, the hint for Q10 suggests that you draw a diagram.

Why is this a good strategy for working out the perimeter of a shape?

What other kinds of diagrams have you used to help you do maths?

4.7 Area

- Find areas of irregular shapes by counting squares
- Calculate the areas of shapes made from rectangles
- Solve problems involving area

Active Learn
Homework

Warm up

1 Fluency What is
 a 8×2 **b** 2^2 **c** 5^2
 d $\sqrt{16}$ **e** $\sqrt{81}$ **f** $\sqrt{64}$

2 Which of these rectangles have area $16\,\text{cm}^2$?
 Which have a perimeter of $16\,\text{cm}$?
 The first one is drawn on centimetre squared paper.

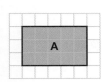

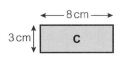
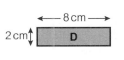
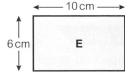

3 Simplify
 a $5 \times y$ **b** $2x \times 4$ **c** $3y \times 3$ **d** $0.5 \times 18x$

Key point

The **area** is the total space covered by a shape. You can find the area of a shape drawn on squared paper by counting the squares inside it.
The **units** used for area are square units, such as mm^2, cm^2, m^2 and km^2.

 has area $1\,\text{cm}^2$.

 is half a square. It has an area of $0.5\,\text{cm}^2$.

Read cm^2 as 'square centimetres'.

4 Find the area of each shape by counting squares.

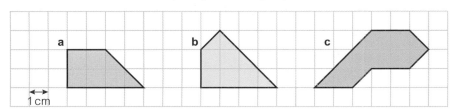

Key point To work out the area of a rectangle or square, use area = length × width.

$A = lw$

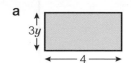

5 Write an expression for the area of each rectangle. Write each answer in its simplest form. All measurements are in centimetres.

a

$3y$ / 4

b

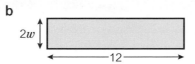

$2w$ / 12

c

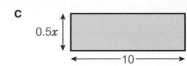

$0.5x$ / 10

6 Work out the area of each square.

a 1 cm / 1 cm

b 2 cm / 2 cm

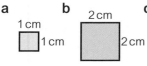

c 3 cm / 3 cm

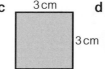

d 4 cm / 4 cm

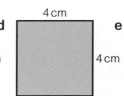

e 5 cm / 5 cm

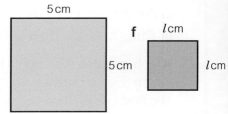

f l cm / l cm

7 Reasoning What type of numbers are your answers to Q6?

Key point To work out the area of a shape made from rectangles, it helps to split the shape into smaller rectangles.

8 Copy and complete the workings to find the area of this shape.

area A = 5 × 3 = 15 cm²
area B = 8 × 7 = ☐ cm²
total area = area A + area B
= ☐ + ☐
= ☐ cm²

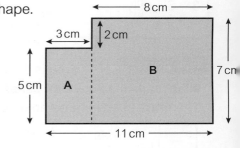

9 These shapes are made from rectangles.

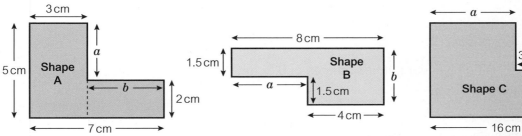

For each shape
a work out the missing length a
b work out the missing length b
c work out the area of the shape.

Q9b and c hint Split the shape into smaller rectangles. Find the area of each rectangle.

10 The diagram shows the dimensions of a café that is going to have new flooring.

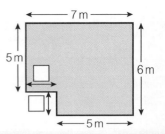

   a Work out each missing length marked with a □.

   b Work out the area of flooring required.

   c **Reasoning** How many different ways are there to work out the area of the floor?

> **Key point** When finding the area of a shape, all dimensions must be in the same units of measure.

11 Work out the area of this shape

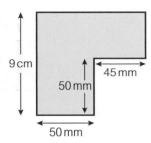

> **Q11 hint** Change the units of measure so that they are all cm. Find the area. Then change them so they are all mm. Find the area.

   a in cm^2 b in mm^2

12 **Problem-solving** A square has area 81 cm^2.

   a What is its side length?

   b What is its perimeter?

13 **Problem-solving** A rectangle has a side length of 12 cm and a perimeter of 25 cm. What is its area?

14 **Problem-solving** A square has area 64 cm^2. What is its perimeter?

15 **Problem-solving** Gold leaf can be bought in 8 cm square sheets. There are 25 sheets in one booklet.

   a Work out the total area that can be covered using one booklet of gold leaf.

   b Aiden needs to cover an area of 5000 cm^2 with gold leaf. How many booklets of gold leaf does Aiden need to buy?

> **Q15b hint** He can only buy whole numbers of booklets.

> **Challenge** On squared paper, draw a square, a rectangle, an L-shape, and one more different shape, each with the same area.

> **Reflect** Here is a list of some jobs where you have to calculate and use area: estate agent, gardener, festival organiser.
> Copy the list and write one more job where calculating and using area is important.
> Describe how you would use area in each job.
> What career are you interested in? Do you think you will need to use area in your job? How?

4.8 More units of measure

- Choose suitable units to measure length and area
- Use units of measure to solve problems
- Use metric and imperial units

Active Learn
Homework

Warm up

1 Fluency

× 10	÷ 10	× 100	÷ 100	× 1000	÷ 1000

Which of these calculations do you use when converting

a m to cm   b cm to mm   c km to m   d g to kg   e litres to ml

2 Work out
a 6 × 30   b 6 × 10 000   c 6.1 × 1000   d 6.1 × 100
e 0.61 × 1000   f 1.6 × 2   g 1.6 × 4

3 Write < or > between each pair of measurements.
a 230 mm … 9.5 cm   b 0.05 kg … 400 g   c 250 ml … 1.5 litres

---

Key point   It is important to be able to choose the most suitable **units** for measuring length, capacity and area. The metric units you already know are
Length: mm, cm, m, km   Mass: g, kg
Capacity: ml, litres   Area: mm², cm², m², km²

---

4 Copy and complete these sentences with the most suitable metric units.
e.g. I would use metres to measure the length of a football field.
a I would use _____ to measure the length of a beetle.
b I would use _____ to measure the capacity of a car fuel tank.
c I would use _____ to measure the mass of a wedding ring.

5 Which unit of area would be sensible for measuring these?
a the area of Wales   b the area of a mobile phone screen
c the area of the deck of a container ship

Key point   Some more metric units that you need to know are
Mass: 1 tonne (t) = 1000 kg   Area: 1 hectare (ha) = 10 000 m²
Capacity: 1 millilitre (ml) = 1 cm³   Volume: 1 litre (l) = 1000 ml

---

6 Copy and complete.
a 3 t = 3 × ☐ = ☐ kg
b 9000 kg = 9000 ÷ ☐ = ☐ t
c 4.6 t = 4.6 × ☐ = ☐ kg
d 5 ha = 5 × ☐ = ☐ m²
e 120 000 m² = 120 000 ÷ ☐ = ☐ ha
f 2 litres = ☐ ml = ☐ cm³
g 75 cm³ = ☐ ml
h 3500 cm³ = ☐ ml = ☐ litres

> **Q6 hint**
> × 1000
> 1 t = 1000 kg
> 1 t = 1000 kg
> ÷ 1000

7 **Reasoning** Do you multiply or divide when converting from
   a a bigger unit of measure to a smaller unit of measure?
   b a smaller unit of measure to a bigger unit of measure?

8 A lorry can take a load of 2300 kg. How many tonnes is this?

9 For each pair of measurements, which is smaller?
   a 8.6 tonnes or 6800 kg   b 86 litres or 8600 ml

10 **Problem-solving** A farm has 5 hectares of land for grazing animals.
   It also has 5000 m² of woodland.
   Which is greater: the land for grazing animals or the woodland?

11 A construction company is building a car park.
   The plot of land is a rectangle 300 m long by 150 m wide.
   a Work out the area of the land   i in m²   ii in hectares
   b **Problem-solving** A standard car park space for one car is 2.4 m by 5 m.
     How many spaces fit in this car park?
   c **Reasoning** Is this a good model? What other space is needed in a car park?

12 **Problem-solving** Automotive engineers calculate the kerb weight of a car by adding the
   dry weight of the car to the mass of oil, water and petrol a car needs. They also add the
   mass of a 75 kg driver.
   Here is some information about a particular model of car.
      dry weight = 915 kg   oil = 5 kg   water = 5 kg   petrol = 40 kg
   Calculate the kerb weight of this car in tonnes.

> **Key point**   You need to know these conversions between **metric** and **imperial units**.
> 1 foot (ft) ≈ 30 cm   1 mile ≈ 1.6 km

13 **Reasoning** Will you walk further if you walk 1 mile or 1 kilometre? Explain.

14 Copy and complete.
   a 6 ft = 6 × ☐ = ☐ cm   b 150 cm = 150 ÷ ☐ = ☐ ft
   c 4 miles = 4 × ☐ = ☐ km   d 160 km = 160 ÷ ☐ = ☐ miles

15 **Problem-solving** This is a formula to estimate how far away a thunderstorm is.
      Distance (in km) = time between lightning and thunder (in s) ÷ 3
   After a flash of lightning, Mary counts 24 seconds before the thunder.
   How far away is the thunderstorm in miles?

> **Challenge**   Write down something you could measure with each unit of measure.
> centimetres   litres   feet   millilitres   kilograms   grams

> **Reflect**   Write two headings, 'metric' and 'imperial'.
> Write down as many different measures as you can under each heading. Do not use this
> book to help you.
> When you were trying to remember all the measures, what were you thinking or imagining?
> Discuss your way of remembering with other students.

4 Check up

Scales and measures

1 On a scale drawing, 1 cm represents 1 m. Use a ruler to draw a line to represent 5 m.

2 Write the measure shown on each scale.

a

b

3 Copy and complete.
   a 8 cm = ☐ mm   **b** 2 kg = ☐ g   **c** 600 cm = ☐ m

4 Copy and complete.
   a 3.2 m = ☐ cm   **b** 8.7 litres = ☐ ml   **c** 2400 m = ☐ km

5 1 tonne = 1000 kg
   Which is heavier: 1.5 tonnes or 15 000 kg?
   Show your working.

Decimals

6 Which number is bigger, 23.45 or 23.8?

7 Write these numbers in order of size, starting with the smallest.
   8.9, 8.47, 8.95, 8.165, 8.3, 8.35

8 Round each number to 1 decimal place.
   a 4.63   **b** 13.96

9 Work out these calculations. Use an estimate to check your answers.
   a 12.6 + 3.9   **b** 32.8 − 17.5

10 A carpenter has a 3.5 m length of wood.
    He cuts a piece that is 0.85 m long.
    What length of wood does he have left?

11 Work out
    a 2 × 0.3   **b** 0.08 × 4   **c** 0.3 × 0.6

12 Work out these calculations. Use an estimate to check your answers.
    a 2.6 × 8   **b** 39.6 ÷ 3

13 A potter uses 2.25 kg of clay per bowl.
    How much clay does she use for 5 bowls?

14 0.27 × 12 = 3.24
    Use this fact to work out these multiplications. Check your answers using an approximate calculation.
    a 2.7 × 12   **b** 0.27 × 120   **c** 0.027 × 12

Perimeter and area

15 Calculate the perimeter of this regular hexagon.

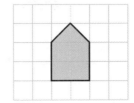
7 cm

16 This shape is drawn on centimetre squared paper. Find its area.

17 A square has area 9 m². What is its perimeter?

18 The diagram shows the dimensions of an office.
   a Work out the perimeter.
   b Work out the floor area.

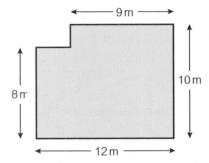
9 m
10 m
8 m
12 m

Challenge

1 Craig has these cards.

| 3 | 2 | 5 | 0 | . |

   a Make five different decimal numbers.
   Use all the cards in each number.
   You must place the decimal point between two of the digits.
   b Write your five numbers in order of size, starting with the smallest.
   c What are the largest and the smallest numbers you can make using all these cards?

2 This rectangle has an area of 24 cm².
   The length and width of the rectangle are
   each a whole number of centimetres.

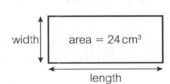
width
area = 24 cm³
length

   a Write three possible lengths and widths
   for the rectangle.
   b Work out the perimeter of each rectangle you wrote in part **a**.

Reflect

How sure are you of your answers? Were you mostly

🙁 **Just guessing**   😐 **Feeling doubtful**   🙂 **Confident**

What next? Use your results to decide whether to strengthen or extend your learning.

4 Strengthen

Scales and measures

1 **a** Copy this scale onto centimetre squared paper.

1 m

b Draw lines on your squared paper to represent
   i 2 m   **ii** 3 m   **iii** 5 m

Q1b hint

1 m 1 m

2 Write the number shown by each arrow.

a

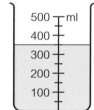

300   400

b

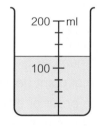

500   600

Q2 hint Try counting up in steps of 1s, 2s, 5s, 10s etc. What size steps take you from 300 to 400? In Q2b, what size steps take you from 500 to 600?

3 Write this measurement. Write the number and the units.

```
|||||||||||||||||||||||||||||||||||||||||||||||||||||||||||||||||||||||||
              80                           85
  cm
```

4 Write down the amount of water.

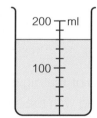

5 Copy and complete the place value tables.

× 10, 100, 1000

Th	H	T	O	.	$\frac{1}{10}$	$\frac{1}{100}$	$\frac{1}{1000}$
			7	.	9		
		7	☐				
	7	☐	0				
7	☐	0	0				

a **i** 7.9 × 10 =
   ii 7.9 × 100 =
   iii 7.9 × 1000 =

× 10, 100, 1000

Th	H	T	O	.	$\frac{1}{10}$	$\frac{1}{100}$	$\frac{1}{100}$
			2	.	1	8	

b **i** 2.18 × 10 =
   ii 2.18 × 100 =
   iii 2.18 × 1000 =

6 Copy and complete the place value tables.

÷ 10, 100, 1000

Th	H	T	O	.	$\frac{1}{10}$	$\frac{1}{100}$	$\frac{1}{1000}$
		8	2				
			8	.	☐		
			0	.	☐	☐	
			0	.	0	☐	☐

a **i** 82 ÷ 10 =
   ii 82 ÷ 100 =
   iii 82 ÷ 1000 =

÷ 10, 100, 1000

Th	H	T	O	.	$\frac{1}{10}$	$\frac{1}{100}$	$\frac{1}{1000}$
	3	2	9				

b **i** 329 ÷ 10 =
   ii 329 ÷ 100 =
   iii 329 ÷ 1000 =

7   Use the 'measure machines' to change millimetres to centimetres and centimetres
    to millimetres.

a smaller   larger   **b** smaller   larger
   measure (mm)   measure (cm)   measure (mm)   measure (cm)

73 mm →   → ☐ cm   ☐ mm ←   ← 4 cm
462 mm →   ÷10   → ☐ cm   ☐ mm ←   ×10   ← 3.2 cm
59.1 mm →   → ☐ cm   ☐ mm ←   ← 5.71 cm

8   Use the 'measure machines' to change centimetres to metres and metres to centimetres.

a smaller   larger   **b** smaller   larger
   measure (cm)   measure (m)   measure (cm)   measure (m)

800 cm →   → ☐ m   ☐ cm ←   ← 5 m
350 cm →   ÷100   → ☐ m   ☐ cm ←   ×100   ← 4.5 m
17 cm →   → ☐ m   ☐ cm ←   ← 6.2 m

9   Use the 'measure machines' to change grams to kilograms and kilograms to grams.

a smaller   larger   **b** smaller   larger
   measure (g)   measure (kg)   measure (g)   measure (kg)

6000 g →   → ☐ kg   ☐ g ←   ← 4 kg
826 g →   ÷1000   → ☐ kg   ☐ g ←   ×1000   ← 6.3 kg
91 g →   → ☐ kg   ☐ g ←   ← 7.62 kg

10 To convert kg into tonnes, divide by 1000.
    To convert tonnes into kg, multiply by 1000.
    Copy and complete.
    a 6 kg = ☐ tonnes   **b** 9.2 kg = ☐ tonnes
    c 3.85 kg = ☐ tonnes   **d** 4000 tonnes = ☐ kg
    e 372 tonnes = ☐ kg   **f** 73 tonnes = ☐ kg

Decimals

1   Write the decimal shown by each diagram.
    The first one is done for you.

a 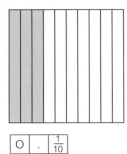   **b**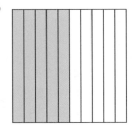

O	.	$\frac{1}{10}$
0	.	3

0.3

2 Write the decimal shown by each diagram.
The first two are done for you.

a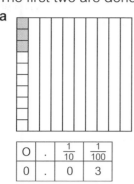

O	.	$\frac{1}{10}$	$\frac{1}{100}$
0	.	0	3

0.03

b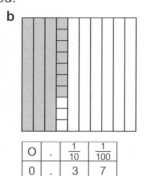

O	.	$\frac{1}{10}$	$\frac{1}{100}$
0	.	3	7

0.37

c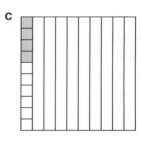

d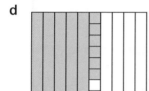

e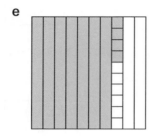

f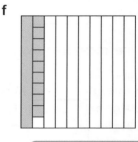

3 Which is larger?
a 0.37 or 0.5   b 0.3 or 0.19
c 0.28 or 0.7   d 4.6 or 4.41   e 4.6 or 4.412

> **Q3a and b hint** Use the diagrams in Q2 to help you.

4 Which of these numbers have only one digit after the decimal point, and so are written to 1 decimal place?
4.6, 3.24, 0.559, 12.8, 5.25, 0.8, 2.44, 156.0

5 **a** Copy and complete this number line.

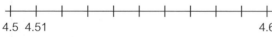
4.5 4.51   4.6

b Round 4.52 to 1 decimal place.
c Round 4.57 to 1 decimal place.
d Round 4.55 to 1 decimal place.

> **Q5 hint**
>
> round   round
> down   up
>
> 4.5   4.55   4.6

6 Round each number to 1 decimal place.
Begin by drawing a number line.
a 6.56   b 6.51   c 6.59
d 6.16   e 6.11   f 6.15
g 14.32   h 141.35   i 14.38

7 Work out these additions.
Part **a** has been started for you.
a 38.6 + 24.6
b 18.58 + 4.67
c 26.14 + 3.8

T	O	$\frac{1}{10}$
	3 8	. 6
+	2 4	. 6
		3 . 2
	1	1

> **Q7 hint** Write one number under the other in a place value table. Line up the decimal points and fill empty spaces with zeros.

8 **Problem-solving** Cosmo buys a tennis racket for £28.65 and some tennis balls for £4.85.
What is the total amount he spends?

9 Work out these subtractions. Part **b** has been started for you.

Q9c hint Write one number under the other, filling empty spaces with zeros.

a 45.9 − 32.7   **b** 8.71 − 6.38   **c** 39.7 − 21.54

$$
\begin{array}{ccccc}
 & & \bigcirc & \frac{1}{10} & \frac{1}{100} \\
 & 8 & . & {}^{6}\!\!\!/\!7 & {}^{1}1 \\
- & 6 & . & 3 & 8 \\
\hline
 & & . & & 3
\end{array}
$$

10 Problem-solving Maya cuts 0.55 m of ribbon from a piece that is 1.8 m long. How much of the ribbon does she have left?

11 Copy and complete these number patterns.

a 2 × 30 = 60   2 × 3 = 6   2 × 0.3 = 0.6   2 × 0.03 = ☐
b 5 × 80 = 400   5 × 8 = ☐   5 × 0.8 = ☐   5 × 0.08 = ☐

12 Use number patterns to work out

a 4 × 0.2   **b** 9 × 0.4   **c** 0.02 × 3   **d** 0.08 × 4

13 Work out these multiplications. Use an estimate to check your answers.

a 4.6 × 8   **b** 3.2 × 6
c 7 × 2.25   **d** 4 × 3.24

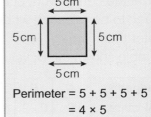

Q13a hint

	40	6
8	☐	☐

☐ + ☐ = ☐

40 × 8 = ☐, 4.6 × 8 = ☐

14 Work out these divisions. Parts **a** and **b** have been started for you.

a 62.8 ÷ 2
$$\begin{array}{r} 3\,☐.☐ \\ 2\overline{)6\,2.8} \end{array}$$

b 37.2 ÷ 3
$$\begin{array}{r} 1\,2.☐ \\ 3\overline{)3\,7.{}^{1}2} \end{array}$$

c 31.2 ÷ 6   **d** 30.1 ÷ 7

Perimeter and area

1 Calculate the perimeter of each regular polygon.

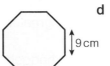

a 5 cm   **b**   **c** 9 cm   **d** 11 cm, 7 cm

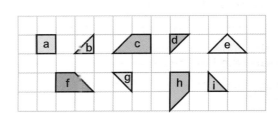

Q1a hint In a regular polygon all the sides are the same length. Sketch the polygon with all sides labelled:

5 cm, 5 cm, 5 cm, 5 cm

Perimeter = 5 + 5 + 5 + 5
= 4 × 5
= ☐ cm

2 These shapes are drawn on centimetre squared paper.
Match each shape to its area.

$\frac{1}{2}$ cm²   1 cm² $1\frac{1}{2}$ cm²

3 Here are two rectangles drawn on centimetre squared grids.

a

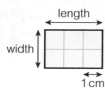

b

For each rectangle, copy and complete.

i The length of the rectangle = ☐ cm

ii The width of the rectangle = ☐ cm

iii The perimeter of the rectangle = all sides added together
$$= ☐ + ☐ + ☐ + ☐ = ☐ \text{ cm}$$

iv The area of the rectangle = length × width
$$= ☐ × ☐ = ☐ \text{ cm}^2$$

4 Work out the perimeter and area of each shape.

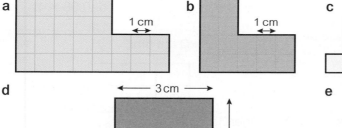

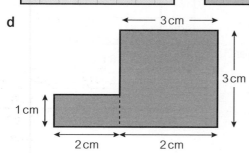

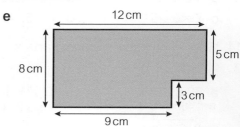

5 A square has area 4 cm². Sketch the square on centimetre squared paper.
  a What is its side length?   **b** What is its perimeter?

Challenge   When a hedgehog arrives at a rescue centre it is weighed and measured. Staff at the centre work out a Body Index (BI) value for the hedgehog. A hedgehog is well enough to be released when it has a BI value that is greater than 0.8 and its weight is greater than 0.65 kg.

The table shows the BI values and weights of some hedgehogs at the rescue centre.

Hedgehog	A	B	C	D	E	F	G	H	I
BI value	0.9	0.82	0.88	0.79	0.95	0.7	0.85	0.76	0.92
Weight (kg)	0.68	0.7	0.78	0.6	0.8	0.58	0.62	0.71	0.85

a Which hedgehogs are well enough to be released?

b Which hedgehog only needs to increase its weight to be released?

c Which hedgehog only needs to increase its BI value to be released?

Reflect   These lessons used place value tables to help solve problems with decimals and measures. Did the place value tables help you? Explain why or why not.

Will you use place value tables to help you solve mathematics problems in future?

4 Extend

1 Copy and complete these number patterns.

a $2080 \div 32 = 65$ b $714 \div 84 = 8.5$
   $208 \div 32 = 6.5$ $714 \div 8.4 = 85$
   $20.8 \div 32 = 0.65$ $714 \div 0.84 = 850$
   $2.08 \div 32 = \square$ $714 \div 0.084 = \square$
   $0.208 \div 32 = \square$ $714 \div 0.0084 = \square$

2 **Reasoning** $150 \div 2.5 = 60$
   Use this fact to work out these calculations.
   Check your answers using an approximate calculation.

   a $15 \div 2.5$ b $1.5 \div 2.5$ c $1500 \div 2.5$
   d $150 \div 0.25$ e $150 \div 0.025$ f $150 \div 25$

> **Q2 hint** Follow similar number patterns to those in Q1 to help you.

> **Q2a hint** Check by estimation: $15 \div 3 = \square$

3 Fynn compares the costs of holiday insurance from two companies.

   a Work out the cost of insurance per day with Diamond. Give your answer to the nearest penny.

   b **Reasoning**   Which company offers the cheaper rate per day?

Company	Price
Diamond	£11.80 for 7 days
Bell	£1.72 per day

4 A boa constrictor is 45 cm long at birth.
   This table shows its increase in length over its first 4 years.

   a Work out the mean increase in length over the first 4 years.

   b Use your mean increase in length to estimate the length of a boa constrictor at the end of the 5th year. Give your answer in metres.

   c **Reasoning**   Is the model used in part b a good model? Explain your answer.

End of year	Increase in length
1	70 cm
2	60 cm
3	33 cm
4	13 cm

5 **Problem-solving**   Faisal is looking for wood to make a shelter. He finds a 3 m tree branch lying on the ground.
   He cuts a 1.1 m piece and a 0.76 m piece from the branch.
   Each of his two saw cuts uses 0.002 m.
   What length is left?

6 **Problem-solving**   The diagram shows four regular polygons.
   The sum of the perimeters of the triangle and hexagon is equal to the sum of the perimeters of the pentagon and square.
   Work out the side length of the square.

5.3 cm   3.6 cm   4.7 cm

7 **Problem-solving**   a A regular volleyball court has a perimeter of 54 m. The length of the court is 18 m. What is the area of the court?

   b A rugby pitch has an area of 9800 m². The width of the pitch is 70 m. What is its perimeter?

> **Q7a hint** Draw a diagram to help you.

8 **Problem-solving**   A square board game has an area of 2500 cm².
   What is the perimeter of the board?

9 For this shape, work out
   a the perimeter   **b** the area

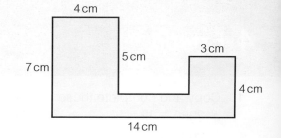

10 Problem-solving The Great Pyramid of Giza has a square base.
The perimeter of the base is 921.6 m.
What is the area of the base?
Give your answer to the nearest square metre.

11 A 2 km long stretch of road needs to be resurfaced. The width of the road is 8.3 m.
   a Model this stretch of road as a rectangle. Work out the area of road that needs to be resurfaced in m².

Resurfacing this road costs £30 per square metre.

   b What is the total cost to resurface this road?

12 The table shows the heights and lengths of roller coasters around the world.

Roller coaster	Country	Height (feet)	Length (feet)
Kingda Ka	USA	456	3118
Leviathan	Canada	306	5486
Steel Dragon 2000	Japan	318	8133
The Ultimate	UK	107	7442
Tower of Terror II	Australia	377	1235

Q12 hint
1 foot ≈ 30 cm
1 mile ≈ 1.6 km

   a Work out the height and length of Kingda Ka in metres.

   b Which roller coasters are more than 2 km in length?
      Give the names of these roller coasters and their length in km, correct to one decimal place.

   c What is the range in heights of the five roller coasters, in metres?

13 A hydro engineer collected data on the smallest and largest amounts of water flowing in a river from 2007 to 2011.
   a Work out the mean of the smallest amount of water flowing in the river.

1 tonne of water is the same as 1000 litres.

   b Convert your answer from part **a** to litres.

   c Work out the range of the largest amount of water flowing in the river in litres.

	Tonnes of water every secor	
Year	**Smallest**	**Largest**
2007	7	387
2008	8	893
2009	6	420
2010	5	462
2011	7	614

Challenge

1 Draw a 1 cm by 1 cm square. Copy and complete.
   area of 1 cm by 1 cm square = ☐ cm²
   area of 10 mm by 10 mm square = ☐ mm²
   1 cm² = ☐ mm²

2 Use the fact you found in Q1 to copy and complete these conversions.
   a 5 cm² = ☐ mm²   **b** 9 cm² = ☐ mm²   **c** 4.2 cm² = ☐ mm²
   d 300 mm² = ☐ cm² **e** 800 mm² = ☐ cm²   **f** 360 mm² = ☐ cm²

Reflect   Write a definition of a decimal, in your own words. Be as accurate as you can.

4 Unit test

1   Calculate the perimeter of this regular hexagon.

3 cm

2   Write > or < between each pair of numbers.
    a 12.39 ... 12.55   **b** 8.6 ... 8.07   **c** 29.8 ... 29.37

3   Write these numbers in order of size, starting with the smallest.
    3.8   3.85   3.35   3.09   3.3   3.41

4   Work out
    a 2 × 0.4   **b** 0.8 × 0.3   **c** 0.09 × 4

5   A recipe for ice cream uses 0.2 litres of cream per person.
    How much cream is needed to make ice cream for 12 people?

6   Write the correct units from the box for each of the statements.

    | cm   mm² | kg | m³ |

    a area of a square = 16☐   **b** perimeter of a rectangle = 18☐

7   Write the value that the arrows on each scale point to.

    a   **b**

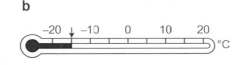

8   Round 55.25 to 1 decimal place.

9   Olga says, 'If I round 27.02 to 1 decimal place, I get the answer 27.'
    Is she correct? Explain your answer.

10 Copy and complete.
    a 800 cm = ☐ m   **b** 2 litres = ☐ ml   **c** 9000 g = ☐ kg

11 Copy and complete.
    a 550 cm = ☐ m   **b** 400 g = ☐ kg   **c** 4.5 km = ☐ m

12 The calculator displays an answer in metres.
    Write the answer in metres and centimetres.

    `532.7`

13 Dave has a mass of 95.45 kg. Caz has a mass of 62.8 kg.
    a What is their total mass?
    Dave is 1.8 m tall and Caz is 1.55 m tall.
    b What is the difference in their heights?

14 Work out these multiplications.
Use an estimate to check your answers.

a 8.7 × 4   **b** 5 × 4.23   **c** 21 × 5.1

15 A baker uses 4.32 kg of dough per tray of rolls.
How much dough does he use for 4 trays of rolls?

16 A square has perimeter 20 cm. What is the area of the square?

17 For this shape, work out
  a the perimeter
  b the area

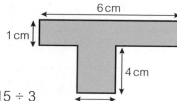

18 Work out these divisions.
  a 39.6 ÷ 4   **b** 84.15 ÷ 3

19 18 × 4.7 = 84.6

Use this fact to work out these multiplications. Check your answers using an approximate calculation.

a 18 × 0.47   **b** 180 × 4.7   **c** 0.18 × 4.7

20 1 mile ≈ 1.6 km

Larry says, '8 miles is approximately 5 km.'
Is Larry correct? Show working to explain.

21 a What is the area of the square
  i in square millimetres
  ii in square metres?
The square and rectangle have the same perimeter.
  b Work out the length of the side of the rectangle marked ☐.

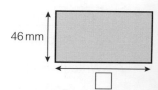
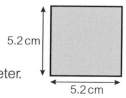

Challenge

1 Work out the missing digits.

a
```
  2 3 . ☐ 2
+ 1 ☐ . 9 ☐
─────────────
  ☐ 2 . 6 8
```

b
```
  4 ☐ . 8 1
− ☐ 5 . ☐ 3
─────────────
  1 7 . 0 ☐
```

2 A rectangle has perimeter 22 cm and area 24 cm².
Work out the side lengths of the rectangle.

Reflect   Put these topics in order, from easiest to hardest.
(You could just write the letters.)

  A Writing decimals in order   **B** Converting measures
  C Rounding decimals   **D** Multiplying and dividing decimals
  E Adding and subtracting decimals   **F** Perimeter
  G Area

Think about the two topics you said were hardest.
What made them hard?
Write at least one hint to help you for each topic.

5 Fractions and percentages

Master　　　Check up p140　　　Strengthen p142　　　Extend p147　　　Unit test p149

5.1 Comparing fractions

- Use fraction notation to describe parts of a shape
- Compare simple fractions
- Use a diagram to compare two or more simple fractions
- Order fractions

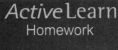

Active Learn
Homework

Warm up

1 Fluency Which of these shapes are $\frac{1}{2}$ shaded?

A　　　　　　　B　　　　　　　C　　　　　　　D

2 a Match each **fraction** to the correct shaded bar.

$\frac{2}{5}$　A　

$\frac{4}{5}$　B　

$\frac{3}{5}$　C　

b Now write the fractions $\frac{2}{5}$, $\frac{4}{5}$ and $\frac{3}{5}$ in order, smallest first.

3 Which bar shows

a $\frac{1}{2}$　A　

b $\frac{1}{4}$　B

c $\frac{6}{10}$　C　

d $\frac{2}{3}$　D

Key point
A **fraction** is part of a whole.
The number above the line in a fraction is the **numerator**.
The number below the line is the **denominator**.

$\dfrac{1}{2}$ ← numerator
← denominator

4 What fraction of each shape is shaded?

a　　　**b**　　　**c**　　　**d**

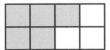

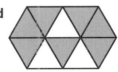

5 What fraction of each shape is
   i shaded   **ii** unshaded?

a   **b**    **c**    **d**

6 Make three copies of this rectangle on squared paper.

Using a new rectangle for each part, shade

a $\frac{1}{6}$ **b** $\frac{1}{2}$ **c** $\frac{1}{3}$

7 **Problem-solving** What fraction of this shape is unshaded?

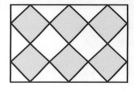

8 Write each set of fractions in **ascending** (smallest to largest) order.

a $\frac{3}{7}$ $\frac{1}{7}$ $\frac{6}{7}$ **b** $\frac{4}{9}$ $\frac{7}{9}$ $\frac{2}{9}$ **c** $\frac{1}{6}$ $\frac{4}{6}$ $\frac{3}{6}$

d $\frac{4}{11}$ $\frac{5}{11}$ $\frac{10}{11}$ $\frac{3}{11}$ **e** $\frac{7}{15}$ $\frac{3}{15}$ $\frac{6}{15}$ $\frac{2}{15}$

9 Write the correct sign, > or <, between each pair of fractions.

a $\frac{3}{4}$ ... $\frac{1}{4}$ **b** $\frac{4}{5}$ ... $\frac{2}{5}$ **c** $\frac{3}{8}$ ... $\frac{5}{8}$ **d** $\frac{4}{9}$ ... $\frac{7}{9}$

> **Key point**   A **unit fraction** has numerator 1. For example, $\frac{1}{2}$, $\frac{1}{3}$ and $\frac{1}{4}$ are unit fractions.

Worked example

Write the correct sign, > or <, between these fractions: $\frac{1}{2}$ ... $\frac{1}{3}$ | Draw two bars the same length. |

| Shade $\frac{1}{2}$ of the first bar. |

| Shade $\frac{1}{3}$ of the second bar. |

$\frac{1}{2} > \frac{1}{3}$

| $\frac{1}{2}$ is greater than $\frac{1}{3}$, so write the 'greater than' symbol, > , between them. |

---

10 a Copy this bar. Shade $\frac{1}{10}$

b Copy this bar. Shade $\frac{1}{8}$

c Write the correct sign, > or <, between this pair of fractions.

   $\frac{1}{8}$ ... $\frac{1}{10}$

11 Write the correct sign, > or <, between each pair of fractions.

a $\frac{1}{2}$ ... $\frac{1}{4}$ **b** $\frac{1}{5}$ ... $\frac{1}{3}$ **c** $\frac{1}{8}$ ... $\frac{1}{6}$ **d** $\frac{1}{7}$ ... $\frac{1}{10}$

12 Reasoning Two shops sell the same pair of flip-flops at the same original price.
Both shops have a sale.
Which pair of flip-flops is cheaper? Explain your answer.

$\frac{1}{3}$ OFF!

BEACH BUOYS

$\frac{1}{2}$ OFF!

Pebble shoes

13 Use the fraction wall to work out which
fraction is larger in each pair.

a $\frac{2}{3}$ or $\frac{3}{4}$ **b** $\frac{3}{7}$ or $\frac{2}{5}$

c $\frac{1}{4}$ or $\frac{2}{7}$ **d** $\frac{4}{5}$ or $\frac{7}{8}$

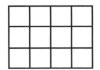

14 Write these fractions in **descending**
(largest to smallest) order.

$\frac{5}{7}$ $\frac{3}{5}$ $\frac{2}{3}$

15 Problem-solving Use the grid to decide which is larger $\frac{3}{4}$ or $\frac{7}{12}$

Challenge

a $\frac{1}{3}$ of this shape is shaded.
How many different ways can you find to shade $\frac{1}{3}$ of the shape?

b Make a copy of this shape and shade in $\frac{1}{2}$
How many different ways can you find to shade $\frac{1}{2}$ of the shape?

Reflect   This may be the first time you have studied fractions since primary school.
Choose A, B or C to complete each statement.

In this lesson, I did …   A well   B OK   C not very well
So far, I think fractions are …   A easy   B OK   C difficult
When I think about the next lesson, I feel …   A confident   B OK   C unsure

If you answered mostly As and Bs, did your experience surprise you? Why?
If you answered mostly Cs, look back at the questions you found most tricky.
Ask a friend or your teacher to explain them to you.
Then complete the statements above again.

5.2 Simplifying fractions

- Change an improper fraction to a mixed number
- Identify equivalent fractions
- Simplify fractions by dividing numerator and denominator by common factors

*Active*Learn
Homework

Warm up

1 Fluency
  a How many thirds are there in one whole?
  b How many quarters are there in one whole?
  c How many fifths are there in one whole?

2 a How many thirds in 2 wholes?
  b How many halves in 3 wholes?

3 Work out the highest common factor (HCF) of
  a 6 and 8   **b** 5 and 15   **c** 12 and 20

Key point
Equivalent fractions have the same value.

4 Reasoning For each pair of diagrams, are the shaded fractions equivalent?

  a   **b**

  c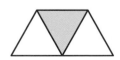

5 Simplify

  a $\frac{3}{3}$ **b** $\frac{4}{4}$ **c** $\frac{6}{6}$ **d** $\frac{10}{10}$

Key point
An **improper fraction** has a numerator that is bigger than its denominator, for example $\frac{3}{2}$
A **mixed number** has a whole number part and a fraction part, for example $1\frac{1}{2}$

Worked example

Write the improper fraction $\frac{4}{3}$ as a mixed number.

$$\frac{4}{3} = 1\frac{1}{3}$$

3 thirds make 1 whole.
There is 1 third left over.

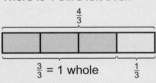

$\frac{3}{3}$ = 1 whole $\frac{1}{3}$

6 Convert these improper fractions to mixed numbers.

a $\frac{5}{3}$ b $\frac{8}{3}$ c $\frac{12}{5}$ d $\frac{7}{5}$ e $\frac{11}{6}$

f $\frac{13}{6}$ g $\frac{7}{4}$ h $\frac{11}{4}$ i $\frac{15}{8}$ j $\frac{21}{8}$

7 **Problem-solving** Rectangular chocolate bars are divided into 10 pieces.

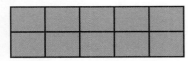

Ossian eats 17 pieces of chocolate.

Write the number of bars of chocolate he has eaten, as a mixed number.

8 A chef calculates that he needs $\frac{17}{4}$ litres of vegetable stock.

How much is this as a mixed number?

Key point You can find equivalent fractions by multiplying or dividing the numerator and denominator by the same number.

Worked example

Complete the equivalent fraction $\frac{2}{3} = \frac{8}{\square}$

$$\frac{2}{3} \overset{\times 4}{\underset{\times 4}{=}} \frac{8}{12}$$

2 has been multiplied by 4 to give 8.
Multiply 3 by 4 to give 12.

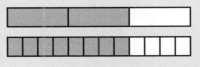

---

9 Copy and complete these equivalent fractions.

a $\frac{5}{20} \overset{\times 2}{\underset{\times 2}{=}} \frac{10}{\square}$ b $\frac{1}{5} \overset{\times 5}{\underset{\times 5}{=}} \frac{\square}{25}$ c $\frac{5}{7} = \frac{15}{\square}$ d $\frac{2}{5} = \frac{\square}{40}$

e $\frac{8}{10} \overset{\div 2}{\underset{\div 2}{=}} \frac{4}{\square}$ f $\frac{4}{12} \overset{\div 4}{\underset{\div 4}{=}} \frac{\square}{3}$ g $\frac{30}{36} = \frac{\square}{6}$ h $\frac{12}{21} = \frac{4}{\square}$

Key point You can **simplify** fractions by dividing the numerator and denominator by a common factor. To write a fraction in its **simplest form**, divide the numerator and denominator by their highest common factor (HCF).

---

10 Simplify these fractions. Copy and complete.

a $\frac{10}{20} \overset{\div 10}{\underset{\div \square}{=}} \frac{\square}{\square}$ b $\frac{50}{60} = \frac{\square}{\square}$ c $\frac{90}{110} = \frac{\square}{\square}$

11 Simplify these fractions. Copy and complete.

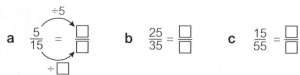

a $\frac{5}{15} = \frac{\square}{\square}$ b $\frac{25}{35} = \frac{\square}{\square}$ c $\frac{15}{55} = \frac{\square}{\square}$

12 Simplify these fractions. Copy and complete.

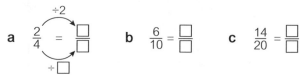

a $\frac{2}{4} = \frac{\square}{\square}$ b $\frac{6}{10} = \frac{\square}{\square}$ c $\frac{14}{20} = \frac{\square}{\square}$

13 Write each fraction in its **simplest form**.

a $\frac{12}{14}$ b $\frac{15}{20}$ c $\frac{20}{30}$ d $\frac{16}{24}$ e $\frac{18}{36}$ f $\frac{14}{21}$

14 Write each fraction in its simplest form.

a $\frac{30}{100}$ b $\frac{70}{100}$ c $\frac{50}{100}$ d $\frac{40}{100}$ e $\frac{20}{100}$

f $\frac{25}{100}$ g $\frac{75}{100}$ h $\frac{22}{100}$ i $\frac{32}{100}$ j $\frac{84}{100}$

15 **Reasoning**   This is how Gary and Lowri cancelled the fraction $\frac{16}{20}$ to its simplest form.

Gary

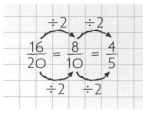

Lowri

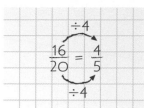

They have cancelled differently but arrived at the same answer. Explain why.

16 Convert each improper fraction to a mixed number.
Write the fraction part in its simplest form.

a $\frac{8}{6}$ b $\frac{10}{6}$ c $\frac{10}{8}$ d $\frac{14}{8}$ e $\frac{14}{10}$ f $\frac{25}{10}$

Challenge

a   Write four fractions to describe the shaded part of this shape.

b   How many possible fractions could you write for part **a**?

Reflect   After this lesson, Lucy said, 'Fractions are not really like whole numbers.'
Kala said, 'Yes! Fractions can be written in many different ways.'
Look back at your work on fractions.

a   What do you think Lucy means?

b   What do you think Kala means?

5.3 Working with fractions

- Add and subtract simple fractions
- Calculate simple fractions of quantities

Active Learn
Homework

Warm up

1 Fluency Work out

a $30 \div 6$ **b** $14 \div 7$ **c** $45 \div 5$ **d** $18 \div 2$

2 Work out

a $\frac{1}{2}$ of £12 **b** $\frac{1}{4}$ of 20 m

3 Write these improper fractions as mixed numbers.

a $\frac{9}{5}$ **b** $\frac{8}{3}$ **c** $\frac{3}{2}$ **d** $\frac{9}{7}$ **e** $\frac{7}{4}$ **f** $\frac{11}{6}$

Key point

When you add or subtract fractions with the same denominator, you add or subtract the numerators. Then write the result over the same denominator.

Worked example

Work out $\frac{1}{5} + \frac{2}{5}$

$$\frac{1}{5} + \frac{2}{5} = \frac{3}{5}$$

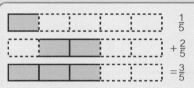

$\frac{1}{5}$

$+ \frac{2}{5}$

$= \frac{3}{5}$

The denominators are the same, so you can add the numerators.

4 Work out

a $\frac{1}{5} + \frac{1}{5}$ **b** $\frac{3}{7} + \frac{2}{7}$ **c** $\frac{2}{9} + \frac{5}{9}$ **d** $\frac{1}{11} + \frac{6}{11}$

5 Work out

a $\frac{4}{5} - \frac{3}{5}$ **b** $\frac{4}{7} - \frac{2}{7}$ **c** $\frac{3}{5} - \frac{2}{5}$ **d** $\frac{8}{9} - \frac{1}{9}$

6 a Work out the missing numbers in these calculations.

i $\frac{3}{5} + \frac{\square}{5} = \frac{5}{5} = \square$ **ii** $\frac{4}{7} + \frac{\square}{7} = \frac{7}{7} = \square$ **iii** $\frac{5}{11} + \frac{\square}{11} = 1$

b Problem-solving Write five pairs of fractions that each add to 1.

7 Copy and complete these subtractions.

a $1 - \frac{11}{12} = \frac{12}{12} - \frac{11}{12} = \frac{\square}{12}$ **b** $1 - \frac{3}{7} = \frac{\square}{7} - \frac{3}{7} = \frac{\square}{\square}$ **c** $1 - \frac{2}{5}$

d $1 - \frac{4}{9}$ **e** $1 - \frac{2}{3}$ **f** $1 - \frac{7}{8}$

8 Work out these calculations. Give each answer in its simplest form.
The first one has been started for you.

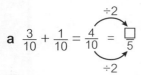

a $\frac{3}{10} + \frac{1}{10} = \frac{4}{10} = \frac{\square}{5}$ ÷2

b $\frac{1}{4} + \frac{1}{4}$ **c** $\frac{1}{8} + \frac{5}{8}$ **d** $\frac{2}{9} + \frac{4}{9}$ **e** $\frac{3}{4} - \frac{1}{4}$

f $\frac{3}{8} - \frac{1}{8}$ **g** $\frac{9}{10} - \frac{3}{10}$ **h** $\frac{11}{12} - \frac{7}{12}$ **i** $\frac{5}{9} - \frac{2}{9}$

9 Work out these additions. Write your answers as mixed numbers.

a $\frac{7}{9} + \frac{3}{9} = \frac{\square}{9} = 1\frac{\square}{9}$ **b** $\frac{6}{9} + \frac{5}{9}$ **c** $\frac{6}{7} + \frac{3}{7}$ **d** $\frac{3}{5} + \frac{4}{5}$

e $\frac{4}{9} + \frac{7}{9}$ **f** $\frac{7}{11} + \frac{7}{11}$ **g** $\frac{2}{3} + \frac{1}{3} + \frac{1}{3}$ **h** $\frac{2}{5} + \frac{3}{5} + \frac{4}{5}$

10 Problem-solving Samyr adds together two different fractions with the same denominator.
He gets the answer $\frac{1}{4}$. Write down two fractions that Samyr might have added.

11 Problem-solving $\frac{1}{9}$ of the guests at a party drink cola and $\frac{4}{9}$ drink lemonade.
The rest drink tea.
a What fraction drink cola or lemonade?
b What fraction drink tea?

Worked example

Find $\frac{1}{3}$ of 18

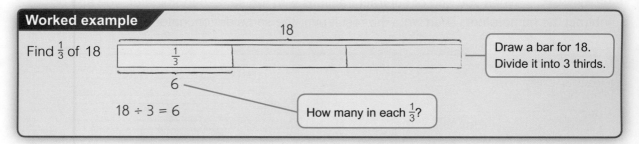

Draw a bar for 18.
Divide it into 3 thirds.

$18 \div 3 = 6$

How many in each $\frac{1}{3}$?

12 Work out

a $\frac{1}{3}$ of 15 kg **b** $\frac{1}{5}$ of 30 kg **c** $\frac{1}{7}$ of £14

d $\frac{1}{9}$ of 36 cm **e** $\frac{1}{8}$ of 24 t **f** $\frac{1}{10}$ of 250 ml

13 Find $\frac{1}{10}$ of each number. Some of your answers will be decimals.
a 40 **b** 700 **c** 55 **d** 99
e 123 **f** 150 **g** 274 **h** 1250

Worked example

Work out $\frac{2}{3}$ of 12

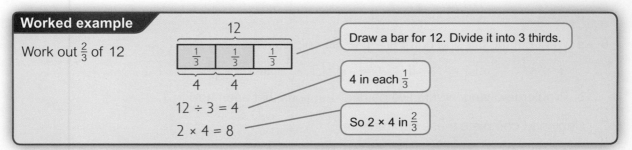

Draw a bar for 12. Divide it into 3 thirds.

4 in each $\frac{1}{3}$

$12 \div 3 = 4$

So 2×4 in $\frac{2}{3}$

$2 \times 4 = 8$

Key point To work out a fraction of a quantity, divide the quantity by the denominator, and then multiply the answer by the numerator.

14 Work out

a $\frac{2}{3}$ of $27

b $\frac{3}{4}$ of 20 m

c $\frac{5}{8}$ of 16 km

d $\frac{4}{5}$ of 30 kg

e $\frac{7}{10}$ of £50

f $\frac{3}{10}$ of £25

15 Problem-solving Red gold is made from $\frac{3}{4}$ gold and $\frac{1}{4}$ copper.
In 24 g of red gold, how many grams are there of
a gold
b copper?

16 Problem-solving The diagram shows the petrol gauge of a car.
The petrol tank holds 56 litres when full.
How much petrol is in the tank?

17 Problem-solving The formula to convert a distance in kilometres to a distance in miles is

distance in miles = $\frac{5}{8}$ of distance in kilometres

Tanya sees this sign on holiday in France.
How far, in miles, is she from Paris?

Paris 60km

18 Problem-solving Hannah carries out a
science experiment.
The table shows her results for the values of A and B.
Hannah says, 'The value of A is always $\frac{2}{3}$ of B.' Is she correct?

A	14	41	66	240
B	21	60	110	380

19 Problem-solving When Sally works on a Sunday she is paid her normal wage plus half again.
Sally is normally paid £9 per hour.
How much is she paid per hour on Sundays?

Challenge Here is a set of fraction dominoes.

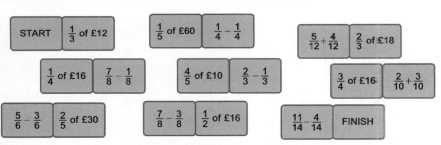

Work out a way to link the dominoes together. A domino can only link with another domino
that gives the same answer. The first two dominoes are

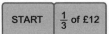

Is there more than one way to link the dominoes?

Reflect

a Write down an easy fraction.
b Write down a difficult fraction.
c What makes one fraction easier or harder than the other?

5.4 Fractions and decimals

ActiveLearn
Homework

- Work with equivalent fractions and decimals
- Write one quantity as a fraction of another

Warm up

1 Fluency What is the fraction that is equivalent to
  a 0.5   **b** 0.25   **c** 0.75

2 What is the value of the digit 7 in each of these numbers?
  a 274.25   **b** 14.75   **c** 112.97

3 Write each fraction in its simplest form.
  a $\frac{2}{10}$ **b** $\frac{18}{100}$ **c** $\frac{45}{100}$

Key point   You can write fractions as decimals.
Three important examples are $\frac{1}{4} = 0.25$, $\frac{1}{2} = 0.5$, $\frac{3}{4} = 0.75$.

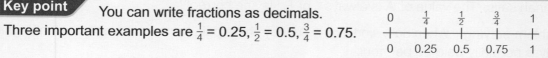

4 a Copy and complete this number line for tenths.

  b Copy and complete this number line for fifths.

Key point   You can convert a decimal to a fraction by looking at the place value.

Worked example

a Write 0.32 as a fraction in its simplest form.

$0.32 = \frac{32}{100}$

$\frac{32}{100} \overset{\div 4}{\underset{\div 4}{=}} \frac{8}{25}$

Look at 0.32 in a place value table.

...	H	T	O	.	$\frac{1}{10}$	$\frac{1}{100}$	...
			0	.	3	2	

0.32 is the same as $\frac{32}{100}$

b Write $\frac{61}{100}$ as a decimal.

$\frac{61}{100} = 0.61$

...	H	T	O	.	$\frac{1}{10}$	$\frac{1}{100}$	...
			0	.	6	1	

5 Write each decimal as a fraction in its simplest form.

    a 0.9   **b** 0.6   **c** 0.5   **d** 0.13

    e 0.34   **f** 0.62   **g** 0.25   **h** 0.75

    i 0.35   **j** 0.06   **k** 0.02   **l** 0.05

6 Write each fraction as a decimal.

    a $\frac{3}{10}$ **b** $\frac{33}{100}$ **c** $\frac{3}{100}$ **d** $\frac{7}{100}$

7 In the 2013 Wimbledon final:
- 0.64 of Andy Murray's first serves were in.
- He won 0.72 of his first serve points.

    Write each decimal as a fraction in its simplest form.

    a 0.64   **b** 0.72

> **Key point**   You can convert a fraction to a decimal by writing an equivalent fraction with a denominator of 10 or 100 and then using place value.

Worked example

Write $\frac{3}{5}$ as a decimal.

$$\frac{3}{5} = \frac{6}{10} \quad (\times 2)$$

Convert to a fraction with denominator 10.

$$\frac{6}{10} = 0.6$$

...	H	T	C	.	$\frac{1}{10}$	...
			C	.	6	

---

8 Write each fraction as a decimal.

    a $\frac{1}{5} = \frac{\square}{10} =$ **b** $\frac{4}{5} = \frac{\square}{10} =$ **c** $\frac{3}{20} = \frac{\square}{100} =$ **d** $\frac{9}{20} = \frac{\square}{100} =$

    e $\frac{2}{25} = \frac{\square}{100} =$ **f** $\frac{12}{25}$ **g** $\frac{7}{50}$ **h** $\frac{17}{50}$

9 In a rugby match, the British Lions won 12 out of the 20 line-outs.

    a Write $\frac{12}{20}$ as a decimal.

    b Is there more than one way to change $\frac{12}{20}$ to a decimal?

10 Convert these fractions to decimals.

    a $\frac{16}{20}$ **b** $\frac{8}{20}$ **c** $\frac{8}{40}$ **d** $\frac{12}{40}$

    e $\frac{12}{60}$ **f** $\frac{21}{60}$ **g** $\frac{21}{75}$ **h** $\frac{60}{75}$

> **Key point**   You can write one quantity as a fraction of another.

$$\frac{\text{number of black squares}}{\text{total number of squares}} = \frac{4}{9}$$

There are 10 apples. 4 of the apples are red.

What fraction of the apples are red?

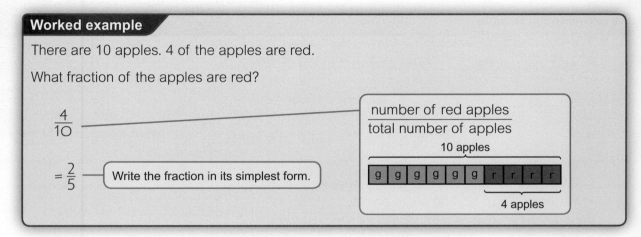

$\frac{4}{10}$

number of red apples / total number of apples

$= \frac{2}{5}$ — Write the fraction in its simplest form.

11 **Problem-solving** There are 17 members in a scout group.
Seven of the members go caving.
What fraction of the group go caving? Give your answer in its simplest form.

12 **Problem-solving** A shop sold 30 pairs of flip-flops in one day.
Five of the pairs were pink.
What fraction of the flip-flops sold were pink? Give your answer in its simplest form.

13 **Problem-solving** There are 3 red, 2 white and 5 blue balls in a bag.
What fraction of the balls are white?
Give your answer in its simplest form.

14 Tom had £8.
He spent £3.
What fraction of £8 did he spend?

15 **Problem-solving** There were 70 000 spectators at a football match.
42 000 of them supported Manchester United.
What fraction of the spectators supported Manchester United?
Write the fraction in its simplest form.

16 **Problem-solving** In the 2012 Tour de France, there were:
9 flat stages   4 medium mountain stages   5 mountain stages   2 individual time-trial stages
a What fraction of the stages were *not* medium mountain stages?
b Write your answer to part a as a decimal.

a Use a calculator to work out $\frac{1}{9}$, $\frac{2}{9}$, and $\frac{3}{9}$ as decimals.
For $\frac{1}{9}$ work out $1 \div 9$, for $\frac{2}{9}$ work out $2 \div 9$, ...

b Describe the pattern you see. What do you think $\frac{7}{9}$ is, as a decimal?

c Check your prediction with a calculator.

d What happens to the pattern when you reach $\frac{9}{9}$, $\frac{10}{9}$ and beyond?

After this lesson, Faiz says, 'Decimals are just another way to write fractions.'
Do you agree with Faiz? Explain.

5.5 Understanding percentages

- Understand percentage as 'the number of parts per 100'
- Convert a percentage to a fraction or decimal
- Work with equivalent percentages, fractions and decimals

Active Learn
Homework

Warm up

1   **Fluency** What fraction of £100 is £9?

2   Write each fraction in its simplest form.

a $\frac{4}{10}$ b $\frac{50}{100}$ c $\frac{8}{100}$ d $\frac{25}{100}$

3   Write each fraction in Q2 as a decimal.

Key point

Per cent means 'out of 100'.

% stands for 'per cent'.

50% means '50 out of 100', which is $\frac{50}{100}$

4   What **percentage** of each block is
    i shaded   **ii** unshaded?

a 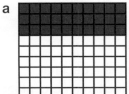 b   c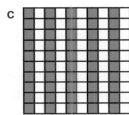

d   What should each pair of answers in parts **a** to **c** add up to?

5   **Reasoning** The manager of Model Fashions has designed a new shop layout.

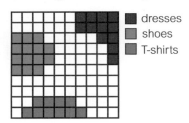
■ dresses
□ shoes
■ T-shirts

a   What percentage of the layout is for
    i shoes **ii** dresses **iii** T-shirts?

b   The manager says, '74% of the shop is empty floor space'.
    Without counting the white squares, how can you tell that the manager is wrong?

6   For every £100 Tim earns, he pays £20 in tax.
    What percentage does Tim pay in tax?

7   A bank account pays £2 interest on savings of £100.
    What percentage is this?

Key point You can write any **percentage** as a fraction with denominator 100.

Worked example

Write 70% as a fraction.

$70\% = \dfrac{70}{100}$ ———— Write as a fraction of 100.

$\overset{\div 10}{\underset{\div 10}{\dfrac{70}{100} = \dfrac{7}{10}}}$ ———— Then write the fraction in its simplest form.

---

8 Write these percentages as fractions.
Write each fraction in its simplest form.
   a 27% b 99% c 30% d 50%
   e 10% f 15% g 25% h 75%

9 A clothes shop makes a different percentage profit on each brand of clothing.
Write each percentage profit as a fraction.
   a 37% b 79% c 61% d 119%

10 A shoe shop gives different percentage discounts on shoes in a sale.
Write each percentage discount as a fraction in its simplest form.
   a 20% b 60% c 40% d 90%

11 **Problem-solving** Is it possible to give a 150% discount?

Key point You can write a percentage as a fraction and then convert to a decimal.
You can write a decimal as a fraction with denominator 100 and then convert to a percentage.

---

12 Convert these percentages to decimals.
The first one has been started for you.

   a $35\% = \dfrac{35}{100} = \square$ b 81% c 9%
   d 1% e 40% f 110%

13 Convert these decimals to percentages.
The first one has been started for you.

   a $0.45 = \dfrac{45}{100} = \square\%$ b 0.72 c 0.03
   d 0.8 e 1.2

14 Convert these fractions to percentages.
Parts **b** and **c** have been started for you.

   a $\dfrac{27}{100}$ b $\dfrac{9}{10} = \dfrac{\square}{100} = \square$ c $\dfrac{11}{50} = \dfrac{\square}{100} = \square$
   d $\dfrac{4}{25}$ e $\dfrac{13}{20}$ f $\dfrac{3}{5}$

15 Problem-solving Match each percentage to its equivalent fraction.

70%	$\frac{21}{100}$
51%	$\frac{4}{5}$
80%	$\frac{39}{100}$
21%	$\frac{7}{10}$
10%	$\frac{1}{10}$
39%	$\frac{51}{100}$

16 Rewrite these statements as fractions, then as percentages.

 a 20 out of 25 students like drawing.

 b 43 out of 50 people play a sport.

 c 7 out of 10 people have a passport.

17 Problem-solving In a rugby match, the British Lions missed 18 out of the 40 missed tackles.

Write $\frac{18}{40}$ as a percentage.

> **Q17 hint** Work out how to change 40 into 100.

Challenge

a Make three copies of this rectangle on squared paper.

b Shade in 65% of the first rectangle. How many squares is this?

c Shade in 0.3 of the second rectangle. How many squares is this?

d Ask a partner to shade, on the third copy, a percentage or decimal that you choose. If it is not possible to do this with a whole number of squares, explain why.

e Explain how you can work out how to choose a percentage or decimal that is a whole number of squares.

Reflect   After this lesson, Alex says, 'Percentages are just another way to write fractions.' Do you agree with Alex? Explain.

5.6 Percentages of amounts

- Use different strategies to calculate with percentages
- Express one quantity as a percentage of another

Active Learn
Homework

Warm up

1 Fluency What fraction is equivalent to
  a 50%   **b** 25%   **c** 75%

2 Work out

  a $\frac{1}{2}$ of £18   **b** $\frac{1}{4}$ of 12 cm   **c** $\frac{3}{4}$ of 12 cm

3 Change each fraction to a percentage.

  a $\frac{4}{100}$ **b** $\frac{7}{10}$ **c** $\frac{11}{50}$ **d** $\frac{9}{20}$

4 Work out
  a 0.1×20 **b** 0.1×32 **c** 0.2×3
  d 0.2×30 **e** 0.2×300 **f** 0.3×300

Key point

50% is the same as $\frac{1}{2}$
To find 50% of an amount, divide by 2.

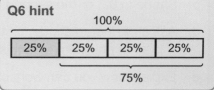

5 Find 50% of
  a 30   **b** 90   **c** 160
  d 54   **e** 18   **f** 19

6 Work out
  a 25% of £40   **b** 75% of £40
  c 25% of 300 kg   **d** 75% of 300 kg
  e 25% of 200 mm   **f** 75% of 200 mm

> **Q6 hint**
>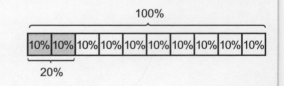

 7 Problem-solving Charities can claim an extra 25% of
the value of a donation back from the Government. This is called Gift Aid.
Work out how much Gift Aid a charity can claim on a donation of £273.50.
Give your answer to the nearest penny.

Key point

$10\% = \frac{10}{100} = \frac{1}{10}$
To find 10% of an amount, divide by 10.
You can then use 10% to find other percentages.

100%
|10%|10%|10%|10%|10%|10%|10%|10%|10%|10%|
20%

8 Work out
   a 10% of 80 kg   **b** 10% of 150 ml
   c 10% of 1500 m   **d** 10% of £45

9 A bank pays 10% interest on all its accounts each year.
Six people have different amounts in their accounts.
How much interest would each person earn in a year?
   a £3000   **b** £1200   **c** £180
   d £445   **e** £183   **f** £963

10 Work out
   a 10% of £50   **b** 20% of £50
   c 80% of £50   **d** 40% of 350 g
   e 30% of 25 km   **f** 20% of £36
   g 60% of £15   **h** 90% of £15

11 There were 500 tickets for a charity concert. 90% were sold.
   a How many tickets were sold?
The tickets sold for £12 each.
   b What was the total amount of money taken in ticket sales?

12 Work out
   a 10% of 70 cm   **b** 5% of 70 cm   **c** 1% of 70 cm
   d 10% of £25   **e** 5% of £25   **f** 1% of £25

Worked example

Find 1% of 300.

$300 \div 100 = 3$

$1\% = \frac{1}{100}$
So to find 1%, divide by 100.

---

13 Find 1% of
   a 600   **b** 920   **c** 3000
   d 50   **e** 75   **f** 199

Key point   When you are working out more complicated percentages of amounts, you can make notes or use **jottings** to help you.

Worked example

Work out 26% of 60 m.

$10\% \rightarrow 60\,m \div 10 = 6\,m$

$20\% \rightarrow 2 \times 6\,m = 12\,m$

$5\% \rightarrow 6\,m \div 2 = 3\,m$

$1\% \rightarrow 6\,m \div 10 = 0.6\,m$

$26\% \rightarrow 12\,m + 3\,m + 0.6\,m = 15.6\,m$

Break down the 26% into 20% + 5% + 1%.
Start by finding 10% of 60 m, then use this to find 20%, 5% and 1%.
Write down the individual parts as you find them.

Finally add the parts together to give 26%.

---

14 Work out these percentages. Use **jottings** to help.
   a 15% of £40   **b** 35% of 90 kg
   c 21% of 50 m   **d** 85% of 120 km

Key point You can use a **multiplier** to work out a percentage, by using the decimal equivalent of the percentage.

$$100\% = 1$$

| 0.1 | 0.1 | 0.1 | 0.1 | 0.1 | 0.1 | 0.1 | 0.1 | 0.1 | 0.1 |

20% = 0.2   10%

$10\% = \frac{10}{100} = \frac{1}{10} = 0.1$

To find 10% you multiply by 0.1, to find 20% you multiply by 0.2, ...

15 Write a calculation using a **multiplier** to work out
   a 20% of £8
   b 30% of 35 kg
   c 40% of 32 litres
   d 60% of 7 t

16 Each month, Karen gets £50 pocket money.
   She spends 24% of her pocket money each month on her phone.
   How much does she pay each month for her phone?

17 A dog agility competition makes £854.87 profit.
   The organisers give 20% of the profit to charity.
   How much do they give to charity?
   Round your answer to a suitable amount.

Challenge A savings account pays 1% interest each month.
At the beginning of January, Shane put £10 000 into the account.

a i How much interest will he receive at the end of January?
   ii How much will he now have in his account?

b i How much interest will he receive at the end of February?
   ii How much will he now have in his account?

c Work out how much he will have in his account at the end of March.

Reflect In this lesson, you found 20% of a quantity in two different ways.
Strategy 1: finding 10%, then multiplying by 2
Strategy 2: using the multiplier 0.2

a Look back at the questions where you used these different strategies.

b Which strategy did you like better? Explain.

c Which strategy would you use if an item was 20% off in a shop? Explain.

d Write down any advantages or disadvantages of each strategy.

5 Check up

Fractions

1 What fraction of each shape is shaded?

a

b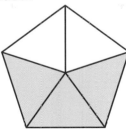

2 Choose one fraction from the cloud to complete this statement:
'$\frac{1}{4}$ is larger than ☐'.

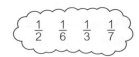

3 Copy and complete each statement using > or <.

a $\frac{4}{9} \ldots \frac{8}{9}$ **b** $\frac{3}{8} \ldots \frac{1}{4}$ **c** $\frac{1}{3} \ldots \frac{1}{5}$

4 Copy and complete these equivalent fractions.

a $\frac{3}{7} = \frac{12}{\square}$ **b** $\frac{30}{40} = \frac{\square}{8}$

5 Convert these improper fractions to mixed numbers.

a $\frac{5}{4}$ **b** $\frac{23}{6}$

6 Work out

a $\frac{2}{5} + \frac{1}{5}$ **b** $\frac{8}{9} - \frac{3}{9}$ **c** $1 - \frac{2}{5}$

7 Work out these calculations. Give each answer in its simplest form.

a $\frac{1}{12} + \frac{5}{12}$ **b** $\frac{13}{20} - \frac{7}{20}$

8 Write each fraction in its simplest form.

a $\frac{6}{10}$ **b** $\frac{18}{24}$

9 Write each fraction in its simplest form.

a $\frac{12}{16}$ **b** $\frac{90}{100}$

10 Work out

a $\frac{1}{6}$ of £18   **b** $\frac{3}{4}$ of 12 km   **c** $\frac{5}{7}$ of 21 kg   **d** $\frac{4}{9}$ of 54 mm

Fractions, decimals and percentages

11 Write each decimal as a fraction in its simplest form.

    a 0.13   **b** 0.7   **c** 0.2   **d** 0.42

12 Write each fraction as a decimal.

    a $\frac{9}{10}$ **b** $\frac{49}{100}$ **c** $\frac{7}{20}$ **d** $\frac{2}{5}$

13 Write each percentage as a fraction in its simplest form.

    a 23%   **b** 60%   **c** 8%   **d** 75%

14 Write each fraction as a percentage.

    a $\frac{42}{100}$ **b** $\frac{3}{10}$ **c** $\frac{41}{50}$ **d** $\frac{11}{25}$

15 Copy and complete this table.

Fraction	Decimal	Percentage
$\frac{1}{2}$		
		70%
	0.25	
		6%

16 There are 12 dogs in a dog training class.
Five of them are spaniels.
What fraction of the dogs are spaniels?

17 There are 8 men and 6 women members in a diving club.
What fraction of the members are women?

Percentages

18 Work out

    a 10% of £40   **b** 50% of 18 cm   **c** 30% of 60 km   **d** 25% of 200 kg

19 Seven out of 10 people own a pet.
Write this number as a percentage.

20 Work out 16% of £30. Use jottings to help.

21 Choose a decimal number from the circle to complete
each of these statements.

    a To find 10% of an amount, you multiply by …

    b To find 20% of an amount, you multiply by …

    c To find 70% of an amount, you multiply by …

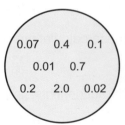

0.07   0.4   0.1
0.01   0.7
0.2   2.0   0.02

Challenge

a Make two copies of this rectangle.
Show two different ways to shade in $\frac{3}{8}$ of the rectangle.

b Write down three different pairs of fractions that add together to give $\frac{1}{2}$

c $\frac{3}{8}$ of £16 = £6

Write down three different 'fraction of an amount' questions that have the answer £6.

Reflect   How sure are you of your answers? Were you mostly

😞 Just guessing   😐 Feeling doubtful   🙂 Confident

What next? Use your results to decide whether to strengthen or extend your learning.

5 Strengthen

Fractions

1 Copy each shape and shade the fraction shown.

a $\frac{3}{5}$

b $\frac{7}{10}$

> **Q1a hint** To shade $\frac{3}{5}$, shade 3 out of the 5 equal parts.

c $\frac{5}{8}$

2

a How many parts are shaded?
b How many parts are in the whole bar?
c What fraction of the bar is shaded?
d What fraction of the bar is unshaded?

> **Q2c hint** $\frac{\text{number of parts shaded}}{\text{total number of equal parts}} = \frac{\square}{\square}$

3 What fraction of each bar is shaded?

a

b

c

d

4 Which is the largest fraction in each set of three bars?

a $\frac{3}{7}$

$\frac{6}{7}$

$\frac{2}{7}$

b $\frac{5}{12}$

$\frac{7}{12}$

$\frac{1}{12}$

5 Which is larger?

a three quarters or two quarters

b three fifths or two fifths

c $\frac{7}{9}$ or $\frac{2}{9}$

d $\frac{11}{12}$ or $\frac{6}{12}$

e $\frac{7}{15}$ or $\frac{11}{15}$

6 Look at the bars. Which is larger, $\frac{1}{5}$ or $\frac{1}{12}$?

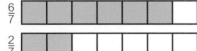

$\frac{1}{5}$

$\frac{1}{12}$

7 Which is larger?

a $\frac{1}{2}$ or $\frac{1}{5}$

b $\frac{1}{6}$ or $\frac{1}{3}$

c $\frac{1}{7}$ or $\frac{1}{8}$

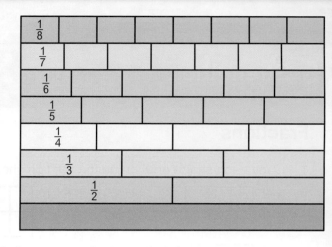

8 How many
   a sixths in a whole   **b** quarters in a whole   **c** tenths in a whole?

9 This diagram represents $\frac{9}{4}$ or nine quarters.

Convert $\frac{9}{4}$ to a mixed number.

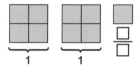

10 Convert each improper fraction to a mixed number.

  a $\frac{6}{5}$ **b** $\frac{12}{5}$ **c** $\frac{13}{6}$ **d** $\frac{7}{2}$

11 Work out

  a $\frac{1}{5} + \frac{2}{5}$

  b $\frac{4}{8} + \frac{1}{8}$

  c $\frac{3}{9} + \frac{2}{9}$

  d $\frac{4}{10} + \frac{3}{10}$

  e $\frac{4}{5} - \frac{1}{5}$

  f $\frac{3}{7} - \frac{1}{7}$

  g $\frac{7}{9} - \frac{2}{9}$

> **Q11a hint**
>
> $\frac{1}{5}$ $\frac{1}{5}$ $\frac{1}{5}$ $\frac{1}{5}$ $\frac{1}{5}$ $\frac{1}{5} + \frac{2}{5}$
>
> $= \frac{3}{5}$

> **Q11e hint** $\frac{4}{5} - \frac{1}{5}$
>
> $= \frac{3}{5}$

12 Work out

  a $1 - \frac{1}{5}$ **b** $1 - \frac{1}{7}$

  c $1 - \frac{1}{3}$ **d** $1 - \frac{1}{4}$

> **Q12a hint**
>
> $1 = \frac{5}{5}$
>
> $\frac{1}{5}$ $\frac{1}{5}$ $\frac{1}{5}$ $\frac{1}{5}$ $\frac{1}{5}$

13 Work out

  a $1 - \frac{3}{5}$ **b** $1 - \frac{5}{6}$ **c** $1 - \frac{2}{3}$ **d** $1 - \frac{7}{9}$

14 Write each fraction in its simplest form.
The first two have been started for you.

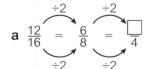

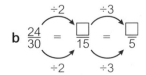

  c $\frac{18}{20}$ **d** $\frac{9}{15}$ **e** $\frac{16}{24}$ **f** $\frac{40}{48}$ **g** $\frac{30}{45}$

15 Copy and complete these simplified fractions.

a
$\frac{4}{10} = \frac{2}{\square}$

b
$\frac{5}{20} = \frac{\square}{4}$

c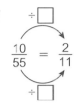
$\frac{30}{40} = \frac{\square}{4}$

d $\div\square$
$\frac{4}{18} = \frac{2}{9}$
$\div\square$

e $\div\square$
$\frac{2}{14} = \frac{1}{7}$
$\div\square$

f $\div\square$
$\frac{10}{55} = \frac{2}{11}$
$\div\square$

16 a Find $\frac{1}{3}$ of 15

b Find $\frac{1}{3}$ of 12   **c** Find $\frac{1}{3}$ of 27   **d** Find $\frac{1}{3}$ of 18

17 a Find $\frac{1}{4}$ of 20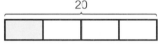

b Find $\frac{1}{5}$ of 10   **c** Find $\frac{1}{6}$ of 12   **d** Find $\frac{1}{10}$ of 30

18 What should you divide by to find these fractions of an amount?

a $\frac{1}{4}$ **b** $\frac{1}{5}$ **c** $\frac{1}{7}$ **d** $\frac{1}{9}$

19 a Find $\frac{1}{5}$ of 20   **b** Find $\frac{2}{5}$ of 20   **c** Find $\frac{3}{5}$ of 20   **d** Find $\frac{4}{5}$ of 20

Fractions, decimals and percentages

1 Copy and complete this diagram.

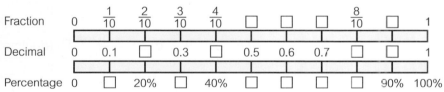

2 Copy and complete. Use the diagram below to help.

Fraction	0		$\frac{1}{4}$		$\frac{1}{2}$		$\frac{3}{4}$		1
Decimal	0		0.25		0.5		0.75		1
Percentage	0		25%		50%		75%		100%

a $0.5 = \frac{\square}{\square}$ (fraction) $= \square\%$ (percentage)

b $\frac{1}{4} = \square$ (decimal) $= \square\%$ (percentage)

c $75\% = \frac{\square}{\square}$ (fraction) $= \square$ (decimal)

d $0.25 = \frac{\square}{\square}$ (fraction) $= \square\%$ (percentage)

3 Copy and complete. Use the diagram below to help.

Fraction   0 $\frac{1}{5}$ $\frac{2}{5}$ $\frac{3}{5}$ $\frac{4}{5}$ 1

Decimal   0   0.2   0.4   0.6   0.8   1

Percentage 0   20%   40%   60%   80%   100%

a $80\% = \frac{\square}{\square}$ (fraction) $= \square$ (decimal)

b $0.6 = \frac{\square}{\square}$ (fraction) $= \square\%$ (percentage)

c $20\% = \frac{\square}{\square}$ (fraction) $= \square$ (decimal)

d $\frac{4}{5} = \square$ (decimal) $= \square\%$ (percentage)

4 **a** There are 9 cats in a rescue centre.
7 of them are black.
What fraction of the cats are black?

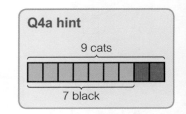

Q4a hint

9 cats

7 black

    b There are 12 dogs in the rescue centre.
5 of them are male.
What fraction of the dogs are male?

5 Copy and complete.

a $21\% = \frac{\square}{100}$

b $66\% = \frac{\square}{100} = \frac{\square}{50}$

c $44\% = \frac{\square}{100} = \frac{\square}{50} = \frac{\square}{25}$

d $\frac{35}{100} = \square\%$

e $\frac{3}{10} \overset{\times 10}{\underset{\times 10}{=}} \frac{\square}{100} = \square\%$

f $\frac{9}{20} \overset{\times 5}{\underset{\times 5}{=}} \frac{\square}{100} = \square\%$

g $\frac{7}{50} \overset{\times 2}{\underset{\times 2}{=}} \frac{\square}{100} = \square\%$

h $\frac{3}{25} \overset{\times 4}{\underset{\times 4}{=}} \frac{\square}{100} = \square\%$

i $0.71 = \frac{\square}{100} = \square\%$

j $0.07 = \frac{\square}{100} = \square\%$

k $0.7 = \frac{\square}{10} = \frac{\square}{100} = \square\%$

l $0.01 = \frac{\square}{100} = \square\%$

m $59\% = \frac{\square}{100} = 0.59$

n $31\% = \frac{\square}{100} = \square$ (decimal)

o $5\% = \frac{\square}{100} = \square$ (decimal)

Percentages

1 **a** What do you divide by to find 10%?
    b Work out 10% of £60.

2 Find 10% of
    a 30   **b** 500
    c 420   **d** 88

Q1a hint

100%

| 10% | 10% | 10% | 10% | 10% | 10% | 10% | 10% | 10% | 10% |

$\square$ parts

3 **a** What do you divide by to find 50%?
    b Work out 50% of 34.

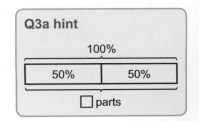

Q3a hint

100%

| 50% | 50% |

$\square$ parts

4 Find 50% of

  a £60   **b** £320   **c** £450   **d** £⁻7

5 Copy and complete.

  a $25\% = \dfrac{\square}{\square}$ of 50%   **b** $75\% = 50\% + \square\%$

6 Work out

    i 50%   **ii** 25%   **iii** 75% of each of these amounts.

  a £20

  b 30 kg

  c 50 ml

  d 80 m

  e £120

7 10% of 45 kg = 4.5 kg

  Work out these percentages of 45 kg.

  a 20% = 2 × 10% of 45 kg =

  b 30%

  c 60%

  d $5\% = \frac{1}{2}$ of 10% of 45 kg =

8 4 out of 10 children like broccoli.

  Copy and complete.

  $\dfrac{\square}{100}$ of children like broccoli.

  $\dfrac{\square}{100} = \square\%$ of children like broccoli.

9 Rewrite these statements as fractions.

  Convert the fractions to percentages.

  a 40 out of 50 students like chocolate cake.

  b 7 out of 25 people go to the gym.

  c 12 out of 20 people have a pet.

  d 8 out of 10 children like fruit.

  e 2 out of 5 students play sport regularly.

> **Q9a hint**
>
> $\dfrac{\square}{\square} = \square\%$ like chocolate cake.

> **Q9b hint**
>
> $\dfrac{\square}{\square} = \square\%$ go to the gym.

Challenge   Shelly thinks that the same fraction of each of these shapes is shaded.

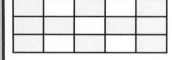 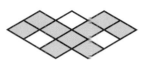

Is Shelly correct? Explain your answer.

Reflect   Caspar says 'In the lesson I am doing lots of division.'

Look back at your work in this lesson. Where did you use division?

How did you use it?

5 Extend

1 Write down how much of each shape is shaded
   i as a fraction   **ii** as a percentage

 a   b   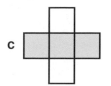 c   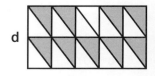 d

2 a Write these fractions in ascending order.

$\frac{1}{7}$ $\frac{1}{2}$ $\frac{1}{11}$

   b Write these fractions in descending order.

$\frac{1}{9}$ $\frac{1}{3}$ $\frac{1}{8}$

3 Problem-solving Sort these cards into groups of equivalent values.
Which card does not belong in any of the groups?

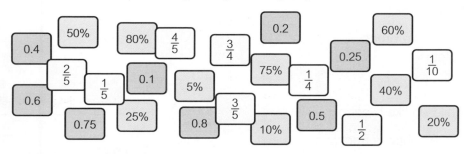

4 Work out these additions. Give each answer as a mixed number in its simplest form.
The first one has been done for you.

a $\frac{3}{4} + \frac{3}{4} = \frac{6}{4} = 1\frac{2}{4} = 1\frac{1}{2}$ **b** $\frac{5}{6} + \frac{5}{6}$

c $\frac{7}{8} + \frac{5}{8}$ **d** $\frac{11}{12} + \frac{7}{12}$

e $\frac{7}{9} + \frac{8}{9} + \frac{5}{9}$ **f** $\frac{9}{10} + \frac{7}{10} + \frac{6}{10}$

5 Reasoning Anil cancels the fraction $\frac{12}{18}$ to its simplest form.
This is what he writes.

Is Anil correct? Explain your answer.

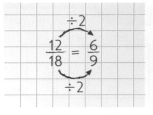

6 Reasoning A blood test shows that 11 ml out of 20 ml is plasma.
The rest of the blood is a mix of red blood cells, white blood cells and platelets.
   a Write the amount of plasma in the blood
     i as a fraction   **ii** as a percentage
   b What percentage of the blood is red blood cells, white blood cells
     and platelets combined?
     Explain how you worked out your answer.

7 **Problem-solving** The bar chart shows the number of T-shirts sold in a zoo shop on one day.
Two-fifths of the T-shirts sold were child sizes.
The rest were adult sizes.
How many adult size T-shirts were sold on that day?

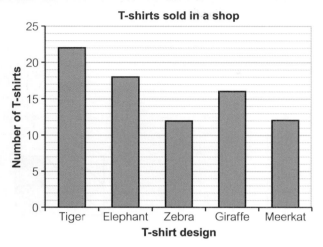

8 **a** Write these test marks as percentages.
   b Who got the highest percentage?

Alex	42 out of 50
Wei Yen	17 out of 25
Nicki	13 out of 20
Robin	32 out of 40

9 **Problem-solving** A square field has a perimeter of 1.2 km.
Two-thirds of the field is to be grazed by sheep, and the rest
by cattle. What area of the field is to be grazed by cattle?
Give your answer in square metres.

> **Q9 hint**
> Draw a diagram to help.

Challenge   To convert a temperature from Fahrenheit to Celsius:
* Subtract 32 from the temperature in Fahrenheit.
* Find $\frac{5}{9}$ of that value.
   a Copy and complete the table to convert these temperatures to Celsius.

Temperature (°F)	59	77	95	167	212
− 32	27				
× $\frac{5}{9}$					
Temperature (°C)					

   b Put these temperatures in order, lowest first.
      77 °F   59 °F   10 °C   20 °C   212 °F   84 °C   167 °F

Reflect   Look back at the questions you have answered in these lessons.
Which question(s) did you find easiest? What made them easy?
Which question(s) did you find most difficult? What made them difficult?
Are there particular kinds of questions that you need more practice with?
If so, which kinds?

5 Unit test

1 Work out
   a 10% of 60 m   **b** 50% of 80 kg   **c** 20% of £90

2 What fraction of each shape is shaded?
   a   **b**

3 Sharon has started to shade this rectangle.
   How many *more* triangles must she shade
   so that $\frac{7}{12}$ of the rectangle is shaded?

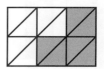

4 **a** Which is larger, $\frac{1}{8}$ or $\frac{1}{9}$?

   b Write these fractions in order of size starting with the smallest.

   $\frac{5}{7}$ $\frac{2}{7}$ $\frac{4}{7}$

5 Copy and complete these percentage and fraction conversions.
   a $47\% = \frac{\square}{100}$ **b** $3\% = \frac{\square}{100}$ **c** $70\% = \frac{\square}{100} = \frac{\square}{10}$

   d $\frac{38}{100} = \square\%$ **e** $\frac{9}{100} = \square\%$ **f** $\frac{3}{10} = \square\%$

6 Copy and complete these decimal and percentage conversions.
   a $0.75 = \square\%$ **b** $0.4 = \square\%$ **c** $0.05 = \square\%$
   d $\square = 50\%$ **e** $\square = 4\%$ **f** $\square = 25\%$

7 **a** Write $\frac{1}{10}$ as a decimal.

   b Use your answer to part **a** to write $\frac{2}{10}$ and $\frac{3}{10}$ as decimals.

8 Write the fraction $\frac{9}{20}$ as a percentage and as a decimal.
   Show all your working.

9 Convert these improper fractions to mixed numbers.

   a $\frac{6}{5}$ **b** $\frac{19}{4}$

10 Copy and complete these equivalent fractions.

   $\frac{7}{9} = \frac{21}{\square}$

11 Which is the correct answer for each question: A, B, C or D?
   a 20% of £32   **A** £16   **B** £6.40   **C** £1.60   **D** £0.64
   b 40% of 8 kg   **A** 32 kg   **B** 5 kg   **C** 3.2 kg   **D** 0.5 kg

12 Work out

   a $\frac{1}{7} + \frac{3}{7}$ **b** $\frac{4}{5} - \frac{3}{5}$

13 Use the diagram to work out which is
   larger, $\frac{3}{5}$ or $\frac{4}{7}$

14 Write the fraction $\frac{24}{32}$ in its simplest form.

15 Work out

    a $\frac{1}{3}$ of 21 cm   **b** $\frac{3}{5}$ of 60 km

16 Nine out of 10 people have a mobile phone.
Write this number as a percentage.

17 Some students were asked the name of their
favourite author.
The pictogram shows the results.
What percentage of the students said Michael
Morpurgo was their favourite author?

Malorie Blackman	📘 📘 📘 📘
Roald Dahl	📘 📘
Michael Morpurgo	📘 📘 📘 📘
J K Rowling	📘 📘 📘 📘 📘
Jacqueline Wilson	📘 📘 📘 📘 📘

Key: 📘 represents 2 students

18 Write whether each of these statements is
true (T) or false (F).
Give a reason for each of your answers.
    a To find 30% of an amount, you multiply by 0.3
    b To find 1% of an amount, you multiply by 0.1

Challenge

a Copy this code box.

I																						?
8	21		15		7	12	9	6	27	12	36		15		5	18	15	5		7	13	4	4	12	13	

Work out the answer to each of the calculations below.
Then use your answers to fill in the letters in the code box and find the secret message.
The first one has been done for you.

$\frac{2}{3}$ of 12 = 8, so 8 = **I**

I	$\frac{2}{3}$ of 12	D	$\frac{1}{4}$ of 20	S	$\frac{3}{5}$ of 35	P	$\frac{1}{7}$ of 49	O	$\frac{4}{9}$ of 27	A	$\frac{5}{6}$ of 18		
Y	10% of 60	G	30% of 90	N	90% of 40	L	25% of 36	T	50% of 26	E	75% of 24	R	1% of 400

b Write your own message and questions for encoding it.

Reflect   Use what you have learned in this unit to work out whether any of these three
statements are true:

A 9% is the same as $\frac{1}{9}$

B 9% and $\frac{1}{9}$ are both the same as 0.9

C 0.9 is the same as 'remainder 9'

Explain your answers.

6 Probability

6.1 The language of probability

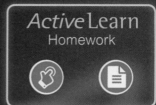

*Active*Learn
Homework

- Use the language of probability
- Use a probability scale with words
- Understand the probability scale from 0 to 1

Warm up

1 **Fluency** What do these words mean?

     possible   impossible   certain

     predict   likely   unlikely

2 What are the missing values on these scales?

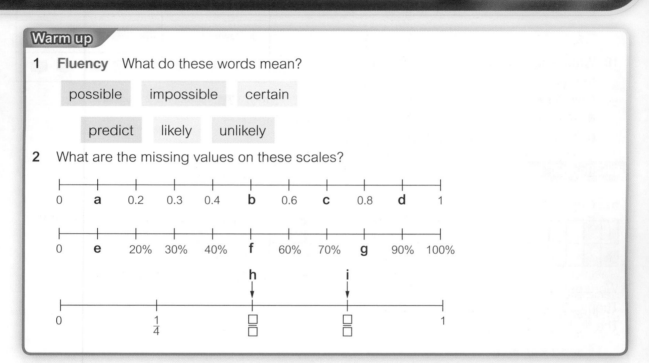

Key point

Probability is the chance that something will happen.
Even chance means that something is as likely to happen as not.
In probability, an **event** is something that might happen.

3 Choose a word to describe the **probability** of each **event**.

     likely   unlikely   very likely   very unlikely

       impossible   certain   even chance

   a getting a 5 when you roll an ordinary dice
   b scoring between 0% and 100% on a maths test
   c picking a card with a number from an ordinary pack of playing cards
   d a domino landing on its end when you drop it on the floor.
   e when you flip a coin it will land heads up
   f the day after Saturday will be Sunday
   g you can run a mile in 60 seconds

4 Copy the **probability scale** below. Mark each event from Q3 on your scale.

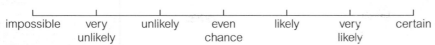

impossible · very unlikely · unlikely · even chance · likely · very likely · certain

5 Write down an event which is
   a impossible   b unlikely   c even chance
   d likely   e certain

6 **Reasoning** Look at these **fair** spinners.

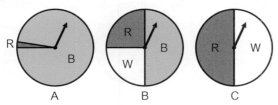

A   B   C

> **Q6 hint Fair** means
> that the pointer on the
> spinner is equally likely
> to stop at any position.

   a Which spinner has an even chance of stopping on red?
   b In which spinner is stopping on white impossible?
   c Which spinner is very likely to stop on blue?
   d Which is more likely – Spinner A stopping on red or Spinner B stopping on white?
      Explain your answer.
   e **Problem-solving** Draw a spinner where red is unlikely and where blue and white are
      equally likely.

7 a Copy the probability scale shown below and mark these probabilities on it.
   Use their capital letters.

impossible   even chance   certain

0     $\frac{1}{2}$ 1

0   0.5   1

0%   50%   100%

   A The probability that someone will grow taller than their father is 50%.
   B The probability of someone born in 2012 living to 100 is about 33%.
   C The probability that a person is right-handed is 0.9.
   D The probability that a train will be delayed is $\frac{1}{10}$
   E The probability of having twins is 1%.
   F The probability of an identical twin having twins is still 1%.
   G The probability of a non-identical twin having twins is 6%.

   b **Reasoning** Who is more likely to have twins – someone who is an identical twin or
      someone who is a non-identical twin? Explain.

8 **Problem-solving** The probabilities of four spinners landing on red are shown on the scale.

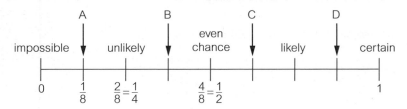

a Write the fraction values of the probabilities A, B, C and D.

b The probabilities A, B, C and D in part **a** are the probabilities of these spinners landing on red.

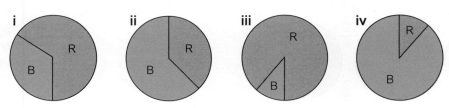

Match each probability to a spinner.

9 **Reasoning** A manufacturer calculates that the probability of one of its dishwashers developing a fault in the first 3 years is 0.07. The manufacturer writes this statement in an advert:

> *Highly unlikely to need repairing in the first three years.*

Do you think this is a reasonable statement? Explain your answer.

10 **Reasoning** The weather forecast says the probability of rain tomorrow is 45%. Do you think you should cancel an outdoor party planned for tomorrow? Explain your answer.

11 **Problem-solving / Reasoning** An expert predicted that there was a 40% chance that Ferrari would win the next Grand Prix. If Ferrari win, what could the expert say to justify her prediction?

Challenge An insurance company estimates the probability of a person having a car accident in the next year. The higher the probability, the more they charge to insure the person.

They group drivers by age: 17–25, 26–40, 41–65, 66–80, over 80.

a   i Which group do you think are most likely to have a car accident?
  ii Which group do you think are least likely to have a car accident?
  iii Show the probability for each age group on a probability scale.

b Do your classmates agree?

Reflect This lesson used a probability scale labelled in different ways.

a Did the scale help you to understand probability, or not? Explain.

b List at least two other areas of maths where you have used a scale.

c Do these scales help you to understand the maths?

6.2 Calculating probability

Active Learn
Homework

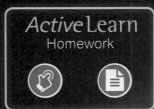

* Identify outcomes and equally likely outcomes
* Calculate probabilities
* Use a probability scale from 0 to 1

Warm up

1 **Fluency**   Here is a fair spinner.
   a Describe the probability that the spinner will land on
      i red   ii blue
   b Which colour is the spinner more likely to land on – red or white?

2 There are five ice creams. One of them is strawberry. Write this as a fraction.

3 Three out of six plums are rotten. Write this as a percentage.

Key point   **Outcomes** are all the possible results of an event. The possible outcomes of flipping a coin are 'heads' and 'tails'.

4 Each of these fair dice is rolled once.
   For each dice
   a write all the possible **outcomes**
   b write the total number of possible outcomes.
   c **Reasoning**   With which dice is the outcome 1 most likely?

6 sides   4 sides   10 sides

Key point   **Successful outcomes** are the outcomes you want.

5 This fair spinner is spun once.
   a Write down all the possible outcomes.
   b What are the successful outcomes for the event 'lands on an even number'?
   c What are the successful outcomes for the event 'lands on a multiple of 3'?
   d What are the successful outcomes for the event 'lands on a prime number'?

Key point   Probability of an event happening = $\dfrac{\text{number of successful outcomes}}{\text{total number of possible outcomes}}$

Worked example

Work out the probability that this fair spinner will land on blue.

Probability that spinner lands on blue = $\dfrac{3}{5}$

There are three successful outcomes: blue, blue, blue.
The total number of possible outcomes is 5.

6 Dewi spins this fair spinner once.
Work out the probability that it lands on
a pink **b** green **c** blue **d** white.

7 Jamie picks one of these number cards **at random**.
  a How many possible outcomes are there?
  b He wants to pick a square number.
     Write down the successful outcomes.
  c Work out the probability that Jamie picks a
     square number.

8 Sara picks one of the cards from Q7 at random. What is the probability that she picks
  a a prime number **b** a number less than 5?

9 The numbers on the faces of a 10-sided dice are

    5   100   20   10   1   1000   15   50   20   1

  Leah rolls the dice once.
  a Work out the probability (as a fraction) that it will land on
     i the number 1 **ii** a 2-digit number
     iii a number greater than 1 **iv** a multiple of 10
  b Write your answers to part **a** as decimal probabilities.
  c Write your answers to part **a** as percentage probabilities.

10 Work out these probabilities, giving your answers as fractions.
  a A fair six-sided dice lands on an odd number.
  b The first ball drawn in a game of Bingo is over 60.
     (There are 100 balls, numbered from 1 to 100.)
  c New Year's Day falls on a Tuesday.
  d You pick a picture card at random from an ordinary shuffled pack of 52 playing cards.
     (The picture cards are the Jack, Queen and King of each of the four suits.)

11 Reasoning 100 raffle tickets are sold.
  One ticket is picked at random to win the prize.
  Frederique has 20 tickets.
  Which letter on this scale shows the probability that Frederique wins the prize?

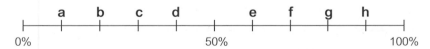

12 A pack of cards contains 13 clubs ♣, 13 spades ♠, 13 hearts ♥ and 13 diamonds ♦.
  Dhevan picks a card at random.
  a Find P(heart). Write your answer as a fraction in its simplest form.
  b Work out P(black card).

13 Reasoning Which of these fair spinners is most likely to land on red?
Write the probability P(red) for each spinner as a percentage to explain.

a   b   c

14 Jarvis rolls a fair 12-sided dice numbered from 1 to 12.

  a Work out the probability of each event:

    A an even number **B** a number greater than 0

    C a multiple of 5   **D** a number less than 10

    E a prime number **F** a square number

> **Q14a hint** P(A) =

  b Draw a probability scale. Number the scale in twelfths: $\frac{0}{12}, \frac{1}{12}, ..., \frac{12}{12}$

    Mark the probabilities from part **a** on it. Use their capital letters.

15 A bag contains 2 chocolates, 3 toffees and 5 chews.
Lin takes a sweet from the bag without looking.
What is the decimal probability that the sweet is

  a a chew   **b** a toffee   **c** a chocolate?

16 Problem-solving

  a A single-wheel lock (with numbers 0–9) can be opened by the number 2.
What is the probability that a stranger can open the lock in one attempt?

  b A different lock has two wheels of numbers.

    i How many possible combinations are there?

    ii What is the probability that a stranger can open this lock at the first attempt?

  c Another lock has three wheels. What is the probability that a stranger can open it at the first attempt?

17 Problem-solving A drawer contains 7 red socks and some black socks.
The probability of picking a red sock is $\frac{1}{4}$

  a How many socks are there altogether in the drawer?

  b How many black socks are in the drawer?

Challenge   Kayla puts red, green, yellow and black counters in a bag.
She picks a counter at random.
The probabilities of getting each of the colours are:

$$P(green) = \frac{1}{4} \qquad P(red) = \frac{1}{2} \qquad P(yellow) = \frac{1}{5} \qquad P(black) = \frac{1}{20}$$

a What is the smallest number of counters that there could be in the bag?

b Use your answer from part **a** to work out how many counters of each colour are in the bag.

Reflect   Andy and Kofi are playing a game with a fair six-sided dice.
Andy needs a 6 to win. He rolls a 2!
Andy says, 'It's not fair. It's harder to roll a 6 than a 2.'
Use what you have learned in this lesson to decide whether Andy is correct. Explain.

6.3 More probability calculations

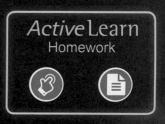
*Active*Learn
Homework

- Calculate more complex probabilities
- Calculate the probability of an event *not* happening

Warm up

1 **Fluency** Work out

    a $1 - \frac{1}{4}$ **b** $1 - 0.6$ **c** $100\% - 20\%$

    d $1 - \frac{2}{5}$ **e** $1 - 0.9$ **f** $1 - \frac{3}{10}$

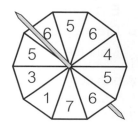

2 Roopa spins this fair spinner once.

    a What is P(5)?

    b What is P(odd number)? Write your answer as a decimal.

    c What is P(even number)? Write your answer as a percentage.

Key point   P(green or blue) means 'the probability of landing on green or blue'.

3   **a** Work out the probability that this spinner will land on blue or red.

      b **Reasoning** Louis says, 'P(blue or red) is the same as P(not yellow).'
        Is Louis correct? Explain.

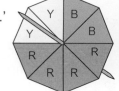

4   At a disco, there are 3 red, 2 green and 1 yellow laser lights.
     A computer turns one on at random.

     a How many possible outcomes are there?

     b Work out

        i P(yellow)   **ii** P(red or yellow)

      iii P(green or red)   **iv** P(red or green or yellow)

       v P(not red)   **vi** P(not red or yellow)

5   Claire takes a card from the top of a shuffled pack of 52 playing cards.
     What is the probability that it is

     a the 3 of clubs or the 2 of diamonds   **b** a 3 or a 2

     c a club, a diamond or a heart   **d** a Jack or an even number

6   **Problem-solving** Olivia's mp3 player has 100 music tracks and is set to play on random.
     There are 20 rock tracks, 40 pop songs, 10 classical tracks and 30 folk songs.

     a Work out the percentage probability that the next track played is

        i rock or pop   **ii** classical or folk

      iii rock, pop or classical   **iv** not rock or folk

     Olivia adds 20 more classical tracks to her mp3 player.

     b Work out the percentage probability that the next track played is classical.

     c Work out the decimal probability that the next track played is folk.

The probability that Alastair wins his tennis match is $\frac{5}{8}$
What is the probability that Alastair does not win his
tennis match?

P(does not win) = 1 − P(wins)

$$= 1 - \frac{5}{8} = \frac{3}{8}$$

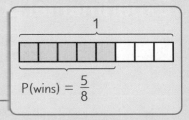

P(wins) = $\frac{5}{8}$

Key point   P(event not happening) = 1 − P(event happening)

7   Gordon is a goalkeeper. The probability that he saves a penalty is 0.3.
     What is the probability that Gordon does *not* save a penalty?

8   The probability that Kalinda makes a basketball shot is $\frac{7}{12}$
     Work out the probability that she does *not* make the basketball shot.

9   In a river, P(catching a fish) = 0.27. Calculate P(not catching a fish).

Key point   Percentage probability of something *not* happening =
100% − percentage probability of event happening

10   A baby has a 5% chance of being born on its due date
     What is the probability that a baby is *not* born on its due date?

11   A software company calculates that the probability its software will crash is 2%.
     Copy and complete this statement for an advert for the company:
     'The probability that our software will not crash is _____ %.'

12   The probability that Haroon does *not* hit the bullseye of a dartboard is 0.8.
     What is the probability that he hits the bullseye with the next dart?

Challenge

a   What is the probability of flipping heads with a fair coin?
b   What is the probability of *not* flipping heads with a fair coin?
c   Add the two probabilities together.
d   What is the probability of rolling a 3 with a fair six-sided dice?
e   What is the probability of *not* rolling a 3 with a fair six-sided dice?
f   Add the two probabilities together.
g   Write a general rule for adding together the probability of an event happening and the
    probability of it *not* happening.

Reflect   Would you find it easier to use a fraction, a decimal or a percentage to write the
probability for
a   rolling 1 on the dice
b   landing on 1 on the spinner
c   flipping the coin and getting a head?
    Explain your answers.

6.4 Experimental probability

- Record data from a simple experiment
- Estimate probability based on experimental data
- Make conclusions based on the results of an experiment

Active Learn
Homework

Warm up

1 **Fluency** In a group of 20 students, 5 are left-handed.
Write this as
   a a fraction   **b** a percentage

2 **a** The tally chart shows the colours of flowers that grew from a mixed packet of seeds.
Copy and complete the table.
   b What is the total frequency?
   c What fraction of the flowers are red?

Colour	Tally	Frequency
Red	ⅢⅣ ⅢⅣ Ⅱ	
Blue		7
White	ⅢⅣ Ⅰ	

3 Describe each of these probabilities in words.
Choose from: impossible, unlikely, even chance, likely, certain.

   a 0.4   **b** $\frac{19}{20}$ **c** 0.5   **d** $\frac{1}{50}$ **e** 60%   **f** 0

Key point

You can use the results of an **experiment** to estimate probabilities.
This is called **experimental probability**.

Experimental probability = $\frac{\text{frequency of event}}{\text{total frequency}}$

Worked example

Andrew dropped a drawing pin lots of times. It could land point up or down.
He recorded the results in a frequency table.
Work out the experimental probability that
a the pin will land point up
b the pin will land point down

Total frequency = 83 + 17 = 100 ——————————— Work out the total frequency first

Position	Frequency	Experimental probability
Point up	83	$\frac{83}{100}$
Point down	17	$\frac{17}{100}$
Total frequency	100	

Experimental probability
= $\frac{\text{number of times pin lands point up}}{\text{total frequency}}$
= $\frac{83}{100}$ = 83% or 0.83

Notice that $\frac{83}{100} + \frac{17}{100} = \frac{100}{100} = 1$

4 Chandak spins this spinner 100 times. He records his results in a frequency table.

Colour	Frequency	Experimental probability
red	17	$\frac{17}{\square} = \underline{\quad}$ %
green	28	
yellow	55	
Total frequency	$\square$	

Copy the table. Calculate the experimental probability of landing on each colour.

5 **Reasoning** A biologist records the number of leaves on some clover plants. He records his results in a frequency table.

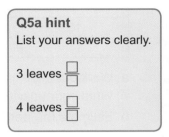

Number of leaves	Frequency
3	156
4	26
5	18
Total frequency	

Q5a hint
List your answers clearly.

3 leaves $\frac{\square}{\square}$

4 leaves $\frac{\square}{\square}$

a Work out the experimental probability of each number of leaves.

b The biologist concludes that it is very unlikely that a clover plant will have more than four leaves. Do you agree with this statement? Explain.

6 **Reasoning** A hospital tried out a new kind of knee surgery on some patients.
After two years, patients were asked how they felt.
The results are shown in the frequency table.

Outcome	Frequency	Experimental probability
symptom free	60	
some improvement	15	
no improvement	5	
Total frequency		

a Work out the total frequency.

b Calculate the experimental probabilities as fractions.

c The hospital claims that patients undergoing the new surgery are very likely to improve. Comment on this claim.

> **Key point** You can estimate a probability from results, using experimental probability.
> Probability can be used to **predict** what may happen in the future.

7 A manufacturer of chocolate biscuit bars tests 500 bars.
17 of these bars have no biscuit inside.
Estimate the probability that a chocolate biscuit bar has no biscuit.

8   Hal counted the passengers in the first 100 cars passing his school.
    He found that 38 of the cars had no passengers.
    Estimate the probability that the next car will have
    a no passengers   b some passengers

9   **Problem-solving**   A skateboard manufacturer gave 100 customers a set of newly designed
    wheels to try out. The table shows for how long the wheels performed well.

Time (months)	Frequency	Experimental probability
5	7	
6	14	
7	35	
8	25	
9	15	
10	3	
11	1	

    a Work out the experimental probabilities for the different times the wheels performed well.
      Write your answers as percentages.
    b Estimate the percentage probability that a randomly chosen set of wheels will perform
      well for longer than 8 months.

10 **Reasoning**   Records show that more than 10 cm of rain fell in Orkney during 415 of the last
    1000 months.
    a Estimate the probability that there will be more than 10 cm of rain in Orkney next month.
    b Is this a good model for predicting the rainfall in the month of July?
      Give a reason for your answer.

11 a **Reasoning**   For which of these events can you work out the exact probability?
      A The next train being late.
      B Picking a particular coloured counter from a bag.
      C A piece of toast falling on the floor butter-side down.
      D The price of your favourite magazine increasing next year.
    b How could you estimate the probabilities of the other events?

> **Key point**   The more times you repeat an experiment, the more accurate the
> experimental probability.

12 **Problem-solving / Reasoning**   Ali is playing computer solitaire.
    He tallies the number of times he wins and loses.

Win	||
Lose	Жℋ	

    a Use Ali's results to estimate the probability of him
      winning the next solitaire game.
    b How could Ali improve the accuracy of his estimate?

13 Problem-solving Paul, Surinda and Amy are investigating how likely it is that a drawing pin will land point up or point down when you drop it. Here are the results of each of their experiments.

	Paul	Surinda	Amy
Point up	8	41	17
Point down	2	9	3

    a Calculate the experimental probability of the drawing pin landing point up for each person.

    b Which result do you think is the most accurate? Give a reason for your answer.

    c Combine the data from all three experiments to work out

      i how many times the pin was dropped

      ii how many times the pin landed point up

      iii a more accurate estimate for the probability that the drawing pin will land point up

14 Reasoning A manufacturer tested a new kind of mobile phone battery.
They claim that there is a 95% experimental probability that the battery will last for 30 hours with average use.

    a Can you tell from the probability how many batteries they tested?

    b What would make you confident that their claim was correct?

Challenge You will need a coin and a piece of lined paper. Work in pairs or small groups.
You will be investigating the number of lines that a coin flipped onto lined paper will cross.

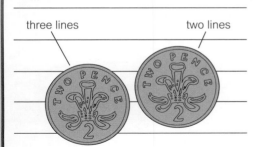

three lines   two lines

a Design a table to record your results.

b Flip your coin onto the lined paper 20 times and record your results in your table. (You will need to decide what to do if the coin just touches a line. Whatever you decide, make sure you use the same rule for every trial.)

c Calculate the experimental probabilities from your results and write a conclusion.

d If you repeated the experiment, would you get exactly the same results?

Reflect In the Challenge in this lesson, you collected your own data and worked out the experimental probability.
Other questions gave you data.
Which was easier? Explain.

6.5 Expected outcomes

- Use probability to estimate the expected number of times an outcome will occur
- Apply probabilities from experimental data in simple situations

Warm up

1 **Fluency** What is the probability of
   a a fair coin landing heads up
   b rolling 'three' on a fair dice?

2 A fair spinner has the numbers 1 to 8.
   If it is spun, what is the probability that it will land on
   a an odd number   b a prime number?

3 A fair ten-sided dice is numbered 0 to 9. At a school fair, you win £1 if you
   roll a multiple of 3. What is the probability of winning £1?

4 At a charity fête, there are two spinner games.
   Both games cost £1 to enter and have the same prize.

Game 1   Game 2

   a Calculate the probability of winning the prize in game 1.
   b Calculate the probability of winning the prize in game 2.
   c **Reasoning** Which game would you play? Give a reason for your answer.

Key point You can use probability to estimate the **expected number** of times an outcome will occur.

Worked example

A fair dice is rolled 60 times.
Estimate the expected number of 'ones'.

$P(\text{rolling 'one'}) = \dfrac{1}{6}$ ———
> Calculate the probability of rolling 'one'.

$\dfrac{1}{6}$ of 60 rolls = 10

Expected number of 'ones' in 60 rolls is 10.

5 A fair coin is flipped 80 times.
Estimate the expected number of 'tails'.

6 Problem-solving A number card is picked at random from this selection.

| 9 | 13 | 3 | 2 | 18 | 7 | 10 | 5 |

a Calculate the probability of picking an even number.
b Each time a card is picked, it is replaced.
Estimate the expected number of even numbers picked in 24 picks.

7 In a biscuit factory, 3 out of 50 biscuits are broken.
a Estimate the probability that a biscuit is broken.
b Estimate the expected number of broken biscuits in a box of 250 biscuits.

8 a How many tails do you expect if a fair coin is flipped 50 times?
b How many 2s do you expect if a fair six-sided dice is rolled 30 times?
c How many even numbers do you expect if a fair ten-sided dice is rolled 80 times?
d You take a card from a shuffled pack of 52 playing cards and put it back.
How many hearts do you expect if you do this 20 times?

Worked example

Davina charges 10p for a raffle ticket.
Davina uses tickets numbered 1 to 100.
Numbers ending in 0 win a 50p prize.
She sells 60 tickets.
How much money can she expect to make?

Money paid for 60 tickets = 60 × 10p = 600p = £6

The probability of winning 50p is $\frac{10}{100} = \frac{1}{10}$ ──── The numbers 10, 20, 30, ..., 100 are the only winning tickets. There are 10 possible winning tickets out of 100.

Expected number of wins = $\frac{1}{10}$ of 60 = 6

Expected prize money = 6 × 50p = 300p = £3

£6 − £3 = £3 ──── Subtract the money paid as prizes from the money paid for the tickets.

Davina can expect to make £3 selling 60 tickets.

---

9 A tombola has paper tickets numbered from 1 to 100.
The tickets are folded and players pick one at random.
Each ticket costs 10p.
If you pick a ticket that ends in 5 or 0 you win 20p.
a How much money could you expect to win from 20 tickets?
b How much money would you pay for 20 tickets?
c **Reasoning** Are you likely to make money if you buy 20 tickets?

10 Edin has made a 'buzzer game'.

Players must carefully move a metal ring along a wire without touching it.

If they are successful they win a prize.

Edin has tested the game with 50 people.

13 completed it successfully and 37 did not.

a What is the estimated probability that a player will be able to complete the game?

Edin charges 10p for a go. He decides to award a prize of 50p.

b How much money should Edin expect to make or lose for 50 players?

c Reasoning Is his game likely to make money?

11 On Steve's stall, players pay 5p to throw a dart at a playing card. He tested the board with his friends, and recorded the results.

	Hit picture card	Hit number card	Missed	Total
Number of outcomes	36	69	95	200

Steve plans to give a prize of 50p for hitting a picture card.

a How many picture card hits should he expect in 50 dart throws?

b Reasoning Explain how much money Steve is likely to make or lose if 100 people play his game.

12 Shaun has this fair spinner.

a What is the probability of **i** winning 20p **ii** winning 10p?

Shaun charges 10p to spin the spinner.

b Problem-solving How much money should Shaun expect to make for every six spins?

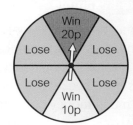

13 A fair spinner has the numbers 1 to 5.

You need to spin an even number to win.

a What is the probability of winning?

b How many wins would you expect in 100 games?

c The game costs 20p a go. How much money do players pay for 100 goes?

d Problem-solving What would be a sensible prize to make sure that this game makes a profit?

Challenge Rhianna has designed a game to raise money.

Two coins are flipped separately, and if they are both heads you win.

Rhianna thinks that this will happen $\frac{1}{3}$ of the time.

Test this with a partner.

How many times should you flip the coins to get a reliable answer? Explain what you find.

Reflect Look back at the questions in this lesson.

List all the mathematics needed to design a fundraising game that raises money.

6 Check up

The language of probability

1   Choose a word to describe the probability of each event.
Choose from: impossible, unlikely, even chance, likely, certain.
A You roll a number greater than 1 with an ordinary dice.
B A stamp falls on the floor sticky side up.
C One of your classmates was born on 30 February.
D A student chosen at random in assembly has a name beginning with Q.
E This maths lesson will end before 6 pm.

2   Copy this probability scale. Mark each event from Q1 on the scale. Use their capital letters.

| impossible | even
chance | certain |

3   Describe each probability using words.
   a 0.8   **b** 0   **c** 0.25   **d** 1   **e** 0.5

4   If divers come up too fast, they can become ill. About 12% of them become ill.
Use words to describe the probability of a diver becoming ill.

Calculating probability

5   Here is a fair spinner.
   a List all the possible outcomes for the spinner.
   b Work out the probability that the spinner lands on white.
   c Work out the probability that the spinner lands on red or white.

6   Tess spins this fair spinner once.
Work out the decimal probability that it lands on
   a 4
   b an odd number
   c a number less than 7

7   These letter cards are shuffled. Jim chooses one of the cards at random.

Work out the percentage probability that he picks
   a the letter S   **b** a vowel
   c not the letter P   **d** a green letter or the letter M
   e not the letter S   **f** the letter J

8   The probability that a new smartphone will develop a fault in the first 12 months is 0.1.
What is the probability that it does *not* develop a fault?

9 Elephant calves have a survival rate of 98%.

    a When an elephant calf is born, is it likely to survive?

    b What is the probability of the elephant calf *not* surviving?

10 On a spinner, P(blue) $= \frac{1}{3}$

    Work out P(not blue).

Experimental probability

11 On a journey, Soujit and Jamie recorded the colour of each
traffic light as they reached it. Here are their results.

    a Calculate the experimental probability for each colour.

    b Soujit says that it is unlikely that a traffic light will be on amber
when he reaches it. Do you agree with Soujit's statement?
Explain.

Colour	Frequency
red	19
amber	7
green	24

12 Here is a spinner for a game.
Isabella spins the spinner 10 times.
Tim spins the spinner 50 times.

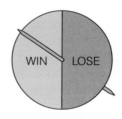

Result	Isabella	Tim
win	6	24
lose	4	26

    a Calculate the experimental probability of winning for each person.

    b Whose results give a more accurate estimate of the probability of winning? Explain.

13 This spinner is spun 100 times.
Estimate the number of times it will land on 'Win'.

14 A quality control inspection of 200 watches found that 20 were faulty.

    a Estimate the probability that a watch will be faulty.

    b Estimate the number of faulty watches in a batch of 500 watches.

Challenge

a Think about your favourite sport. Describe an event that is

    i impossible   **ii** likely   **iii** very unlikely   **iv** certain

b Design a spinner using the colours blue, yellow, red and white.
The probability of the spinner landing on blue must be 25%, on yellow 0.1, on red $\frac{1}{2}$

Reflect   How sure are you of your answers? Were you mostly

    😟 Just guessing   😐 Feeling doubtful   🙂 Confident

What next? Use your results to decide whether to strengthen or extend your learning.

6 Strengthen

The language of probability

1 Match each description to a probability.

An event

| A cannot happen |
| B often happens |
| C rarely happens |
| D happens as many times as it doesn't happen |
| E always happens |

The probability is

| 1 unlikely |
| 2 likely |
| 3 certain |
| 4 impossible |
| 5 even chance |

2 Here is a fair red and blue spinner.
a Which colour is the spinner likely to land on?
b Which colour is the spinner unlikely to land on?

3 Jen spins this red and blue spinner once.
Which of these statements is true?
A The spinner is likely to land on blue.
B The spinner is unlikely to land on blue.
C The spinner has an even chance of landing on blue.

4 The probabilities of three spinners landing on white are shown on the probability scale.
Match each probability to a spinner.

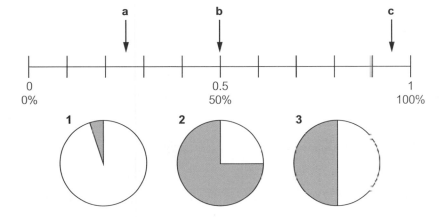

5 a Copy this percentage probability scale.

Mark on it the letter of each of these events.

A There is a 50% chance that the first student through the school gate tomorrow will be female.

B A teacher says there is a 10% probability that the school will be closed due to snow next week.

C There is a 100% probability that the school will close later today.

D If you cheat in a maths exam, the probability of being caught is 90%.

b Describe each of the events in part **a** using probability words.

Calculating probability

1 This four-sided dice has a shape drawn on each side.

a How many possible outcomes are there?

b The dice is rolled once. How many successful outcomes are there for each event?

   A The dice lands on a triangle.

   B The dice lands on a circle.

   C The dice lands on a shape with straight sides.

c Write the probability of each event in part **b**.

$$\frac{\text{number of successful outcomes}}{\text{total number of possible outcomes}} = \frac{\square}{\square}$$

> **Q1b hint** Event A can happen in two ways because there are two triangles. There are two successful outcomes for event A.

2 Alessandra put these counters in a bag and then took one out without looking.

a How many possible outcomes are there?

b What is the probability (as a fraction in tenths) that the counter is

   i blue

   ii red

   iii pink

   iv blue or pink

   v not pink?

> **Q2a hint** How many different ways can Alessandra take a counter out of the bag? Some counters are the same colour but they are still different from one another.

c Mark the probabilities in part **b** on a probability scale like this.

0   $\frac{1}{10}$ $\frac{2}{10}$ $\frac{3}{10}$ $\frac{4}{10}$ $\frac{5}{10}$ $\frac{6}{10}$ $\frac{7}{10}$ $\frac{8}{10}$ $\frac{9}{10}$ 1

impossible   even chance   certain

d Write each of the probabilities in part **b** as a decimal.

e Write each of the probabilities in part **b** as a percentage.

f Alessandra wrote each letter of her name on a counter and picked one at random.

What is the probability the counter is

   i the letter A   **ii** blue and the letter A

  iii blue or the letter A   **iv** not a vowel

   v not red?

3 Due to weather conditions, the probability that there will be a NASA rocket launch tomorrow is $\frac{1}{10}$
Work out the probability that there will *not* be a launch.

Q3 hint

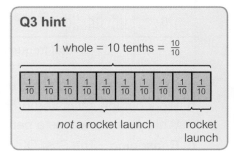

1 whole = 10 tenths = $\frac{10}{10}$

not a rocket launch   rocket launch

4 For a computer game, P(win) is 18%.
Work out P(not win).

5 Astronomers predict that there is a 45% chance of a solar storm tomorrow.
What is the probability that there will *not* be a solar storm tomorrow?

Q4 hint

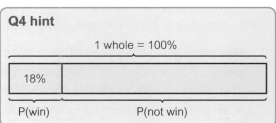

1 whole = 100%

18%

P(win)   P(not win)

Experimental probability

1 Sanchez's teacher secretly put 10 cubes in a bag. Some were blue, some yellow and some black. Sanchez took one out and recorded its colour in the tally chart below. Then he put the cube back into the bag. He repeated this 20 times.

Colour	Tally	Frequency	Experimental probability
blue	卌 卌 |||	13	$\frac{13}{20}$
yellow	卌		
black	||		
	Total frequency		

Copy the table.

a Complete the Frequency column.

b Calculate the total frequency.

c Calculate the experimental probability of picking each colour.

d Which counter is more likely to be picked from the bag, black or yellow? Explain.

e Write the probability of picking each colour in words.

Q1b hint The total frequency is the total number of times Sanchez took a cube from the bag.

2 The tally chart shows the visits to some Post Office cashier desks on a Saturday morning.

Cashier desk	Tally	Frequency	Estimated probability
1	JHT JHT IIII	14	$\frac{14}{100} = 14\%$
2	JHT JHT JHT I		
3	JHT JHT JHT JHT JHT		
4	JHT JHT JHT JHT JHT JHT		
5	JHT JHT JHT		
	Total frequency		

a Copy the table and complete the Frequency column.

b Work out the estimated probability that the next customer will visit cashier desk 3. Write your answer as a percentage.

c **Reasoning** The Post Office manager says that the probability of a customer visiting cashier desk 1 next Monday is 14%. Explain why this might not be true.

Q2c hint Think of a reason why things might be different in the Post Office on Monday compared with on Saturday.

3 **Reasoning** During Euro 2008, Paul the Octopus predicted the results of Germany's matches. This table shows his results for each match.

a How many matches did Germany play in total?

b How many results did Paul correctly predict?

c Calculate the experimental probability that Paul will correctly predict the result of the match.

d Give a reason why your answer to part c might not be very accurate.

Match	Paul's prediction
Germany v Poland	correct
Germany v Croatia	incorrect
Germany v Austria	correct
Germany v Portugal	correct
Germany v Turkey	correct
Germany v Spain	incorrect

Q3d hint More repeats of experiments give more accurate estimates of probability.

Challenge

a Write six numbers to label a blank six-sided dice so that the probability of rolling an even number is unlikely.

b Write ten numbers to label a blank ten-sided dice so that there is an even chance of rolling a number greater than 3.

Reflect

a Copy and complete this sentence with three different endings.
In this unit I learned to ____.

b Copy and complete at least four of these sentences.
I showed I am good at ____ .   I was surprised by ____ .
I found ____ hard.   I was happy that ____ .
I got better at _____ by ____ .   I still need help with ____ .

6 Extend

1 Toni rolls a fair six-sided dice once.
   Describe a possible event with each of these probabilities.
   a impossible   **b** 50%   **c** unlikely

   d $\frac{5}{6}$ **e** 1   **f** $\frac{1}{3}$

2 There is a 40% chance that a car accident is caused by the driver
   using their mobile phone.
   What is the probability that a car accident is *not* caused by a mobile phone?

3 About 200 000 people are chosen to sit on a jury each year in the UK.
   They are chosen from a population of 48 million adults.
   a What is the probability that a particular adult will be chosen next year?
      Write your answer as a fraction in its simplest form.
   b Describe this probability using words.
   c What is the probability that a particular adult will *not* be chosen for a jury?
      Write your answer as a fraction.

4 An astronomer recorded the number of shooting stars she saw each night between
   midnight and 1 am.

Shooting stars	0	1–2	3–5	6–10	11–20	more than 20
Frequency	3	12	20	22	15	8

   a For how many nights did she record the number of shooting stars?
   b Estimate the probability that she will see at least three shooting stars between midnight
      and 1 am the next night.

5 **Reasoning**   The median number of customers visiting Lydia's café each day is 36.
   What is the probability that more than 36 customers will visit the café tomorrow?

6 Adi dropped 30 chocolate eggs. She found that 22 eggs broke and 8 eggs remained whole.
   a Work out the experimental probability of a chocolate egg breaking when dropped.
   Adi repeated the experiment with 40 chocolate eggs and found that 28 broke and 12
   remained whole.
   b Work out the experimental probability of a chocolate egg breaking when dropped in the
      second experiment.
   c **Reasoning**   Why are your answers to parts **a** and **b** not the same?
   d **Problem-solving**   Use the results from both experiments to estimate how many eggs
      would be broken if Adi dropped 175 chocolate eggs.

7 **Problem-solving**   At Christmas, a store lift continuously plays these five songs, one after
   the other, without any gaps between them. The song durations are shown in seconds.

   Jingle Bells (150 sec)   Silent Night (140 sec)   White Christmas (180 sec)
   We Three Kings (120 sec)   Away in a Manger (160 sec)

   Work out the probability that White Christmas will be playing when Izzy enters the lift.
   Write your answer as a decimal.

8 **Reasoning** Morine recorded the darts thrown
by two of her favourite players.

	Tom Sharp	Sneaky Joe
Single	39	8
Double	30	5
Treble	25	7
25 ring	5	3
Bull	1	2

a Draw a suitable graph for the data.

b Estimate the probability of each player
hitting a treble.

c Which player is more likely to hit a treble?
Explain your answer.

d Whose estimated probability is a more reliable model for their future dart throws?
Explain why.

e If Sneaky Joe threw 200 darts tomorrow, estimate the number of trebles he would hit.

9 **Problem-solving**

a A tennis ball is made by covering a rubber ball with two identical yellow patches.
One patch is marked with an X.

   i What is the probability that the ball will land on the marked patch after being hit?

   ii Estimate the number of times it will land on the marked patch after 30 hits.

b A rugby ball is made by stitching four identical patches together.
One patch is marked with an X.

   i What is the probability that the ball will land on the marked patch after being kicked?

   ii Estimate the number of times it will land on the marked patch after 60 kicks.

c About 360 million square kilometres of the Earth's surface is sea.
The other 150 million square kilometres is land.
If a scale model of the Earth rolls across the floor, what is the probability of it stopping
on land?

Challenge   Work in a group of five.

a Each person draws a straight line between 1 cm and 30 cm long,
secretly noting its length.

b Take turns to show your line to the group.

c Each person estimates the length of the line.

d Record each estimate of the length.

e Check whether the estimate is within 10% of the true length. (Work out 10% of the true
length. Add and subtract this to the true length to give the range of estimates
within 10% of the true value.)

f Repeat until each person's line has been estimated.

g Record all of the results in the same tally chart. Use the headings 'good estimate'
and 'poor estimate'. A good estimate is within 10% of the true length.

h Work out the experimental probability of a person making a good estimate.

Reflect   Probability is used in many jobs, for example

   sport   medicine   insurance   forensic science   weather forecasting

For each job, write down one way it uses probability.
What career are you interested in?
Do you think you will need to use probability in your job? If so, how?

6 Unit test

1 **a** Use **impossible**, **unlikely**, **even chance**, **likely** or **certain** to describe the probability of each of these events.

   A You choose the King of hearts from a shuffled pack of 52 playing cards.

   B You get 110% in your next maths exam.

   C A letter is delivered to your home next week.

   D You run a mile in 3 minutes.

   E You get an odd number when you roll an ordinary dice.

   b Copy the probability scale. Mark each event from part **a** on it. Use their capital letters.

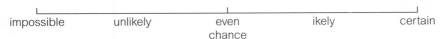

| impossible | unlikely | even chance | ikely | certain |

2 Describe each probability using words.

   a $\frac{3}{4}$ **b** 0.2   **c** 50%   **d** 0   **e** 100%

3 Duane spins this fair spinner.

   a Write the probability that the spinner lands on

   A the number 4   **B** the number 3

   C an even number   **D** the number 6

   E a number less than 10

   b Copy the probability scale. Mark each event from part **b** on it.
   Use their capital letters.

| 0 | $\frac{1}{4}$ | $\frac{1}{2}$ | $\frac{3}{4}$ | 1 |

4 Emily wrote these words on her fridge using magnetic letters.

   a One of the letters falls off the door when it is closed.
   What is the decimal probability that the fallen letter is

   A the letter o   **B** n, o or d   **C** a red letter

   D a black vowel   **E** a blue letter or a vowel?

food inside

   b Copy the probability scale. Mark each event from part **a** on it.
   Use their capital letters.

| 0 | 0.5 | 1 |

5 A fair 10-sided dice numbered 0, 1, 2, 3, 4, 5, 6, 7, 8, 9 is rolled once.
   Work out the percentage probability that it lands on

   a the number 3   **b** not the number 3

   c a square number   **d** a prime number

6 The probability of a stolen car being found is 0.3.
   What is the probability of a stolen car *not* being found?

7 A government website says that the chance of catching flu this year is 20%.
   What is the probability that a person will *not* catch flu?

8 As part of an experiment, Luke cut off part of a rubber ball.

He predicted that the ball would be unlikely to land on the curved surface when dropped. Here are Luke's results.

Outcome	Tally	Frequency	Experimental probability
curved surface	⅂⅂⅂⅂⅂ |||		
flat surface	⅂⅂⅂⅂⅂ ⅂⅂⅂⅂⅂ ||		
	Total frequency		

a How many times did Luke drop the ball?

b Work out the experimental probabilities.

c Is Luke's prediction correct? Explain your answer.

d How can Luke improve his estimates of the experimental probabilities?

9 In a video racing game, an obstacle randomly appears on the race track 3 times every 20 laps.
Estimate the number of times the obstacle will appear in 100 laps.

10 The bar chart shows the A-level mathematics grades achieved by some students one year.

a How many students got a grade of B or C?

b How many students took A-level mathematics?

c Estimate the probability that a student will get a grade of B or C next year.

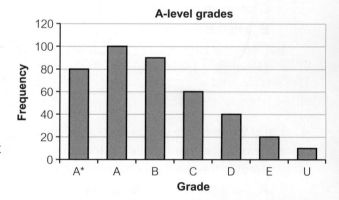

Draw a four-sided and a six-sided fair spinner.
Put the numbers 1, 2 or 3 on the sections of each spinner.
You can use each number as often as you like.
Make up three probability questions about each spinner.
Work out the answers to your questions.

Think carefully about the work you have done in this unit.

a Write down, in your own words, a short definition for
   i probability   **ii** probability scale
   iii possible outcomes **iv** successful outcomes

b The word 'event' is used a lot in probability. How is it used differently in probability compared with everyday life?

7 Ratio and proportion

Master Check up p191 Strengthen p193 Extend p198 Unit test p200

7.1 Direct proportion

- Use direct proportion in simple contexts
- Solve simple problems involving direct proportion
- Use the unitary method to solve simple word problems involving direct proportion

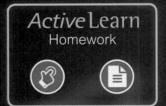

ActiveLearn
Homework

Warm up

1 Fluency What is
   a double 3   **b** double 4   **c** half of 18   **d** half of £5

2 Work out
   a 250g ÷ 5   **b** £1.20 ÷ 6   **c** 35kg ÷ 7   **d** 56m ÷ 8

3 Work out
   a 40 ÷ 5 × 3   **b** 20 ÷ 4 × 6   **c** 16 ÷ 8 × 5   **d** 50 ÷ 10 × 12

> **Key point** When two quantities are in **direct proportion**, as one increases or decreases, the other increases or decreases at the same rate.
> This means that when one quantity is zero, so is the other; when one doubles, so does the other; when one is multiplied by 3, so is the other, etc.

4 A theatre ticket costs £21. How much do 2 tickets cost?

5 2 cinema tickets cost £30. How much is 1 cinema ticket?

> **Key point** In the **unitary method**, you find the value of one item before finding the value of more.

Worked example

3 memory sticks cost £24.
How much do 7 memory sticks cost?

1 memory stick = £24 ÷ 3 = £8

7 memory sticks = £8 × 7 = £56

Draw a bar model to show this. Each equal section represents 1 memory stick.

£24

Use the bar model to help you find the cost of 1 memory stick.

£8

Use the bar model to help you find the cost of 7 memory sticks.

| £8 | £8 | £8 | £8 | £8 | £8 | £8 |

6 A recipe for 4 people needs 100 g of sugar.
How many grams of sugar are needed for
a 1 person **b** 7 people?

7 6 packets of sweets contain 42 sweets.
How many sweets in 10 packets?

8 3 packs of gift tags contain 18 tags.
How many tags in 4 packs?

9 4 notepads cost £12.
How much do 5 notepads cost?

10 Ben buys 7 sheets of coloured card for 56p.
How much would he pay for
a 12 sheets **b** 18 sheets **c** 20 sheets?

11 **Problem-solving** Burning 4 litres of diesel emits about 10 kg of CO_2.
Work out how much CO_2 is emitted from burning
a 1 litre of diesel **b** 5 litres of diesel **c** 7 litres of diesel.

12 5 equal sized coaches can carry a total of 215 people.
How many people can be carried by
a 9 coaches **b** 15 coaches **c** 21 coaches?

13 Mike can make 3 bird tables from 4.5 m of wood.
How many metres of wood will he need to make
a 11 bird tables **b** 17 bird tables **c** 28 bird tables?

14 A 300 g pot of yoghurt contains 75 g of fat.
How much fat is there in a 700 g pot?

15 **Problem-solving** It takes a music examiner 1 hour to test 3 students.
How long will it take the examiner to test 8 students?

16 **Problem-solving** A scout leader orders a mug for each of the
31 scouts in his group. The total value of the order is £93.
4 more scouts join the group, so he orders an extra 4 mugs.
What is the total value of the order now?

> **Q15 Hint** Change the
> hour into minutes first.

17 **Reasoning** A recipe for 4 people uses 120 g of cheese.
How much cheese is needed for 8 people?
Andrew works out the answer like this:

120 g				120 g				120 g			

8 people would need twice as much. 120 g × 2

a What is Andrew's answer?
b Use Andrew's method to answer this question:
A recipe for 3 people uses 90 g of flour.
How much flour is needed for 6 people?

18 6 eggs cost £1.50.
How much are a dozen eggs?

Q18 hint 1 dozen = 12

19 18 oranges cost £5.
How much do 9 oranges cost?

Q19 hint 9 oranges is half as many as 18 oranges.

20 A fairground ride for 4 people is £22.
How much is the ride for 2 people?

21 A recipe for 6 people uses 4 eggs.
How many eggs are needed for
a 12 people
b 3 people
c 9 people
d 15 people?

Q21c hint

6 people			3 people		

22 **Reasoning** 4 tickets to a music concert cost £140.
Show two different ways you can work out the cost of 14 tickets.
Which method is better? Why?

23 **Problem-solving** Which of these are in **direct proportion**? Explain.
a A recipe for 12 muffins uses 120 ml of milk;
a recipe for 24 muffins uses 240 ml of milk.
b A shop offer: 1 pair of jeans costs £25;
2 pairs of jeans cost £45.
c One bunch of flowers costs £6; three bunches
of flowers cost £18.
d One week, Jamie receives £40 for delivering 1000 leaflets; the next week,
he receives £20 for delivering 460 leaflets.

Q23 hint When two quantities are in direct proportion, when one multiplies or divides by an amount, the other multiplies or divides by the same amount.

Challenge This pancake recipe makes 10 pancakes:

100 g plain flour

2 eggs

300 ml milk

How much of each ingredient would you need to make 2 pancakes each for everyone in
your class?

Reflect Alan tells his mum he spotted two examples of direct proportionality at the
petrol station:
A Mum's car: 30 litres of fuel costs £42
Next car: 10 litres of fuel costs £14
B In our queue six people are served in 10 minutes.
In the other queue, three people are served in 6 minutes.

His mum is impressed!
Should his mum be impressed? Explain.

7.2 Writing ratios

- Use ratio notation
- Reduce a ratio to its simplest form
- Reduce a three-part ratio to its simplest form by cancelling

Active Learn
Homework

Warm up

1 Fluency Work out

a 24 ÷ 6   **b** 30 ÷ 5   **c** 49 ÷ 7   **d** 56 ÷ 8

2 What is the highest common factor (HCF) of

a 3 and 6   **b** 6 and 10   **c** 12 and 16   **d** 8, 12 and 24?

3 What are the missing numbers?

a

For every 1 black bead there are ☐ white beads.

b

For every 2 black beads there are ☐ white beads.

Key point

A **ratio** is a way of comparing two or more quantities.

Ratios are written as numbers separated by a colon ':'

For example, in this tile pattern there are 2 blue tiles for 1 red tile.

| B | R | B |

The ratio of blue tiles to red tiles is 2:1.

4 Write the **ratio** of blue beads to yellow beads for each necklace.

a   YBYBY

b   BBYBB

c   YBYBYBY

d   YBBBBBY

5 Draw beads to show the ratios

a blue to yellow 5:1   **b** blue to yellow 1:5

6 Reasoning Is the ratio 5:1 the same as the ratio 1:5? Explain your answer.

7 Write these as ratios.

a There are 5 g of chemical A for every 2 g of chemical B.
Write the ratio of chemical A to chemical B.

b In a sample of blood, there are 1250 red blood cells for every white blood cell.
Write the ratio of red to white blood cells in this sample.

c Ammonia is made of 1 nitrogen atom for every 3 hydrogen atoms.
Write the ratio of nitrogen to hydrogen.

> **Q7a hint**
> A:B
> ☐:☐

8 a Write the ratio of green to white tiles for each pattern.

a ▮☐☐☐ b ▮☐☐☐☐▮☐☐ c ▮☐☐☐☐▮☐☐☐☐▮☐☐

d Copy and complete for the tiles in parts **a** to **c**.
   For every 1 green tile there are ☐ white tiles.

> **Key point**   You can make the numbers in a ratio as small as possible by **simplifying**.
> You simplify a ratio by dividing the numbers in the ratio by the **highest common factor**.

9 Write the ratio of blue beads to yellow beads for each necklace.
   Simplify each ratio if possible. The first one has been started for you.

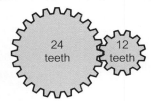

a ⟨YBBBBY⟩   blue : yellow = 4 : 2 = 2 : ☐

b ⟨BBBYYYBBB⟩

c ⟨YYBBBBYY⟩

10 Write each ratio in its simplest form.

  a 2 : 20   **b** 25 : 5   **c** 4 : 24

  d 6 : 30   **e** 8 : 24   **f** 6 : 66

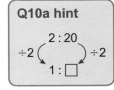
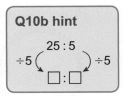

Q10a hint
2 : 20
÷2 ⤵⤴ ÷2
1 : ☐

Q10b hint
25 : 5
÷5 ⤵⤴ ÷5
☐ : ☐

11 A gear has two cogs. There are 24 teeth on the large cog and 12 teeth on the small cog.
   What is the ratio of the number of teeth on the large cog to the small cog?

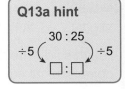

24 teeth   12 teeth

Write your answer as a ratio in its simplest form.

12 A sample of methane has 100 carbon atoms for every
   400 hydrogen atoms.
   Write the ratio of carbon atoms to hydrogen atoms in its simplest form.

13 Write each ratio in its simplest form.

  a 30 : 25   **b** 24 : 10   **c** 16 : 12

  d 40 : 15   **e** 9 : 30   **f** 21 : 28

Q13a hint
30 : 25
÷5 ⤵⤴ ÷5
☐ : ☐

14 A solution is mixed from two different liquids.
   There are 50 ml of liquid A and 125 ml of liquid B.
   Write the ratio of liquid A to liquid B in its simplest form.

15 Problem-solving   At a nursery there are 5 members of staff and 20 children.
   The recommended ratio of staff to children is 1 : 4.
   Has the nursery got the correct ratio?

16 Problem-solving Digital television screens usually have a width:height ratio of 16:9.
This screen has width 80 cm and height 45 cm.
Does the screen have a ratio of 16:9?
Show how you worked out your answer.

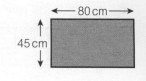

17 Copy and complete the ratio of red to blue to white bunting.

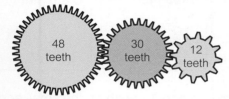

6:☐:☐

18 Write each ratio in its simplest form.
a 6:4:2   b 20:5:10   c 8:4:12
d 8:24:12   e 20:12:16   f 24:30:18
g 14:21:42   h 25:15:35   i 8:32:56

Q18d hint

8:24:12

÷4 (÷4) ÷4

☐:☐:☐

19 Problem-solving This gear has three cogs.

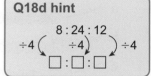

48 teeth   30 teeth   12 teeth

The first cog has 48 teeth, the second has 30 teeth and the third has 12 teeth.
What is the ratio of the number of teeth on the first cog to the second cog to the third cog?
Write your answer as a ratio in its simplest form.

20 Problem-solving A rose gold bracelet is made from 30 g of gold, 8 g of copper and 2 g of silver.
Write the ratio of gold:copper:silver in its simplest form.

Challenge   Write three different ratios that simplify to
a 4:5   b 1:2:3

Reflect   After this lesson, Miguel and Judith discussed what they noticed about ratios.
Miguel said, 'A ratio tells you how much of one thing there is compared to another.'

Judith said, 'The things must always be the same kind. So, they must all be numbers of objects, or all numbers of people, or all lengths.'

Look back at the questions you answered about ratios.
Is Miguel correct?
Is Judith correct?
What other quantities or amounts could you write in a ratio?

7.3 Using ratios

* Find equivalent ratios
* Divide a quantity into two parts in a given ratio
* Solve word problems involving ratio
* Use ratios and measures

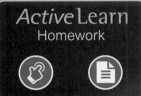

*Active*Learn
Homework

Warm up

1   **Fluency**   Here is a bag containing white and blue balls.
What is the ratio of white : blue balls in the bag?

2   **Fluency**   Work out

   a $28 \div 4$ **b** $36 \div 6$ **c** $60 \div 5$ **d** $44 \div 11$

3   Copy and complete

   a $1\,m = \square\,cm$ **b** $\square\,m = 1\,km$ **c** $1\,kg = \square\,g$ **d** $\square\,ml = 1\,litre$

Key point

Multiplying all the numbers in a ratio by the same number gives an **equivalent ratio**.

4   Write an **equivalent ratio** of white to green beads for each necklace.

   a

   b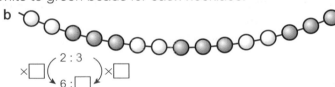

5   Copy and complete these equivalent ratios.

   a $5:2 = \square:4$ **b** $1:2 = 3:\square$ **c** $6:1 = \square:3$

   d $7:3 = 14:\square$ **e** $4:5 = \square:20$ **f** $11:2 = 22:\square$

   g $6:7 = \square:49$ **h** $15:8 = 30:\square$

> **Q5a hint**
>

Worked example

The ratio of cumin to paprika in a recipe is $1:2$.
Nimah uses 3 teaspoons of cumin. How many teaspoons of paprika does she use?

C : P
$\times 3 \left(\begin{array}{c} 1:2 \\ 3:6 \end{array} \right) \times 3$ —— | Multiply each part by the same number to get an equivalent ratio.

Nimah uses 6 teaspoons of paprika.

6   A scarf is made from balls of grey wool and pink wool in the ratio $1:3$.
How many balls of pink wool are needed for 4 balls of grey wool?

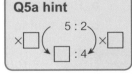

7   Hummingbirds eat nectar made from sugar and water in the ratio $1:4$.
How much water is needed for 3 teaspoons of sugar?

A **8** **Problem-solving** A recipe for Thai chicken uses Thai sauce and fresh ginger in the ratio 2:1. Anna uses 4 tablespoons of Thai sauce. How much ginger does she use?

A **9** **Problem-solving** Harry invests some money in low-risk and high-risk investments in the ratio 7:3.
He invests £1800 into the high-risk investments.
How much money does he invest altogether?

L : H
7 : 3
1800

B **10** In a necklace there is one blue bead for every four yellow beads.
There are 15 beads altogether.

B   Y
1   4

   a How many blue beads are there?
   b How many yellow beads are there?
   c **Reasoning** How can you check that your answers to parts **a** and **b** are correct?

Q10 hint
blue : yellow   total beads
      1 : 4   5
×☐ (⟳) ×☐   ) ×☐
      ☐ : ☐   15

B **11** **Problem-solving** A game requires 18 counters in total.
There must be one white counter for every two black counters.
How many of each colour counter must be included in the game box?

B **12** Share these amounts between Alice and Ben in the ratios given.
Show how you check your answers.
   a £21 in the ratio 2:1   **b** £45 in the ratio 2:3
   c £96 in the ratio 7:5   **d** £28 in the ratio 4:3
   e £72 in the ratio 3:5   **f** £60 in the ratio 11:4

Q12a hint
      2 : 1   £3
×☐ (⟳) ×☐   ) ×☐
      ☐ : ☐   £21

B **13** Red gold is made from copper and gold in the ratio 1:3.
A red gold necklace weighs 24 g.
   a How much copper is in the necklace?   **b** How much gold is in the necklace?

14 In the UK, the ratio of men to women over 80 is approximately 4:7.
In a village in the UK there are 110 people over 80.
   a **Problem-solving** How many would you expect to be men?
   b **Reasoning** Why can't you be sure this is the exact number of men?

> **Key point** You can use **ratios** to convert between **metric units**.

15 Copy and complete.
Every 1 cm is the same as _____ mm. The ratio cm:mm is 1:10.

16 Write these conversions as ratios.
   a mm:cm   **b** cm:m   **c** km:m
   d kg:g   **e** ml:litres   **f** m:cm

17 Complete these conversions.
   a 9 m = ☐ cm   **f** 30 mm = ☐ cm
   b 2 cm = ☐ mm   **g** 12 000 ml = ☐ litres
   c 7 litres = ☐ ml   **h** 10 cm = ☐ mm
   d 5000 m = ☐ km   **i** 100 m = ☐ km
   e 200 cm = ☐ m   **j** 1500 ml = ☐ litres

Q17a hint
   m : cm
   1 : 100
×9 (⟳) ×9
   9 : ☐

Q17d hint
   m : km
   1000 : 1
×5 (⟳) ×5
   5000 : ☐

18 Complete these conversions.

a 3.6 m = ☐ cm   **b** 2.8 kg = ☐ g   **c** 3.1 cm = ☐ mm

d 8.9 kg = ☐ g   **e** 3900 m = ☐ km   **f** 630 cm = ☐ m

g 84 mm = ☐ cm   **h** 8600 ml = ☐ litres   **i** 70 m = ☐ cm

19 A carpenter has a piece of wood 1.2 m long.

a How long is the piece of wood in centimetres?

b He cuts the wood into two pieces in the ratio 2:3.
Work out the length of the shorter piece of wood.

20 Problem-solving The heights of a man and a tree are in the ratio 1:6. The man's height is 180 cm.
What is the height of the tree? Give your answer in metres.

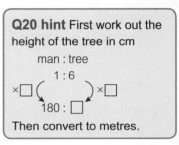

Q20 hint First work out the height of the tree in cm

man : tree

1 : 6

×☐ ⟨ ⟩ ×☐

180 : ☐

Then convert to metres.

21 Problem-solving Jesse makes green paint by mixing blue paint and yellow paint in the ratio 3:5.
He uses 750 ml of yellow paint.
How much green paint does he make in total?
Give your answer in litres.

22 Problem-solving The Wilson family and the Jones family share the cost of a holiday cottage in the ratio of the number in each family. The table shows the cost of the cottage and the number of people in each family for the two years they go away together.

	Number in Wilson family	Number in Jones family	Cost of holiday cottage
2016	2	3	£450
2019	4	5	£630

a How much do the Wilson family pay in

i 2016   **ii** 2019?

b How much do the Jones family pay in

i 2016   **ii** 2019?

c Which family has the biggest increase in price from 2016 to 2019?

Challenge Two people start a business. The first person invests £4000 in the business and the second person invests £6000 in the business.
At the end of the first year, they have both worked the same number of hours.
The profit they have made is £50 000.
Is it fair that they get £25 000 each? Explain.

Reflect Look back at the questions you answered in this lesson.
For type A questions, you were given the ratio and one part and asked to find the other part (for example Q6, Q7). For type B questions, you were given the ratio and total and asked to find the parts (for example Q10, Q11).

Find two other questions for types A and B.

What maths operations (addition, subtraction, multiplication, or division) did you use to solve each type of question?

7.4 Ratios, proportions and fractions

- Use fractions to describe and compare proportions
- Understand and use the relationship between fractions, ratio and proportion

Active Learn
Homework

Warm up

1 Fluency   In each list, which fraction is
    i smallest   **ii** largest?

a $\frac{3}{8}, \frac{7}{8}, \frac{1}{8}$

b $\frac{1}{2}, \frac{1}{5}, \frac{1}{3}$

c $\frac{11}{15}, \frac{4}{15}, \frac{7}{15}$

2 Write Simon's age as a fraction of Debbie's age when
    a Simon is 15 and Debbie is 20   **b** Simon is 20 and Debbie is 25
    Write your answers as fractions in their simplest form.

3 a Write the ratio of blue tiles to white tiles. Give the ratio in its simplest form.

    b What fraction of the tiles are
       i blue   **ii** white?
       Simplify your fractions.

Key point   A **proportion** compares a part with a whole. You can write a proportion as a fraction, a decimal or a percentage.

Worked example

In a biscuit tin, there are 10 chocolate and 4 shortbread biscuits.
What proportion are

a   chocolate   **b** shortbread?

$10 + 4 = 14$ —————— Work out the total number of biscuits.

a $\dfrac{\text{chocolate biscuits}}{\text{total biscuits}} = \dfrac{10}{14} = \dfrac{5}{7}$ ————— Write each amount as a fraction of the total. Simplify.

b $\dfrac{\text{shortbread biscuits}}{\text{total biscuits}} = \dfrac{4}{14} = \dfrac{2}{7}$ ————— Check your answer: $\frac{5}{7} + \frac{2}{7} = 1$

---

4   A zoo has five adult tigers and three tiger cubs.
    What proportion of the zoo's tigers are cubs?

5   In a play there are 4 musicians and 6 actors.
    What proportion of those in the play are actors? Give your answer in its simplest form.

6 In class 7G, 14 students are 11 years old and 16 students are 12 years old. Write down the proportion of the class that are each age.

7 In class 7G, there are 12 boys and 18 girls.
   a Write down the ratio of boys to girls. Give the ratio in its simplest form.
   b Reasoning Sam writes that the proportion of boys is $\frac{2}{3}$
      Is Sam correct? Show working to support your answer.

8 Bell metal is made from copper and tin in the ratio 4:1.
   a What fraction of bell metal is
     i copper   **ii** tin?
   b Show how to check your answers to part **a**.

> **Q8 hint**
>
> | copper | copper | copper | copper | tin |
> |--------|--------|--------|--------|-----|
>
> Proportion as a fraction = $\frac{\square}{5}$

9 Brass is made from copper and zinc in the ratio 7:3.
   a What proportion of brass is
     i copper   **ii** zinc?
   b Show how to check your answers to part **a**.

10 The sauce for a Thai dish uses Thai sauce and oyster sauce in the ratio 5:3.
   a What fraction of the sauce is
     i Thai sauce   **ii** oyster sauce?
   b Show how to check your answers to part **a**.

11 Joshua makes green paint by mixing blue paint and yellow paint in the ratio 4:7.
   a What proportion of the green paint is
     i blue paint   **ii** yellow paint?
   b Show how to check your answers to part **a**.
   c Problem-solving Joshua needs 22 litres of green paint. How many litres of each colour does he need?

12 In a shoe shop, six of the 42 pairs of shoes sold were size 5.
In a different shoe shop, 12 out of the 60 shoes sold were size 5.
Which shop sold the greater proportion of size 5 shoes?

> **Q12 hint**
> Cancel each fraction to its simplest form, then compare the fractions.
>
> $\frac{6}{42} = \frac{\square}{\square}$, $\frac{12}{60} = \frac{\square}{\square}$

13 There are 32 children in class 7H.
20 of them are boys.
There are 24 children in class 7T.
9 of them are boys.
Which class has the greater proportion of boys?

14 Problem-solving The table shows the number of toffee-fudge ice creams and the total number of ice creams sold one weekend.

Day	Number of toffee-fudge ice creams sold	Total number of ice creams sold
Saturday	25	150
Sunday	15	120

Which day had the greater proportion of sales of toffee-fudge ice cream?

15 Problem-solving The table shows the number of seats for season-ticket holders and the total number of seats at two football stadiums.

Stadium	Seats for season-ticket holders	Total number of seats
Manchester City FC	36 000	48 000
Reading FC	18 000	24 000

Which stadium has the greater proportion of seats for season-ticket holders?

16 Problem-solving These are the ages of passengers on two buses.

Bus 5A	12 38 42 25 15 14 61 27 50
Bus 12	26 11 12 11 63 71 58 32 13 68 29 80 41 12 14

Which bus has the greater proportion of children (age under 16) on it?

17 Problem-solving A music website sells singles and albums in the ratio 9:5.
A different website sells singles and albums in the ratio 5:2.
Which website sells the greater proportion of albums?

Q17 hint Use equivalent fractions to compare.

18 A school trip to a theme park has 4 full coaches of staff and students.
One coach seats 52 people. Altogether there are 16 staff.
Write the answer to these questions in their simplest form.
a What fraction of the people on the trip are staff?
b **Problem-solving** What is the ratio of staff to students?

Q18 hint How can you work backwards from a fraction to a ratio?

19 Problem-solving / Reasoning Tim makes orange squash by mixing 50 ml of squash with 450 ml of water.
Peter makes orange squash by mixing 30 ml of squash with 210 ml of water.
Who has made the stronger squash? Explain your answer.

Challenge In about 1490, Leonardo da Vinci made a drawing to show the proportions of the human body.
- The length of a person's ear is one-third of the length of their face.
- A person's arm span is equal to their height.
- The distance from the top of the head to the bottom of the chin is $\frac{1}{8}$ of a person's height.
- The distance from the elbow to the tip of the hand is $\frac{1}{5}$ of a person's height.
- The length of a person's foot is $\frac{1}{7}$ of a person's height.
- The length of a person's hand is $\frac{1}{10}$ of a person's height.

a Write these statements as ratios.

b Are any of these statements true for you?

Reflect Write in your own words the difference between a ratio and a proportion.
Use this bar as an example to write some statements about the ratio of red to yellow and the proportion of red and yellow.

R	R	R	R	Y	Y	Y	Y	Y	Y	Y	Y

7.5 Proportions and percentages

- Use percentages to describe proportions
- Use percentages to compare simple proportions
- Understand and use the relationship between percentages, ratio and proportion

Active Learn
Homework

Warm up

1 **Fluency** What are the missing numbers?
   a $10 \times \square = 100$ **b** $5 \times \square = 100$ **c** $4 \times \square = 100$

2 Copy and complete these equivalent fractions.
   a $\frac{23}{50} = \frac{\square}{100}$ **b** $\frac{7}{25} = \frac{\square}{100}$ **c** $\frac{9}{10} = \frac{\square}{100}$ **d** $\frac{13}{20} = \frac{\square}{100}$ **e** $\frac{126}{200} = \frac{\square}{100}$

3 **a** Write the ratio of red : blue in the bar.

R	R	R	R	R	R	R	B	B	B

   b What fraction of the bar is red?
   c What fraction of the bar is blue?
   d Copy and complete the percentage of the bar that is red. $\frac{\square}{10} = \frac{\square}{100} = \square\%$
   e What percentage of the bar is blue?

4 A rugby team won 9 out of their 10 line-outs.
   What proportion of their line-outs did they win?
   Give your answer as a percentage.

> **Q4 hint** Write the proportion as a fraction.
> $\frac{9}{10} = \frac{\square}{100} = \square\%$
> Convert the fraction to a percentage.

5 After a day's holiday, Dan has 50 new email messages in his inbox.
   13 of them have attachments.
   What proportion of the emails have attachments?
   Give your answer as a percentage.

6 An activity centre offers 30 different activities.
   15 of them are outdoors.
   What proportion of their activities are outdoors?
   Give your answer as a percentage.

Key point
You can compare **proportions** using **percentages**.

7 In a high street bookshop, 14 of the 20 books sold in a day were fiction.
   In an online bookshop, 138 of the 200 books sold in a day were fiction.
   a What proportion of the books sold by the high street bookshop were fiction?
      Give your answer as a percentage.
   b What proportion of the books sold by the online bookshop were fiction?
      Give your answer as a percentage.
   c Which bookshop sold a greater proportion of fiction books?

8 Problem solving In one season, a hockey team scored goals from 18 out of the 20 penalties they were awarded.

The hockey team also scored goals from 42 out of the 50 penalty corners they were awarded.

Did they score a greater proportion of goals from penalties or penalty corners?

9 Problem-solving The diagrams show a rectangle and a square.

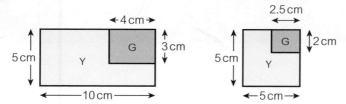

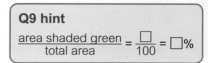

Which shape has the greater proportion shaded green?

Q9 hint

$\dfrac{\text{area shaded green}}{\text{total area}} = \dfrac{\square}{100} = \square\%$

10 Problem-solving A recipe for oatmeal bread uses 350 g flour and 250 g oatmeal.

A recipe for wheatmeal bread uses 550 g flour and 450 g wheatmeal.

Which bread has the greater proportion of flour?

Worked example

The ratio of boys to girls in a swimming club is $3:7$.
What percentage of the children are girls?

$3 + 7 = 10$ parts

$\dfrac{7}{10} = \dfrac{70}{100}$

70% are girls

For every three boys there are seven girls.
The percentage of girls is $\dfrac{7}{10} = \dfrac{70}{100} = 70\%$.

10 children

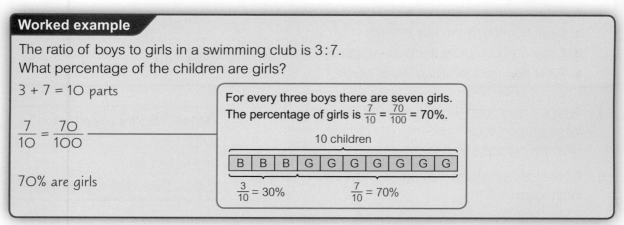

$\dfrac{3}{10} = 30\%$ $\dfrac{7}{10} = 70\%$

---

11 Bell metal is used for making bells.
Bell metal is tin and copper in the ratio $4:1$.
  a What percentage of bell metal is
    i tin   **ii** copper?
  b Reasoning Show how you can check your answers to part **a**.

12 The ratio of girls to boys in a school choir is $7:3$.
  a What percentage of the choir are
    i girls   **ii** boys?
  b Reasoning Show how you checked your answers to part **a**.

13 A sugar solution is made from glucose and fructose in the ratio $7:13$.
  a What percentage of the sugar solution is
    i glucose   **ii** fructose?
  b Reasoning Show how you checked your answers to part **a**.

14 The ratio of brass to string players in a school orchestra is $1:3$.
   The ratio of brass to string players in a professional orchestra is $1:4$.
   a What percentage of the school orchestra play brass instruments?
   b What percentage of the professional orchestra play string instruments?
   c **Reasoning** Luke says, 'There is a bigger proportion of brass players in the school orchestra than in the professional orchestra, therefore the school orchestra has more brass players.'
   Is Luke correct? Explain.

15 **Problem solving** The ratio of adults to children ice skating at a rink is $9:16$.
   The ratio of adults to children watching the ice skaters is $9:1$.
   Kate says there is a bigger proportion of children ice-skating than adults watching the ice skaters.
   Is she correct? Explain.

16 Sandra makes pink paint by mixing 400 ml of red paint with 850 ml of white paint.
   a Write the ratio of red paint to white paint in its simplest form.
   b What percentage of the pink paint is
      i red   ii white?
   c **Reasoning** Show how you checked your answers to part **b**.

17 **Problem solving** A financial advisor suggests that a client invests in low-risk shares and medium-risk shares in the ratio $2:3$.
   The client invests 70% of his money in medium-risk shares and the rest in low-risk shares.
   Has he followed the advisor's suggestion?
   Show your working.

18 **Problem solving** A fitness instructor advises Jeff to do an exercise. She says it should have a work-out to rest ratio of $3:1$.
   Jeff does the exercise, resting for 75% of the time.
   Does Jeff do the exercise as advised?
   Show your working.

Challenge

a Draw three different bar models where the proportion of sections coloured grey is 40%.

b What is the ratio of grey to white for each one?

Reflect   Use what you have learned in this unit to write five sentences that describe the ratio, proportion, fraction and percentage of the different colours in this bar.

6 cm   2 cm

Which word was easiest to write a sentence for? Why?
Which word was hardest to write a sentence for? Why?

7 Check up

Direct proportion

1 Three packets of cashew nuts cost £6.
Work out the cost of
a 1 packet   **b** 2 packets   **c** 5 packets

2 A recipe for four people uses 200 g flour.
How much flour is needed for
a 8 people   **b** 2 people   **c** 6 people?

3 It takes Mary half an hour to clean 4 dog kennels at a rescue centre.
She works from 8 am to 11 30 am.
Has Mary got enough time to clean all 28 kennels?
Show your working.

4 It costs eight people £320 to go horse riding.
How much will it cost five people?

Ratio

5 Write the ratio of green to yellow tiles.

a | Y | Y | G | G | G | Y | Y |
    b | Y | G | G | Y | Y | G | G | G |

6 Write each ratio in its simplest form.
a 2:12   **b** 28:4
c 10:15   **d** 8:18:20

7 Which of these ratios is equivalent to 4:9?
A 1:6   **B** 7:12   **C** 12:27   **D** 18:8

8 Joe uses coriander and turmeric in a curry in the ratio 3:1.
How much coriander does he use with 2 teaspoons of turmeric?

9 The ratio of boys to girls in a volleyball club is 4:3.
12 club members are boys.
a How many members are girls?
b What is the total number of members in the club?

10 Katy keeps chickens. For every one white egg she gets five brown eggs.
There are 30 eggs altogether.
a How many white eggs are there?   **b** How many brown eggs are there?

11 a Share £30 in the ratio 2:3.
b Show how you checked that your answer is correct.

12 The lengths of two rollercoaster tracks are in the ratio 3:10.
When the rollercoaster tracks are joined together, the total length is 2600 m.
What is the length of the longer rollercoaster track? Give your answer in kilometres.

Comparing proportions

13 A dance group has 11 dancers. 4 of them are tap dancers.
   What proportion of the dance group are tap dancers?

14 In an audience, there are 24 adults and 48 children.
   Write down the proportion of
   a adults   **b** children
   Give your answers as fractions in their simplest form.

15 In the matches in their first season, a football team scored no goals in 7 out of the
   20 matches they played. In the matches in their second season, they scored no goals in 9
   out of the 25 matches they played.
   a In what proportion of their matches did they score no goals
      i in their first season   **ii** in their second season?
      Give your answers as percentages.
   b In which season did they score no goals in a greater proportion of their matches?

16 These are the ages of the members of two badminton clubs.

Dragons	12 14 15 15 17 17 19 20 22 22
Swifts	11 11 13 14 14 15 15 16 16 16 17 19 19 20 21 21 22 24 25 25

   Which badminton club has the greater proportion of members over the age of 18?

17 When Gill makes bread, she uses white flour and wholemeal four in the ratio 3 : 1.
   What fraction of the flour is
   a white flour   **b** wholemeal flour?
   c Gill needs 300 g of flour. How much of each type of flour does she use?

18 The ratio of men to women in a quiz team is 2 : 3.
   a What percentage of the quiz team are
      i women   **ii** men?
   b Show how you checked your answers to part **a**.

Challenge

1 Copy this rectangle on to squared paper.
   a Show three different ways that you can divide the rectangle into
      two parts so that the areas of the two parts are in the ratio 1 : 2.
   b Show three different ways that you can divide the rectangle into
      two parts so that the areas of the two parts are in the ratio 5 : 3.

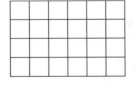

2 The ratio of boys to girls in a swimming club is 4 : 5.
   Write three different numbers of boys and girls that could be in the swimming club.
   In each case, give the total number of children in the swimming club.

Reflect   How sure are you of your answers? Were you mostly

                😞 Just guessing   😐 Feeling doubtful   😊 Confident

   What next? Use your results to decide whether to strengthen or extend your learning.

7 Strengthen

Direct proportion

1 A raffle ticket at a charity event costs £3.
Copy and complete the bar models and working.

a | £3 | £3 | 2 raffle tickets = 2 × £3 = ☐

b | £3 | £3 | £3 | £3 | £3 | 5 raffle tickets = ☐ × £3 = ☐

c | £3 | £3 | £3 | £3 | £3 | £3 | £3 | 7 raffle tickets = ☐ × ☐ = ☐

2 A book of 5 raffle tickets at another charity event costs £20.
Copy and complete the bar models and working.

a £20   1 raffle ticket = £20 ÷ 5 = £☐

b | £☐ | £☐ | £☐ | 3 raffle tickets = 3 × £☐ = £☐

c | £☐ | £☐ | £☐ | £☐ | 4 raffle tickets = 4 × £☐ = £☐

3 It costs £60 for 6 children to go sledging in a snowdome.
How much does it cost for
a 1 child   **b** 2 children
c 4 children   **d** 8 children?

Q3 hint

£60

4 3 packets of rice cost £1.50.
Work out the cost of
a 2 packets   **b** 4 packets
c 5 packets   **d** 10 packets

Q4 hint Find the
cost of 1 packet first.

5 **Problem-solving** There are 100 calories in 5 teaspoons of sugar.
How many calories are there in 8 teaspoons of sugar?

Ratio

1 Copy and colour this bar to show the ratios

| | | | |

a black to white 1 : 3
b black to white 3 : 1
c black to white 2 : 2

Q1a hint
black to white 1 : 3
Colour 1 black. Leave 3 white.

2 a Copy and colour this bar to show the ratios

| | | | | | | | |

   i black to white 5 : 3
  ii black to white 2 : 6
 iii black to white 1 : 3

Q2a iii hint For every 1 black
part there are 3 white parts.

b What do you notice about your answers to parts **ii** and **iii**?

3 Write $2:6$ in its simplest form.

4 Copy and complete to write each ratio in its simplest form.

Q3 hint

a  $\div2 \begin{pmatrix} 2:4 \\ \square:\square \end{pmatrix} \div\square$ $2:4 = \square:\square$

b $\div3 \begin{pmatrix} 6:3 \\ \square:\square \end{pmatrix} \div\square$ $6:3 = \square:\square$

c $\div\square \begin{pmatrix} 3:12 \\ \square:\square \end{pmatrix} \div\square$ $3:12 = \square:\square$

d 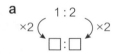 $\div\square \begin{pmatrix} 30:5 \\ \square:\square \end{pmatrix} \div\square$ $30:5 = \square:\square$

e $6:8 = \square:\square$

f $6:27 = \square:\square$

g $16:24 = \square:\square$

5 **Reasoning** This is how Ian simplifies the ratio $36:54$ to its simplest form.
Is Ian correct? Explain your answer.

6 A campsite has pitches for 20 static caravans and 35 touring caravans. Write the ratio of static caravans to touring caravans in its simplest form.

Q6 hint $20:35 = \square:\square$
Try dividing by 2, by 3, by 4, … until you find a factor that works for both numbers.

7 Copy and complete these equivalent ratios.

a $\times2 \begin{pmatrix} 1:2 \\ \square:\square \end{pmatrix} \times2$ **b** $\times3 \begin{pmatrix} 1:2 \\ \square:\square \end{pmatrix} \times3$ **c** $\times4 \begin{pmatrix} 1:2 \\ \square:\square \end{pmatrix} \times\square$ **d** $\times5 \begin{pmatrix} 1:2 \\ \square:\square \end{pmatrix} \times\square$

e $\times2 \begin{pmatrix} 4:5 \\ \square:\square \end{pmatrix} \times\square$ **f** $\times3 \begin{pmatrix} 4:5 \\ \square:\square \end{pmatrix} \times\square$ **g** $\times5 \begin{pmatrix} 4:5 \\ \square:\square \end{pmatrix} \times\square$ **h** $\times10 \begin{pmatrix} 4:5 \\ \square:\square \end{pmatrix} \times\square$

8 Write two equivalent ratios to $2:3$.
Check that your ratios simplify to $2:3$.

9 Jan uses the herbs basil and oregano in a sauce in the ratio

basil : oregano
$1:2$

She uses 2 teaspoons of basil.
Copy and complete to find out how much oregano she uses.

basil : oregano

$\times2 \begin{pmatrix} 1:2 \\ 2:\square \end{pmatrix} \times\square$
teaspoons   teaspoons

10 The ratio of boys to girls in a gym club is

boys : girls
2 : 3

Q10a hint

There are six boys in the club.
a How many girls are in the club?
b What is the total number of children in the club?

11 The ratio of steel to plastic in a washing machine is 10 : 7.
The steel in one washing machine weighs 40 kg.
How much does the plastic in the washing machine weigh?

12 Draw bars to show these ratios.
For each one, write the total number of parts.
The first one has been done for you.

   a 1 : 3   **b** 1 : 2   **c** 4 : 1   **d** 2 : 3

4 parts

13 Share these amounts between Andy and Bern in the
ratios given. The first one has been started for you.
   a £12 in the ratio 1 : 3

Q13a hint

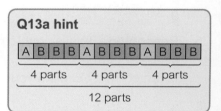

4 parts   4 parts   4 parts

12 parts

A : B
1 : 3
×3   × □
£3 : £□

Check: £3 + □ = £12
   b £15 in the ratio 1 : 2
   c £20 in the ratio 4 : 1
   d £20 in the ratio 2 : 3

14 **Problem-solving** A hairdresser makes hair
colouring by mixing 1 part dye with 2 parts peroxide
solution. He wants 60 ml of hair colouring.
   a How much dye does he use?
   b How much peroxide solution does he use?
   c Show how you checked your answers to
      parts **a** and **b**.

Q14 hint

60 ml

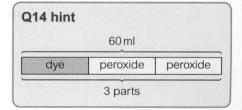

| dye | peroxide | peroxide |

3 parts

Comparing proportions

1 There are 7 chocolate muffins in a pack of 10 muffins.
What proportion of the pack are chocolate muffins?

2 Annie gets 9 spellings out of 10 correct
in a spelling test.
What proportion does she get correct?

Q1 hint

10

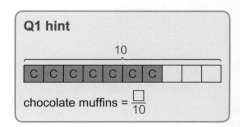

| C | C | C | C | C | C | C | | | |

chocolate muffins = □/10

3 There are 10 flowers in a vase. 3 of them are red.
  a What proportion are red?
  b What proportion are not red?

4 There are 10 meals on a menu. 4 of them are fish dishes.
  Copy and complete.

Proportion of fish dishes =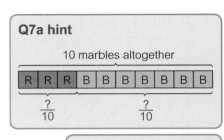

5 8 out of 12 meals on a menu are vegetarian.
  Write the proportion of vegetarian meals as a fraction in its simplest form.

6 40 out of the 120 pupils in Year 8 live less than 5 miles away from school.
  a What fraction is that?
  b Write the fraction in its simplest form.

7 **a** There are 3 red and 7 blue marbles in box A.
    What fraction of the marbles are red?
  b There are 3 red and 2 blue marbles in box B.
    What fraction of the marbles are red?
  c Which box has the greater proportion of
    red marbles?

> **Q7a hint**
>
> 10 marbles altogether
>
> | R | R | R | B | B | B | B | B | B | B |
>
> $\frac{?}{10}$ $\frac{?}{10}$

> **Q7c hint** Compare the
> fractions in parts **a** and **b**.
> Which is greater?

8 **a** There are 5 boys in a group of 8 students.
    What fraction of the group are boys?
  b There are 6 boys in a group of 16 students.
    What fraction of this group are boys?
  c Which group has the greater proportion of boys?

9 There are 100 emails in Carol's inbox.
  73 of them are from friends.
  What percentage of the emails are from friends?

> **Q9 hint**
>
> $\frac{\square}{100} = \square\%$

10 Ellie has 5 songs on a playlist. 3 of them are R&B.
  a What fraction of the songs are R&B?
  b What percentage of the songs on Ellie's playlist are R&B?
  Dev has 25 songs on a playlist. 12 of them are R&B.
  c What percentage of the songs on Dev's playlist are R&B?
  d Whose playlist has the greater proportion of R&B songs?

> **Q10 hint**
>
> $\frac{3}{5} = \frac{6}{10} = \frac{\square}{100} = \square\%$
>
> $\frac{12}{25} = \frac{\square}{100} = \square\%$

11 Penny has 25 pairs of socks. 10 pairs are black.
  Sarah has 20 pairs of socks. 9 pairs are black.
  a What percentage of Penny's socks are black?
  b What percentage of Sarah's socks are black?
  c Who has the greater proportion of black socks?

12 Problem-solving The table shows the gender of the members of two canoe clubs.
G stands for girls; B stands for boys.

Seals	G B B G B B G G G G
Dolphins	B B B G G B G G G B G B G G B B G G G B

  a For each club, count the number of girls and the total number of children.

  b For each club, work out the percentage of girls.

  c Which canoe club has the greater proportion of girls?

13 When Sally bakes a teabread, she makes a topping out
of syrup and water in the ratio 2 : 1.
What fraction of the topping is
  a syrup   **b** water?

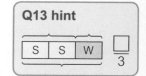

14 There are 3 red and 4 blue cards in a game.
  a What proportion are red?
  b What proportion are blue?
  c What is the ratio of red cards to blue cards?

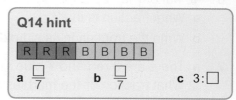

15 There are 11 milk chocolates and 9 dark chocolates in a box.
  a What proportion are milk?
  b What proportion are dark?
  c What is the ratio of milk to dark chocolates?

16 The ratio of silver to gold rings in a jewellery box is 2 : 5.
  a What is the proportion of silver rings?
  b What is the proportion of gold rings?
  c **Reasoning** Harry says, 'The proportion of silver rings is $\frac{2}{5}$.'
    What mistake has Harry made?

Challenge

1 In a surfing lesson $\frac{2}{3}$ of the group are boys and the rest are girls.
    Harry says, 'The ratio of boys to girls is 2 : 3.'
    Lily says, 'The ratio of boys to girls is 2 : 1.'
Who is correct? Explain your answer.

2 **Problem-solving** 21 friends go to the cinema.
They order taxis to take them from the cinema to a restaurant. One taxi carries 4 passengers.
  a What is the smallest number of taxis they need to order?
  b No more than 4 people are allowed in one taxi.
    Write down three ways that the 21 people can fit in the taxis.
  c Each taxi charges £15 for the journey, and the people in each taxi share the cost
    equally between them. Which of the three ways you found in part **b** would be the
    fairest way to travel, so that all the people pay roughly the same?

Reflect These lessons used bar models to help you solve ratio and proportion problems.
Did the bar models help you or not? Explain.

7 Extend

1 **Problem-solving / Reasoning** The bar graphs show the number of recorded delivery and special delivery letters a company sends during one week.

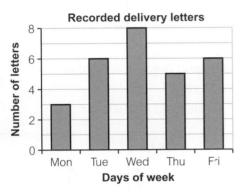

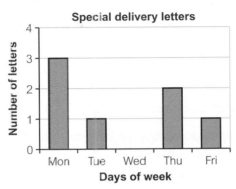

What is the ratio of recorded delivery to special delivery letters?
Write your answer in its simplest form.

> **Q1 hint** What do you need to work out first, before you can work out the ratio?

2 **Problem-solving** A recycling plant takes the glass out of plasma TVs. Half the weight of a plasma TV is glass and iron in the ratio 3:2. What is the weight of the glass in a plasma TV that weighs 30 kg?

3 **Problem-solving** A nurse uses this formula to work out the amount of medicine a patient needs each hour.

$$A = B \times M \times 60$$

> **Q3 hint** Try some easier numbers in the formula first such as $B = 20$ and $M = 60$.

where A is the amount of medicine per hour (in micrograms)
    B is the amount of medicine per kg body mass (in micrograms)
    M is the mass of the patient (in kg)

The ratio for converting from micrograms to milligrams is 1000:1.
Work out the value of A when $B = 18$ and $M = 62$.
Give your answer in milligrams.

4 **Problem-solving** The graph shows the income from souvenirs, photos and refreshments at a children's farm one weekend.

> **Q4 hint** What do you need to work out before you can work out the proportion of the income that came from souvenirs?

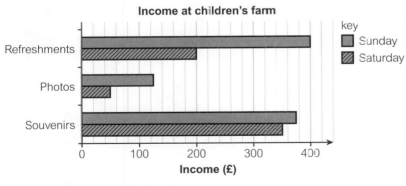

a Which day had the greater income from souvenirs?
b Which day had the greater proportion of income from souvenirs?

5 **Problem-solving** These are the ages of the 5 members of a riding club.
  8, 10, 11, 15, 16
  These are the ages of the 10 members of a diving club.
  15, 16, 16, 18, 20, 22, 22, 25, 37, 39
  In which club are the greater proportion of the members older than the mean age?

6 **Reasoning** Sam makes light blue paint by mixing 250 ml of blue paint with 750 ml of white paint.
  Tony makes light blue paint by mixing 400 ml of blue paint with 1600 ml of white paint.
  Who has made the lighter blue paint? Explain your answer.

7 60 g of green gold contains 45 g of gold, 12 g of silver, and the rest is copper.
  a Write the ratio of gold : silver : copper in its simplest form.
  b A green gold ring weighs 8 g. What is the weight of the gold in the ring?

8 In a class of students, $\frac{2}{5}$ have brown hair and $\frac{3}{5}$ have black hair. Write the ratio of students with brown hair to students with black hair.

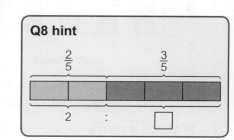

Q8 hint

9 In a gluten-free loaf, $\frac{3}{8}$ of the flour is gram flour, $\frac{1}{8}$ is rye flour and $\frac{4}{8}$ is rice flour. Work out the ratio of gram flour to rye flour to rice flour.

10 Four cans of beans weigh the same as six boxes of chocolates.
  Copy and complete the sentences.
  a Eight cans of beans weigh the same as □ boxes of chocolates.
  b Six cans of beans weigh the same as □ boxes of chocolates.
  c □ cans of beans weigh the same as 30 boxes of chocolates.
  d □ cans of beans weigh the same as 15 boxes of chocolates

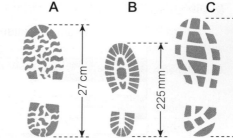

11 Pete invests £6000 into Premium Bonds, Income Bonds and Growth Bonds in the ratio 1 : 5 : 2.
  a How much does he invest in
    i Premium Bonds   ii Income Bonds   iii Growth Bonds?
  b Show how you checked your answers to part **a**.

Challenge   Forensic scientists use the length of a footprint to predict the height of a criminal.
For an adult, the usual ratio of length of foot to height is 3 : 20.

a Predict the height of the criminals who have left these footprints.
  Give your answers in metres.

b Measure your foot length and see if this ratio works for you.

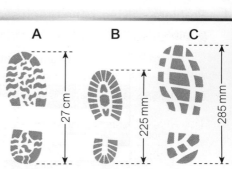

A   B   C

27 cm   225 mm   285 mm

Reflect   In these lessons, you were asked questions about ratios and proportions.

Are ratio and proportion the same thing or different? Explain.

In which other subjects might understanding ratio and proportion be useful to you?

7 Unit test

1 Abbie makes a necklace out of beads.
  For every red bead, she uses three green beads.
  She uses 24 beads altogether.
  a How many red beads does she use?
  b How many green beads does she use?

2 A T-shirt costs £3.
  Work out the cost of 11 T-shirts.

3 There are 4 teaspoons of sugar in 200 ml of cola.
  How many teaspoons of sugar are there in these amounts of cola?
  a 400 ml   b 100 ml   c 1 litre?

4 It takes a kitchen assistant 4 minutes to make 6 slices of toast.
  The kitchen assistant has 20 minutes to make 35 slices of toast.
  Has the kitchen assistant got enough time to complete the order?
  Show your working and explain your answer.

5 a Share 27 kg in the ratio 4 : 5.
  b Show how you checked that your answer is correct.

6 Write each ratio in its simplest form.
  a 3 : 15   b 30 : 5   c 15 : 25   d 80 : 60

7 There are 4 blue lights in a row of 10 disco lights.
  a What fraction of the disco lights are blue?
  There are 7 blue lights in a row of 20 disco lights.
  b What fraction of this row of disco lights are blue?
  c Which row of disco lights has the greater proportion of blue lights?
    Show how you worked out your answer.

8 The first time Abu played darts he missed the dartboard in 3 out of 10 throws.
  The second time Abu played darts he missed the dartboard once in 5 throws.
  a What proportion of Abu's throws missed the dartboard
    i the first time   ii the second time?
    Write your answers as percentages.
  b Did Abu miss the dartboard on a greater proportion of throws the first or the second time?

9 These are the times, in minutes, for members of two clubs to swim 100 lengths.

Dolphins	62 74 75 75 87 87 89 90 102 116
Sharks	51 71 73 74 74 75 75 76 76 78 78 79 89 90 91 91 92 94 95 95

  Which swimming club has the greater proportion of members with times under 80 minutes?

10 3 mugs cost £12. Work out the cost of
   a 1 mug   b 2 mugs   c 8 mugs

11 It costs £420 for 6 tickets to watch a football match.
How much does it cost for 5 tickets?

12 Greg uses petrol and oil in his chainsaw in the ratio of 50 : 1.
How much oil does he use with 1500 ml of petrol?

13 The ratio of girl to boy members in a club is 2 : 3.
There are eight girl members.
  a How many boy members are there?
  b What is the total number of children in the club?

14 When Gavin makes a fruit cake, he uses cherries and sultanas in the ratio 1 : 4.
  a What fraction of the fruit is
    i cherries   **ii** sultanas?
  b Gavin uses 250 g of fruit. How much of each type of fruit does he use?

15 The ratio of home to away supporters at a basketball match is 7 : 3.
  a What percentage of the supporters are
    i home supporters   **ii** away supporters?
  b Show how you checked your answers to part **a**.

16 Rob makes a rice pudding using 800 ml of milk and 200 ml of cream.
  a Write the ratio of milk to cream in its simplest form.
  b What percentage of the milk and cream mix is
    i milk   **ii** cream?

17 Debbie makes lemon squash by mixing 100 ml of squash with 900 ml water.
Sian makes lemon squash by mixing 40 ml of squash with 460 ml water.
Who has made the stronger squash? Explain your answer.

Challenge   Here are seven cards with ratios written on them.

| 1 : 2 | 1 : 3 | 1 : 4 | 2 : 3 | 3 : 5 | 3 : 7 | 4 : 5 |

RULES
Use each card once only.
Only use a card if it shares the amount exactly into whole numbers of pounds.

Follow these steps.
1 Choose a card and share £1000 in the ratio written on it.
Circle the larger amount.
2 Choose a different card and share the circled amount in this ratio. Circle the smaller amount.
3 Choose a different card and share the circled amount in this ratio. Circle the larger amount.
4 Continue until you can go no further.
  a What is the smallest final amount you can find?
  b Is there a strategy you can use to try to find the smallest amount? Explain your answer.

EXAMPLE
£1000 shared in the ratio 3 : 5 is £375 : £625
£625 shared in the ratio 2 : 3 is £250 : £375
£250 shared in the ratio 3 : 7 is £75 : £175
£175 shared in the ratio 1 : 4 is £35 : £140
£35 cannot be shared by any of the other ratios so STOP.

Reflect   List five new skills and ideas you have learned in this unit.
What mathematical operations did you use most (addition, subtraction, multiplication or division)? Which lesson in this unit did you like best? Why?

8 Lines and angles

Master   Check up p220   Strengthen p222   Extend p228   Unit test p230

8.1 Measuring and drawing angles

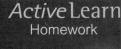

Active Learn
Homework

* Use a protractor to measure and draw angles
* Recognise acute, obtuse and reflex angles

Warm up

1 Fluency Which of these angles are larger than a right angle?

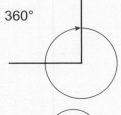

A   B   C   D

2 a What is the size of the angle measured on this protractor?
b What type of angle is it?

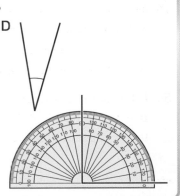

Key point   **Angle** measures a turn. Angles are measured in **degrees** (°).

A whole turn is 360°

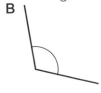

A **right angle** is a quarter turn, or 90°.

A half turn is 180°.

This symbol means an angle is a right angle.

---

3 What angles do these diagrams show?

a

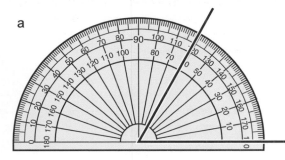

b

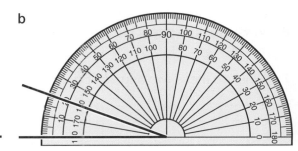

4 Look at these angles.

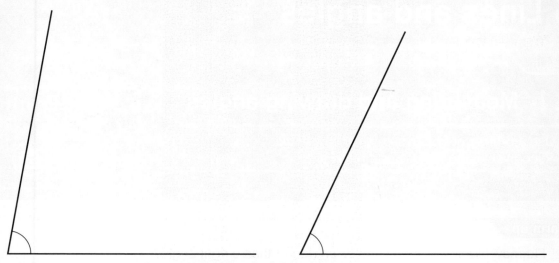

    a Is each angle larger or smaller than 90°?
    b Measure each angle.
       Give your answers to the nearest degree. Check they are sensible.

5 Measure each angle to the nearest degree.
    a

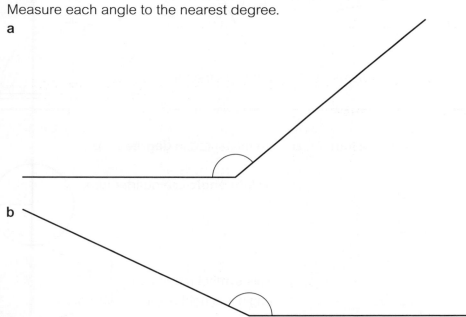

    b

> **Key point** An **acute** angle is smaller than 90°. An **obtuse** angle is between 90° and 180°.

6 What types of angle are the angles in
    a Q4   **b** Q5?

7 What do these two angles add up to?

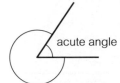

acute angle

8   Work out the size of each unknown angle.

a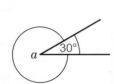

b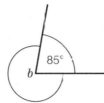

c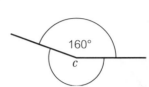

Worked example

Measure the reflex angle shown by the red arc.

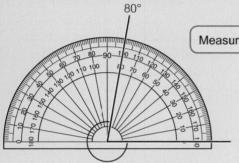

80°

> Measure the smaller angle.

$$360° - 80° = 280°$$

> Work out the size of the **reflex angle**.

---

9   **Reasoning**   Isabel needs to measure these reflex angles to the nearest degree.

     Copy and complete Isabel's work.

a   smaller angle = 120°, so marked angle = 360° − 120° = ☐

b   smaller angle = ☐, so marked angle = ☐

c   smaller angle = ☐, so marked angle = ☐

a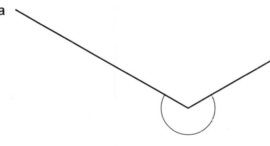

b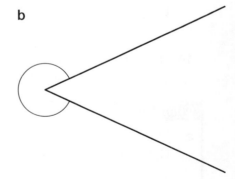

c

Worked example

Use a protractor to draw an angle of 60°.

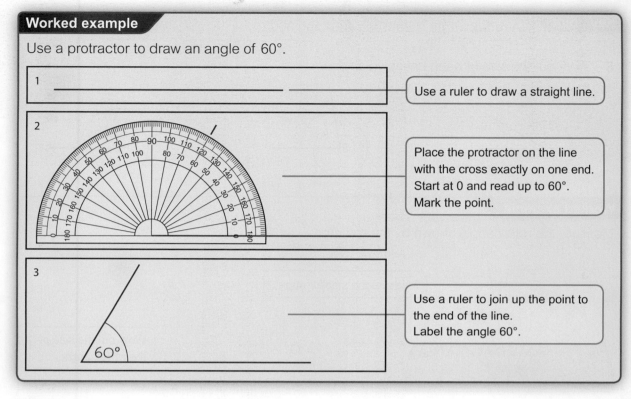

1 — Use a ruler to draw a straight line.

2 — Place the protractor on the line with the cross exactly on one end. Start at 0 and read up to 60°. Mark the point.

3 — Use a ruler to join up the point to the end of the line. Label the angle 60°.

60°

10 Use a protractor to draw these angles.
   a 80°   **b** 37°   **c** 100°   **d** 145°

11 Safety regulations state that ladders should be placed at 75° from the horizontal. Draw a diagram to show this angle accurately.

12 **Problem-solving** Use a protractor to draw two lines at right angles to each other.

13 **a** Use a protractor to draw an angle of 140°.
       Label the 220° angle on your diagram.
   b Use a protractor to draw an angle of 60°.
       Work out the size of the reflex angle.

14 Use a protractor to draw these reflex angles.
   a 320°   **b** 280°   **c** 250°   **d** 200°

Challenge

a The angle between the hour and minute hands on a clock face is 90°.
   Both hands are pointing at a number.
   What time could it be?

   Hint Sketch or use a clock face.

b What if the angle is 60°?
c What if the angle is 120°?

Reflect

Dylan measures this angle as 130°.

What mistake has he made?

50°

8.2 Lines, angles and triangles

- Estimate the size of angles
- Describe and label lines, angles and triangles
- Identify angle and side properties of triangles

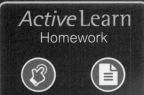
Warm up

1 Fluency How many sides does a triangle have?

2 What type is each of these angles? Choose from acute, obtuse or right angle.

a   b  c   d

Key point

- The number of equal sides and angles can help you identify a **triangle**.
- Equal sides are marked using a dash.
- Equal angles are shown using the same number of arcs.

scalene	isosceles	equilateral	right-angled
all angles and sides different	two equal angles and sides	all angles and sides equal	one angle is a right angle

3 Measure the sides and angles of each triangle.
Write down if it is scalene, isosceles or equilateral.

a 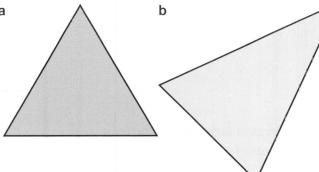 b c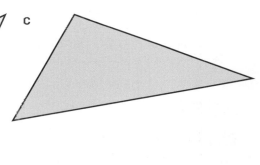

4 Name each triangle. Choose from scalene, isosceles, equilateral and right-angled.

a   b 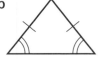 c 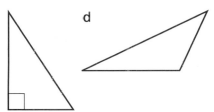 d

e Which triangle has two of these names?

5 The thick line is AB. It joins A and B.
Use letters to write down the names of the lines marked
w, **x**, **y** and **z**.

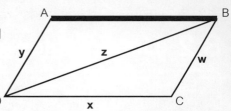

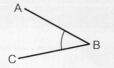

6 Use letters to write down the name of each angle.

a

b
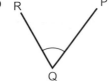

c

d Is there more than one name for each angle?

7 Name the angle where
a line AB meets line BD
b line BD meets line AD
c line AD meets line AB

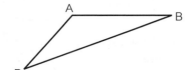

8 This is a Howe truss. It was used in wooden railway bridges.
a Write down a line that is
   i parallel to CD   **ii** perpendicular to DG
b Are CI and CG parallel? Explain your answer.
c What type of angle is $\angle$GCD?
d What type of angle is $\widehat{EFD}$?
e Write down a 90° angle.

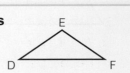

9 The diagram shows triangle ABC.
Write down
a a letter at one vertex of the triangle
b the equal sides
c the type of triangle
d the equal angles

10 Reasoning Look at these triangles.

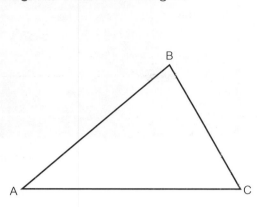
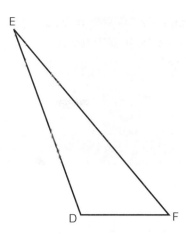

a One of the angles is 40°. Without using a protractor, decide which one.

b Which angles are less than 90°?

c Estimate the size of ∠FDE.

d Estimate the size of ∠BCA.

e Which angle looks less than 40°?

f Use a protractor to check your answers to parts **c** to **e**.

11 Problem-solving Show how you can make an equilateral triangle from two identical right-angled triangles.

12 This is a common truss. It is used to support roofs.

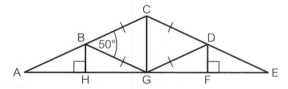

a What type of triangle is BGC?

b How many right-angled triangles are there?

c Write down two lengths that are equal.

d Write down two angles that are equal. Use a protractor to check.

Challenge

a Draw as many *different* triangles as you can on square dotted paper, by joining three dots. Here is one example.

b Label the triangles: scalene, isosceles, equilateral, right-angled.

c Which types of triangle can you draw?

d Repeat parts **a** to **c** using triangle dotted paper.

Reflect 'Notation' means symbols. Mathematics uses a lot of notation.

For example:

= means 'is equal to'   ° means 'degrees'   ⌐ means 'a right angle'

Look back at this lesson on angles, lines and triangles.

Write a list of all the maths notation used.

Why do you think mathematicians use notation?

8.3 Drawing triangles accurately

* Use a ruler and protractor to draw triangles accurately

Active Learn
Homework

Warm up

1 Fluency What can you say about the sides of
    a an equilateral triangle   **b** an isosceles triangle   **c** a scalene triangle?

2 Draw each of these lines accurately.
    a 7 cm   **b** 6.2 cm   **c** 50 mm

3 Use a protractor to draw these angles.
    a 25°   **b** 120°

4 On a scale drawing 1 cm represents 1 m.
    Use a ruler to draw a line to represent 6 m.

Worked example

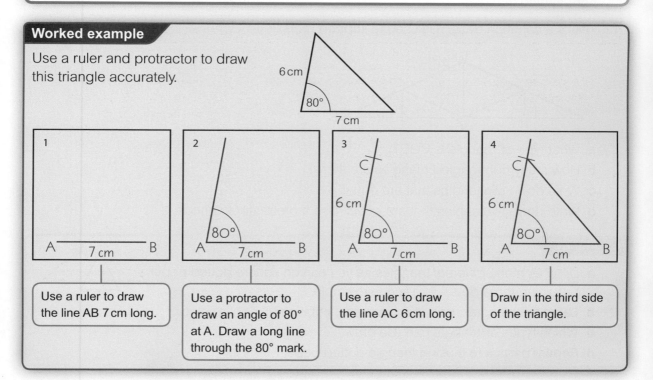

Use a ruler and protractor to draw this triangle accurately.

6 cm   80°   7 cm

1 Use a ruler to draw the line AB 7 cm long.

2 Use a protractor to draw an angle of 80° at A. Draw a long line through the 80° mark.

3 Use a ruler to draw the line AC 6 cm long.

4 Draw in the third side of the triangle.

5 Use a ruler and protractor to draw these triangles accurately.
    Start by drawing a side with a length you know.

a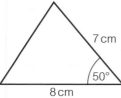
7 cm   50°   8 cm

b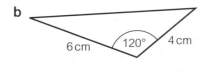
6 cm   120°   4 cm

To make an accurate drawing, you must use a ruler and protractor.

6   **a** Make an accurate drawing of this triangle.
    b Measure the length of AB, to the nearest millimetre.
    c Measure the sizes of angles CAB and CBA.
    d **Reasoning**   What type of triangle is ABC?

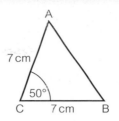

Worked example

Make an accurate drawing of this triangle.

1	2	3
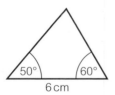 A ——— 8 cm ——— B	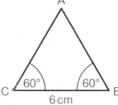	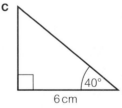

Use a ruler to draw the line AB 8 cm long.

Use a protractor to draw an angle of 110° at A. Draw a long line through the 110° mark.

Use a protractor to draw an angle of 30° at B. Draw a line through the 30° mark until it crosses the 110° line.

7   Use a ruler and protractor to draw these triangles accurately.
    a   **b**   **c**

8   **a** Make an accurate drawing of this triangle.
    b Measure the lengths of AC and AB.
    c Measure the size of ∠CAB.
    d **Reasoning**   What type of triangle is ABC?

9   **Problem-solving/Reasoning**   This is a common roof truss.
    a Make an accurate drawing of the truss. Use 1 cm to represent 1 m in real life.
    b Ray needs a truss with a height of 4 m. Is this truss big enough? Explain your answer.

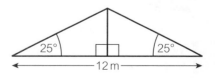

10 Reasoning Triangle XYZ has

∠XYZ = 120° ∠YZX = 30° XZ = 5 cm

a Which is the best sketch of triangle XYZ? Explain.

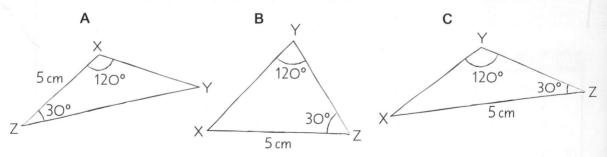

b Draw triangle XYZ accurately.

11 a Sketch triangle ABC with three acute angles.
b On your sketch, label
  i ∠ABC 50° **ii** side AB 5 cm **iii** ∠ACB 60°
c Draw triangle ABC accurately.

12 a Sketch triangle DEF, where DE = 9 cm, DF = 7 cm and ∠FDE = 60°
b Draw triangle DEF accurately.

13 a Sketch an isosceles triangle with base angles 35° and base length 6 cm.
b Draw this isosceles triangle accurately.

14 Problem-solving Draw accurately a right-angled triangle with perpendicular sides 3 cm and 4 cm long.

Challenge

a Use a ruler and protractor to accurately draw an equilateral triangle with side length 8 cm.
b Mark the **midpoint** of each side.
c Use a ruler to join up the three midpoints.
d Measure the side lengths of all the small triangles.
e Measure the angles of all the small triangles.
f What can you say about these triangles?
g How many equilateral triangles are there in your diagram? Remember to count the outer triangle.
h Repeat parts **b** to **g** for the smaller triangles.

Reflect In this lesson, you used two tools to help you draw triangles accurately – a ruler and a protractor.
Which did you find more difficult to use? Why?
Write down a hint or hints, in your own words, to help you use this tool in future.

8.4 Calculating angles

Active Learn
Homework

- Use the rules for angles on a straight line, angles around a point and vertically opposite angles
- Solve problems involving angles

Warm up

1 Fluency Find the missing numbers.
  a $40 + \square = 180$ **b** $180 = 65 + \square$
  c $280 + \square = 360$ **d** $360 = 175 + \square$

2 a What is $360 \div 3$?
  b Divide 360 into nine equal parts.

3 a What is 60×6?
  b What is the missing number? $180 \times \square = 360$

> **Key point**   Angles are sometimes labelled with lower case letters inside the angle.

4

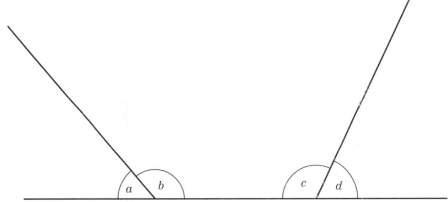

  a Measure the angles marked with letters.
  b Work out $a + b$ and $c + d$.
  c Reasoning   What do you notice about your answers to part **b**?

> **Key point**   The angles on a straight line add up to 180°.
> $a + b = 180°$
>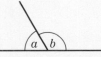

5 Calculate the size of each unknown angle.
  a   **b**   **c**

   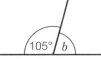   

6 Calculate the size of each unknown angle.

a
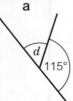

b

c
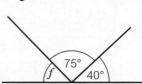

d

> **Key point** In a diagram, angles labelled with the same letter are the same size.

7 **Problem-solving / Reasoning** Work out the size of angle a.

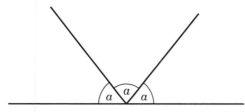

8 **Problem-solving** When light is reflected, the **angle of incidence** (i) equals the **angle of reflection** (r).

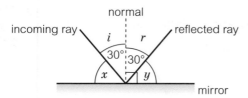

Work out the size of
a angle x **b** angle y

9 **Problem-solving / Reasoning** Angle A is five times the size of angle B. What are the sizes of angle A and angle B?

> **Q9 hint** Use the ratio A : B
> 5 : 1

10 **Problem-solving / Reasoning**
   a Work out the size of angle a.
   b Work out the size of angle b.
   c Work out the sum of angles round a point.

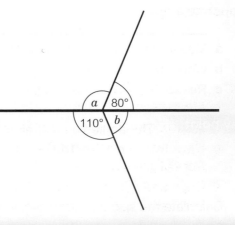

> **Key point** The angles around a point add up to 360°.
> $$a + b = 360°$$

11 Calculate the size of each unknown angle.

a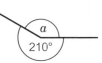

b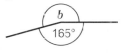

c

12 Work out the size of ∠ABC in each diagram.

a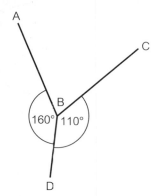

b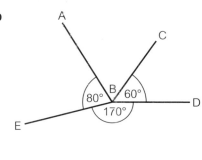

13 **Problem-solving** Mike broke three plates into pieces.
Work out which pieces belong together.

A
35°

B
93°

C
322°

D
169°

E
180°

F
145°

G
38°

H
98°

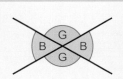

> **Key point** When two lines cross, they make pairs of **vertically opposite angles**.

14 a **Reasoning** Draw two straight lines that cross each other.
  b Measure two vertically opposite angles. What do you notice?
  c Measure the other two vertically opposite angles. What do you notice?
  d Draw two more straight lines that cross.
    Repeat parts **b** and **c**.
  e Copy and complete the sentence.
    Vertically opposite angles are _____

> **Key point** Vertically opposite angles are equal.
> The green angles are equal.
> The blue angles are equal.

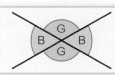

15 Work out the sizes of the unknown angles.

Write down your reason for each angle you find. Choose from:

- Angles on a straight line add up to 180°.
- Angles at a point add up to 360°.
- Vertically opposite angles are equal.

a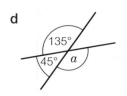

b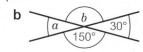

c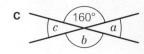

d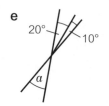

e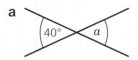

f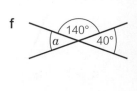

16 Problem-solving The diagram shows the path of light rays through a glass lens. Work out the sizes of angles m and n.

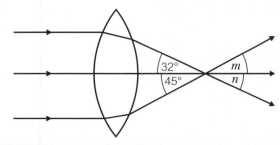

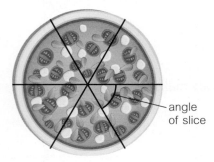

Challenge   A group of friends share a pizza fairly.

angle of slice

a Work out the angle of each slice when there are

i 2 friends   **ii** 3 friends   **iii** 5 friends   **iv** 8 friends

b Write an expression for the angle of each slice when n friends share a pizza.

c Work out how many slices you can cut with

i angle 60°   **ii** angle 90°   **iii** angle 36°   **iv** angle 30°

Reflect   How do you know whether to subtract from 90°, 180° or 360° when calculating angles?

8.5 Angles in a triangle

Active Learn
Homework

- Use the rule for the sum of angles in a triangle
- Calculate interior and exterior angles
- Solve angle problems involving triangles

Warm up

1 **Fluency** Work out
   a 90 + 30   **b** 90 − 55   **c** 65 + 40   **d** 180 − 115   **e** 180 − 65 − 20

2 Copy these triangles.
   Mark equal sides with dashes.
   Mark equal angles with the same number of arcs.

equilateral   isosceles

3 Work out the size of the unknown angles.
   a   **b**   **c**

  a 70°   85° b 145° c

Key point

An **interior** angle is inside a shape.
An **exterior** angle is outside the shape on a straight
line with the interior angle.

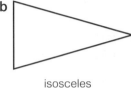

interior
exterior

4 For each diagram, write down the size of
   i the interior angle   **ii** the exterior angle
   a   **b**   **c**

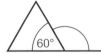 60°   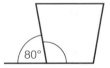 80°

   d What angle fact did you use?

5 **a** Draw a triangle with angles a, b and c.
   b Cut out the triangle and tear off each of the corners.
   c Arrange the three parts in a straight line so that angles a, b and c meet.
   d **Reasoning** What does this tell you about the angles in a triangle?
      Does this work for all triangles?

Key point

The angles in a triangle add up to 180°.
   $a + b + c = 180°$

6 Calculate the size of each unknown angle.

a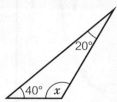

b

c

Q6a hint $x = 180° - 40° - 20°$

7 Sketch each triangle. Mark the angles you know. Work out the size of the unknown angle.
   a ABC where $\angle ABC = 55°$ and $\angle BAC = 35°$
   b PQR where $\angle PQR = 65°$ and $\angle QPR = 90°$
   c DEF where $\angle DFE = 40°$ and $\angle FED = 100°$

8 **Reasoning** A sail-maker used this plan to mark out a sail
   on a sheet of cloth.
   She checked the shape by measuring the angle b and
   found it to be 125°.
   Did she draw the sail correctly? Explain your answer.

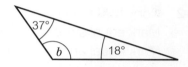

9 **Reasoning** Work out the size of each angle in an equilateral triangle.

> **Key point** In an isosceles triangle, the angles at the base of the
> equal sides are equal.

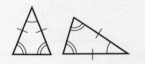

10 Work out the missing angles in these isosceles triangles.
   Write down a reason for each step of your working.

a

b

c

> • Angles in a triangle add up
>   to 180°.
> • Base angles in an isosceles
>   triangle are equal.

11 **Reasoning** Calculate the size of each angle labelled with a letter.
   Give reasons for each angle that you find.

a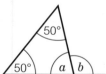

b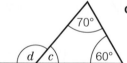

c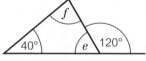

d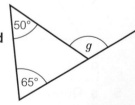

> **Challenge** One of the angles in an isosceles triangle is 50°.
> Sketch two possible isosceles triangles it could be.
> Label the size of each angle.

> **Reflect** In this lesson, you did a lot of subtracting from 180°.
> What strategy did you use to mentally subtract from 180°?
> Would your strategy still work to subtract 46°? If not, what strategy would you use?
> Would your strategy still work to subtract a number from 360°?
> If not, what strategy would you use?

8.6 Quadrilaterals

- Identify and name types of quadrilaterals
- Use the rule for the sum of angles in a quadrilateral
- Solve angle problems involving quadrilaterals

Active Learn
Homework

Warm up

1 Fluency How many sides does a quadrilateral have?

2 Work out
a 30 + 90 + 150   **b** 75 + 135 + 125   **c** 360 − 170 − 90 − 20

Key point
A **trapezium** is a quadrilateral with **one** pair of parallel sides.

An **isosceles** trapezium is a trapezium with one pair of equal sides.

3 Name each of these quadrilaterals.
Choose from square, rectangle, rhombus, parallelogram, trapezium, kite or arrowhead.

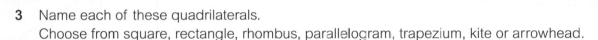

4 Reasoning Which quadrilateral am I?
a I have four right angles. All my sides are equal.
b I have four equal angles and two pairs of equal sides.
c I have one pair of parallel sides.
d I have no right angles. My opposite sides are equal and parallel.

Q4 hint Look at the quadrilaterals in Q3.

Key point
The **properties** of a shape are facts about its angles and sides.

5 Here is a rectangle cut in two.
a Name the two new shapes that are made.
b Write down the properties of each shape.

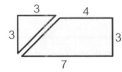

6 Problem-solving Show how two identical isosceles triangles can be joined together to make
a a rhombus   **b** a square

Key point
A **diagonal** is a line joining two opposite vertices.

7 Reasoning In this diagram, the diagonal divides the quadrilateral into two triangles.
a What do the three angles in the top triangle add up to?
b What do the three angles in the bottom triangle add up to?
c What do all the angles in the quadrilateral add up to?
d Will this work for *any* quadrilateral?

The angles in a quadrilateral add up to 360°.
$$a + b + c + d = 360°$$

8   Calculate the size of each unknown angle.

Q8a hint $a = 360° - 135° - \Box° - \Box°$

a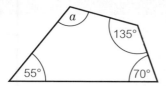

b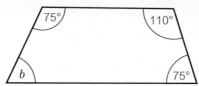

c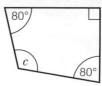

d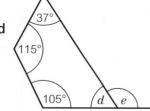

e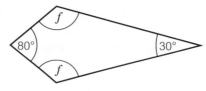

f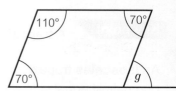

9   **Reasoning**   Here is a rhombus cut in two.
   a Calculate the size of each unknown angle.
      Write down a reason for your answers.
   b What do you notice about opposite angles in this rhombus?

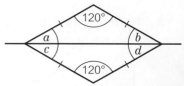

10 a Calculate the size of ∠DAB.
    b Which two sides are parallel?
    c Name this quadrilateral.

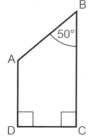

11 **Problem-solving**   The diagram shows a bicycle frame.
    Work out the size of the unknown angle.

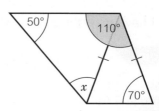

Take two squares of any size and overlap them to make
another shape. For example:
Which of these shapes can be made?

   rectangle, square, rhombus, isosceles triangle, kite, trapezium

Draw diagrams to show how.

Look back at all the lessons you have completed in this unit so far.
Write down at least two new or important mathematics words you have learned. Make sure
you spell them correctly.
Beside each word, write a definition in your own words. You can draw diagrams to help.

8 Check up

Measuring and drawing angles

1 **a** What type of angle is
   i ∠ABC   **ii** ∠DEF?
   b Measure each angle.
   Write your answers to the nearest degree.

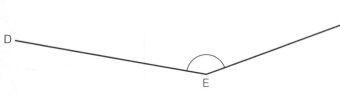

2 Choose the best estimate for each angle.

a

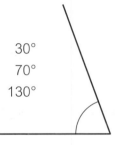

30°
70°
130°

b

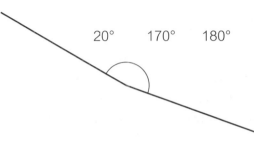

20°   170°   180°

3 Use a protractor to draw an angle of
   a 115°   **b** 310°

4 **a** Make an accurate drawing of this triangle.
   b Measure angle ACB.
   c What type of triangle have you drawn?

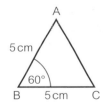

Calculating angles

5 Work out the size of each unknown angle.

a

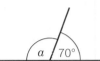

b

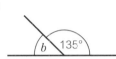

6 Work out the size of angle a. Show your working.

7 Calculate the size of each unknown angle.
   Write down a reason for each angle you find.

a

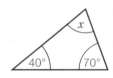

b

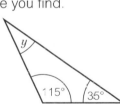

8 The diagram shows two straight lines crossing.
Find the size of
a angle a **b** angle b
Write down a reason for each angle you find.

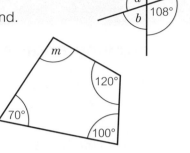

9 The diagram shows a quadrilateral.
Calculate the size of the unknown angle.
Give a reason.

Solving angle problems

10 Write down the name of each of these quadrilaterals.

a 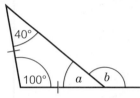   **b**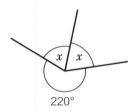

11 Work out the size of each angle labelled with a letter.
Write down a reason for each angle you find.

a   **b**

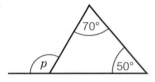

12 Work out the size of each angle labelled x. Give a reason for each angle you find.

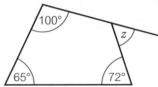

13 Work out the size of each angle labelled with a letter. Give a reason for each angle you find.

a   **b**   **c**

Challenge   Use a 4 × 4 square on square dotted paper.
Draw as many different quadrilaterals as you can.
a Name the quadrilaterals you have drawn.
b Are there types of quadrilateral that you cannot draw on this 4 × 4 square?

Reflect   How sure are you of your answers? Were you mostly

🙁 Just guessing   😐 Feeling doubtful   🙂 Confident

What next? Use your results to decide whether to strengthen or extend your learning.

8 Strengthen

Measuring and drawing angles

1 Read the size of each angle from the protractor.

a

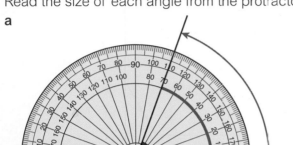

b

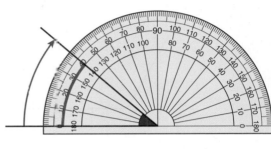

2 Caroline measures an angle.
She says, 'This angle measures 115°.'
Andrew says, 'It can't be.
It is smaller than a right angle.'
What has Caroline done wrong?

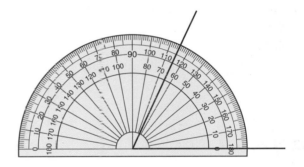

3 Choose the correct measurement for each angle.

a

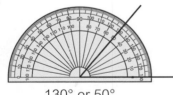

130° or 50°

b

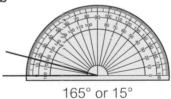

165° or 15°

c

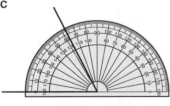

63° or 117°

d

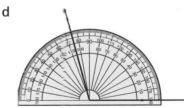

75° or 105°

e

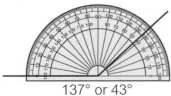

137° or 43°

f

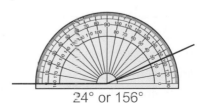

24° or 156°

4 Use a protractor to measure these angles.
Place the cross of the protractor on the point of the angle.
Line up the zero line with one line of the angle.

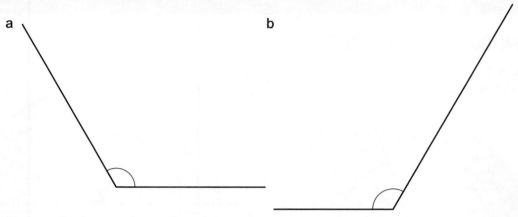

a

b

5 Choose the best estimate for each angle.
Is the angle smaller or larger than 90°? If it is larger, is it closer to 90° or 180°?
If it is smaller, is it closer to 0° or 90°?

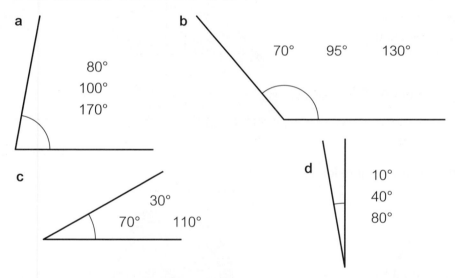

a

80°
100°
170°

b

70°   95°   130°

c

30°
70°   110°

d

10°
40°
80°

6 Three students were asked to draw a triangle ABC where
AB = 8 cm, BC = 7 cm and ∠ABC = 85°.
They each made a sketch of the triangle before drawing it accurately.

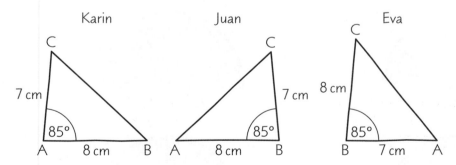

Karin   Juan   Eva

a Only one of the sketches is correct. Who drew the correct sketch?
b Draw a line AB 8 cm long.
c Use a protractor to draw the given angle ∠ABC.
d Draw in the other lines needed to complete the triangle accurately.

Calculating angles

1   There are 360° in a full turn.

   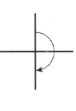

   a How many degrees are there in half a turn?
   b How many degrees are
      there in a quarter turn?

2   Work out the size of each angle marked with a letter.

   a
   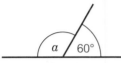

   b

   > **Q2a hint** $a = 180° - \square°$

   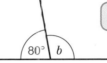

   c
   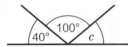

   d
   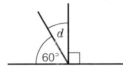

3   What do the angles around a point add up to?

4   Work out the size of each angle marked with a letter.

   a x
      200°

   b
   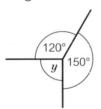

   c

   > **Q4a hint** $x = 360° - \square°$

   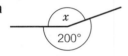

5   The angles in a triangle add up to 180°.
   Calculate the size of each unknown angle.

   > **Q5 hint** $180° - \square° - \square° = \square$

   a
   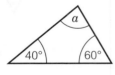

   b
   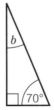

   c
   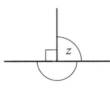
   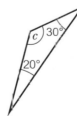

6   **a** Measure the interior angles of this triangle.
      Check that they add up to 180°.
   b Measure the exterior angles.
   c Work out
      i $a + e$ **ii** $c + f$ **iii** $b + d$

   What do you notice?

   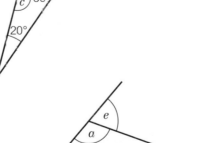

7 Work out the size of angle z.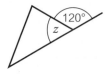

8 The diagram shows two straight lines crossing.

Catrin says, 'Angle a is also 60°.'
Explain why Catrin is correct. Choose from:
   A Angles on a straight line add up to 180°.
   B Vertically opposite angles are equal.
   C The angles look the same size.

9 **a** Complete the working to find the sizes of angles a and b.

angle $a = \square$ Vertically opposite angles are _____.
angle $b = 180° - \square° = \square°$ Angles on a _____ add up to $\square°$.

b Work out the size of each unknown angle. Write down the reasons.

i   **ii**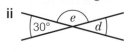

10 **Reasoning** Draw a **quadrilateral** with angles A, B, C and D.
Cut out the quadrilateral and tear off each of the corners.
Arrange the torn-off corners so that angles A, B, C and D meet around a common point.
What does this tell you about the angles in a quadrilateral?

11 Work out the size of each unknown angle.

a 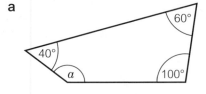   **b**

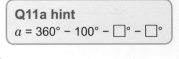
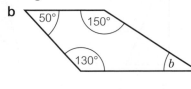

12 Look at this diagram.
   a Work out the size of angle b.
      Use the fact about angles on a straight line.
   b Work out the sizes of angles c, d and e.

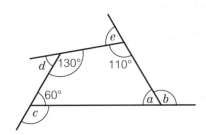

Solving angle problems

1 **Reasoning**

  a Diagram A shows one side of a square.
    Copy the diagram on squared paper and
    draw three more lines to complete the square.

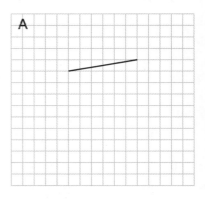

  b Diagram B shows one side of a quadrilateral.
    The opposite sides of this quadrilateral are equal.

    i Copy the diagram on squared paper and draw
      three more lines to show what the quadrilateral
      could be.

    ii Copy the diagram again and draw three more lines
      to show a different quadrilateral that it could be.

   iii What are the names of the shapes that you
      have drawn?

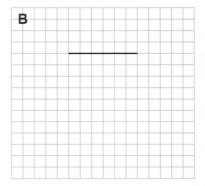

2 **Problem-solving** A cake was cut into equal slices.

> **Q2 hint** Draw a sketch to show how many 60° slices there are in 360°.

These two slices are left.
How many slices have been eaten?

3 **a** Show how two of these scalene triangles can be joined together to make a parallelogram.

  b Show how two right-angled triangles can be joined together to make a parallelogram.

4 The diagram shows a triangle ABC.
  $AB = AC$

> **Q4 hint** $AB = AC$ means that the length of AB is the same as the length of AC.

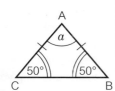

  a How are the equal sides shown on the diagram?
  b What is the name of this type of triangle?
  c What is the size of angle a?

5 The diagram shows a triangle DEF.

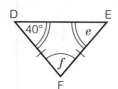

Q5 hint What type of triangle is DEF?

a What is the size of angle *e*? Give a reason for your answer.
b Work out the size of angle *f*.

6 Copy these isosceles triangle diagrams.
Mark the equal angles with double arcs.
Work out the unknown angles.

a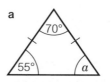
b

c

7 i Work out the interior angle in each triangle.
ii Work out the exterior angle.

a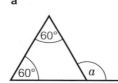
b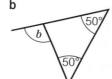
c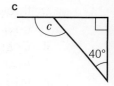

8 Work out the size of angle *x*.

9 The diagram shows a quadrilateral.
Work out the size of
a angle *a*
b angle *b*

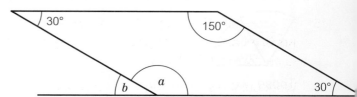

Challenge The diagram shows a tilted platform.
a Work out the size of angle *x*.
b Work out the size of angle *y*.
c As the platform is tilted further, the 70° angle decreases to 65°.
What happens to angle *y*?

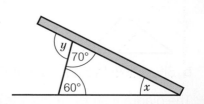

Reflect Write down one thing you find easy and one thing you find difficult when

Hint Look back at the questions in these strengthen lessons to help you.

• drawing angles accurately
• calculating angles
• solving geometric problems

For each thing you find difficult, write a hint in your own words.

8 Extend

1 **Problem-solving**   The mirror in a periscope is set at an angle x.
Work out the size of angle x.

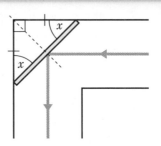

2 If you add two acute angles, do you always get an obtuse angle?

3 Accurately draw triangle ABC with
   AB = 8 cm,
   BC = 6 cm and
   angle ABC = 80°

4 **Problem-solving**
   a Draw a rectangle.
      Draw one straight line to divide the rectangle into two right-angled triangles.
   b Draw another rectangle.
      Draw two straight lines to divide the rectangle into four isosceles triangles.

5 **Reasoning**   The diagram shows an angle.
Kevin measures the angle. He says, 'The angle is 120°.'
Is he correct? Explain your answer.

6 Work out the size of each unknown angle.

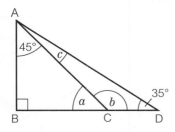

> **Q6 hint** Sketch the diagram. Write on the angles as you work them out.

7 Work out the size of angle x in each diagram.

a   **b** 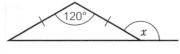   **c**

d   **e**

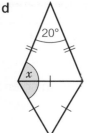   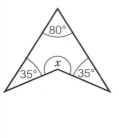

8 **Reasoning** Explain why a triangle can never have a reflex angle but a quadrilateral can.

9 The diagram shows a parallelogram.

a Measure angles ABC, BCD and BAD.

b What do you notice about the opposite angles of a parallelogram?

Q9 hint Draw diagrams to help with your explanations.

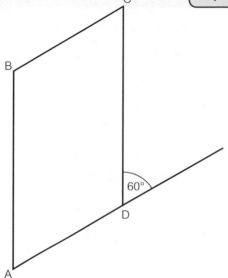

10 **Reasoning** Andre says, 'A rhombus is a parallelogram, but a parallelogram is not a rhombus.' Is he correct? Explain your answer.

11 **Problem-solving** An ash tree is 20 m tall.
It casts a 12 m shadow on the ground.

a Accurately draw a triangle to represent this.

b What angle does the light from the Sun make with the ground?

Q11 hint Use a vertical line to represent the tree.

Challenge Make a large copy of this table.

		Number of pairs of parallel sides		
		0	1	2
Number of pairs of equal sides	0			
	1			
	2			

a Write each of these quadrilaterals in the correct position in the table:
rectangle, rhombus, kite, parallelogram, arrowhead, trapezium

b One of the shapes can fit in more than one box.
Which shape?

Hint Think carefully about the definitions of each shape.

c Draw a quadrilateral to fit in each empty box.
Use markings to show which sides are parallel and which are equal.

Reflect In these extend lessons, you have answered questions about angles and:

i a periscope   ii the shadow from a tree.

Where else do you think it is useful to have an understanding of angles in maths, in science and in real life?

8 Unit test

1   The diagram shows two angles.
    Work out the size of angle m.

2   One of these angles measures 140°.

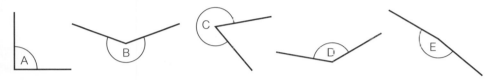

    Without using a protractor, decide which one.

3   Make an accurate drawing of this angle.

70°
5 cm

4   Alan measured the angles in a triangle.
    He said, 'The angles are 40°, 50° and 100°.'
    Is he correct? Explain.

5   Calculate the size of angle x in each diagram. Show your working.
    a   b

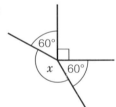

6   A pizza is cut into equal slices like this.
    How many slices is it cut into?

40°

7   a Triangle ABC has three equal sides.
       What are the sizes of the angles in this triangle?
    b The right-angled triangle DEF has two equal sides.
       What are the sizes of the angles in this triangle?

8   Work out the sizes of angles x and y.

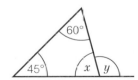

60°
45° x y

9 Make an accurate drawing of this triangle.

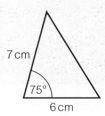

10 a Accurately draw triangle ABC with
$\angle ACB = 50°$
$\angle ABC = 40°$
$BC = 5\,cm$

b What type of triangle is ABC?

11 Here is a list of some types of quadrilaterals.

kite, parallelogram, rectangle, rhombus, square

a Write down the names of the quadrilaterals that have two pairs of parallel sides.

b Write down the names of the quadrilaterals that must have two pairs of equal sides.

12 The diagram shows triangle ABC.
Work out the sizes of angles a, b and c.

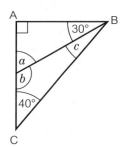

13 ABCD is a quadrilateral.
Work out the size of
a angle c **b** angle d.

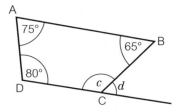

Challenge Start with these two shapes.

a What new shapes can you make by joining them together?

b What are the properties of each new shape?

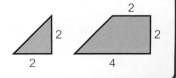

Reflect Look back at the questions you answered in this test.
Find a question that you could not answer straight away, or that you
really had to think about. While you worked on this question how did you feel?

- What were you thinking about? Were you calm? Were you panicky?
- Did you keep trying until you had an answer? Did you give up before reaching an answer, and move on to the next question?
- Did you think you would get the answer correct or incorrect?

Write down any strategies you could use to stay calm and positive when answering tricky maths questions in tests. Compare your strategies with other people's.

9 Sequences and graphs

Master   Check up p249   Strengthen p251   Extend p256   Unit test p258

9.1 Sequences

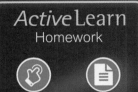

*Active*Learn
Homework

- Recognise, describe and continue number sequences
- Generate terms of a sequence using a one-step term-to-term rule
- Find missing terms in a sequence

Warm up

1 **Fluency** Count up in threes starting at 14. Stop when you get to 26.

2 **Fluency** Find the difference between 7 and 13.

3 Starting at 24, keep halving until you get a mixed number.

4 Count up in fives starting at 8. What do you notice?

Key point
A number **sequence** is a set of numbers that follow a rule.
Each number in a sequence is called a **term**.
The **term-to-term rule** tells you how to get from one term to the next in a sequence.
It can use adding, subtracting, multiplying and dividing.

5 Write the first five terms in each sequence.
- **a** First term 5, term-to-term rule '+3'
- **b** First term 10 000, term-to-term rule '÷10'
- **c** First term 3, term-to-term rule '×2'
- **d** First term 20.5, term-to-term rule '−0.5'
- **e** First term 3.2, term-to-term rule '+0.4'
- **f** First term 10, term-to-term rule '−5'
- **g** First term −7, term-to-term rule '+2'

Worked example

a Write down the next three terms in this sequence.
4, 7, 10, …

4, 7, 10, 13, 16, 19

b Write down the first term and the term-to-term rule.

First term is 4.

Term-to-term rule is 'add 3'.

> Work out how to get from one term to the next. Continue the pattern for the next three terms.

> Write down the first term and the rule to get from one term to the next.

6 Write down the next three terms of each sequence.
- **a** 6, 11, 16, 21, …
- **b** 60, 52, 44, …
- **c** 3, 3.2, 3.4, …
- **d** −12, −10, −8, −6, …

7 Write down the first term, the term-to-term rule and the next three terms of each sequence.

  a 2, 6, 10, 14, ...　　　**b** 2, 6, 18, 54, ...

  c 2, 4, 6, 8, ...　　　　**d** 2, 4, 8, 16, ...

  e 52, 48, 44, ...　　　　**f** 32, 36, 40, ...

  g –15, –12, –9, ...　　　**h** 4, 1, –2, ...

  i 2, –1, –4, ...　　　　**j** 1, 2, 4, 8, ...

  k 160, 80, 40, ...　　　**l** 250 000, 25 000, 2500, ...

> **Q7 hint** What do you need to do to get from one term to the next? Do you add, subtract, multiply or divide?

> **Key point**　Sequences where the numbers **increase** are **ascending** sequences.
> Sequences where the numbers **decrease** are **descending** sequences.

8 Is each sequence ascending or descending?

  a 3, 5, 7, 9, 11, ...　　　　　　**b** 10, 12, 14, 16, 18, ...

  c 20, 18, 16, 14, 12, ...　　　　**d** 9, 8.5, 8, 7.5, 7, ...

  e 1, 5, 25, 125, 625, ...　　　　**f** 1, 1.3, 1.6, 1.9, 2.2, ...

  g –8, –10, –12, –14, –16, ...　　**h** 6.4, 3.2, 1.6, 0.8, 0.4, ...

  i –20, –17, –14, –11, ...　　　　**j** 1.3, 1.25, 1.2, 1.15, 1.1, ...

> **Key point**　A sequence that carries on for ever is **infinite**.
> A sequence with a fixed number of terms or a 'last term' is **finite**.

9 **Reasoning**　Decide whether each sequence is finite or infinite.

  a The odd numbers between 0 and 10　　**b** The even numbers

  c The multiples of 3 between 10 and 20　**d** The multiples of 10 larger than 100

  e The multiples of 10 between 0 and 100　**f** The numbers in the 12 times table

10 A sequence has first term 2.
  The term-to-term rule is 'add 5'.
  The last term is 42.

  a Write down all the terms in the sequence.

  b Is the sequence ascending or descending?
    Is it finite or infinite?

> **Key point**　When you know the first term and the term-to-term rule, you can work out all the terms in the sequence.

11 a Write a sequence with first term 7, term-to-term rule '+0.6'.
    Stop at 13.

  b **Problem-solving**　Write down a first term and a term-to-term rule for another ascending sequence that includes the term 7.6.

12 **Problem-solving**　Ranjit has £450 in his bank account.

  a He spends £50 every month. How much will he have after 4 months?

  b How many months will it be before the account is empty?

13 Rasheed reads on a website that his height should increase by 8 cm a year. His height is 131 cm when he is 10 years old.

    a Write down his predicted heights for the next four years.

    b Problem-solving How many years will it be before he is likely to have a height of 171 cm?

    c Reasoning Could you use the sequence of heights to predict his height when he is 30 years old? Explain.

14 Problem-solving The number of cells in a bacteria sample doubles every 20 minutes. There is one bacteria cell in a dish.

    a How many cells will there be after 40 minutes?

    b How many minutes will it be before there are 16 cells?

> **Q14 hint** You could draw a table like this to help.
>
> | Minutes | 0 | 20 | 40 | ... |
> |---|---|---|---|---|
> | Cells | 1 | | | |

15 Reasoning There was a FIFA World Cup in the years 2010, 2014 and 2018. Will there be a FIFA World Cup in the year 2054 if this pattern continues?

16 a Write down the next four terms of each sequence.

    i 1.2, 1.4, 1.6, ...   **ii** $1\frac{1}{2}$, 2, $2\frac{1}{2}$, 3, ...   **iii** −9, −7, −5, ...

    iv 57.1, 57.8, 58.5, ...   **v** $\frac{1}{2}$, $\frac{1}{4}$, $\frac{1}{8}$, $\frac{1}{16}$, ...   **vi** 2, 0, −2, ...

    b Which of the sequences in part **a** are descending?

17 Problem-solving Work out the missing terms in each sequence.

    a ___ , ___ , ___ , 18, 21, 24   **b** ___ , ___ , ___ , 55, 58, 61

    c ___ , ___ , ___ , 500, 250, 125   **d** ___ , ___ , ___ , 100, 1000, 10 000

18 Problem-solving A sequence begins 1, 4, ...

    a What could the next term be?

    b What is the term-to-term rule?

    c Find another term-to-term rule. What is the next term for this rule?

    d Write as many term-to-term rules as you can for sequences that start 1, 4, ...

    e Answer parts **a** to **d** for the sequence that begins $\frac{1}{2}$, $\frac{1}{4}$, ...

    f Will $\frac{1}{27}$ be a term in any of your sequences? Why, or why not?

19 Problem-solving Work out the missing terms in each sequence.

    a 23 , ___ , 15 , ___ , 7   **b** 23, ___ , 31, ___ , 39

    c −5 , ___ , −15 , ___ , −25   **d** −5, ___ , 5, ___ , 15

    e 7.9, 8.3, ___ , 9.1, ___ , 9.9, ___ , ___   **f** 0.45, ___ , ___ , 0.6, 0.65, ___ , 0.75, ___

    g 3.8, 4.0, 4.2, ___ , ___ , ___ , ___ , 5.2   **h** −1.5, ___ , −0.5, ___ , ___ , 1, ___

> **Challenge** If you write down the sequence of all odd numbers and the sequence of all even numbers, will you have written all the numbers that exist?

> **Reflect** Think carefully about your work on sequences. How would you define a sequence in your own words? Write down your definition. Compare your definition with someone else's in your class.

9.2 Pattern sequences

- Find patterns and rules in sequences
- Describe how a pattern sequence grows
- Write and use number sequences to model real-life problems

Active Learn
Homework

Warm up

1 **Fluency** What is the term-to-term rule of each sequence.
  a 4, 6, 8, 10, …   **b** 12, 18, 24, 30, …   **c** 23, 33, 43, 53, …

2 Write down the next four terms of this sequence: 5, 9, 13, 17, …

3 A sequence has first term 4 and term-to-term rule '+7'.
  Write down the first four terms of the sequence.

Key point   You can draw the next pattern in a sequence by working out how the pattern grows.

4 Draw the next pattern in each sequence.
  a
  b

5 Look at this sequence of patterns made from counters.

Pattern 1   Pattern 2   Pattern 3

  a Draw the next pattern in the sequence.
  b Copy and complete this table for the sequence.

Pattern number	1	2	3	4	5
Number of counters					

Q5c hint Explain how to make the next pattern in the sequence.

  c Describe how the sequence grows.

6 Look at this sequence of patterns made from squares.

Pattern 1   Pattern 2   Pattern 3

  a Draw the next pattern in the sequence.
  b Copy and complete this table for the sequence.

Pattern number	1	2	3	4	5
Number of squares					

  c Describe how the sequence grows.

7 Jack is collecting trading cards. He starts with a gift pack of 15 cards.
He plans to buy 10 cards every week until he has 75.

   a Copy and complete the table to show the number of cards he will
have each week.

Week number	1	2	3	4	5	
Number of cards	15	25				

   b In which week will he reach his target?

8 This is a sequence of growing rectangles.

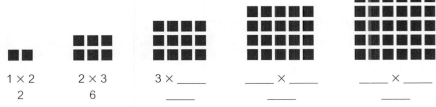

1×2 2×3 $3 \times \underline{}$ $\underline{} \times \underline{}$ $\underline{} \times \underline{}$

2   6 $\underline{}$ $\underline{}$ $\underline{}$

   a Copy and complete the multiplications for the rectangles.

   b Describe how this sequence grows.

   c Reasoning The 2nd rectangle is 2×3. What will the 8th rectangle be?

9 Abi and Ben are both saving up for a computer game that costs £40.
Abi starts with £20 and saves £4 per week.
Ben starts with £10 and saves £8 per week.

   a Problem-solving Who will be first to have enough money for the game?

   b Reasoning Show another way to work this out.

Challenge   A potato in the middle of a tray has gone rotten!
It is making all the potatoes around it rotten too.
Each potato immediately next to a rotten potato goes rotten in a day.

This diagram shows the number of rotten potatoes (including the
original one) after one day.

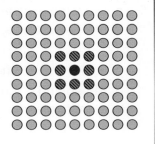

a Copy the diagram onto squared paper.
   Continue the pattern.

The sequence showing the number of new rotten potatoes
each day begins 1, 8, …

b How does it continue?

c Is this a realistic way to predict the spread of a virus in humans?

Reflect   Look back over lessons 9.1 and 9.2. Sequences are shown in three different ways:

A Diagrams   **B** Tables   **C** Lists of numbers

Which of these ways most helped you to understand sequences and answer the questions?

9.3 Coordinates and midpoints

Active Learn
Homework

- Generate and plot coordinates from a rule
- Solve problems and spot patterns in coordinates
- Find the midpoint of a line segment

Warm up

1 Fluency Work out the difference between −3 and 6.

2 Fluency Add 4 and 10, then halve the answer.

3 Write down the coordinates of points I, J, K, L, M, and N.

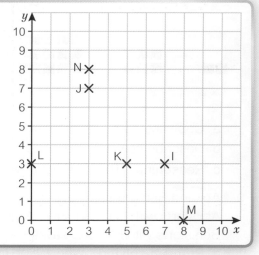

4 Write down the coordinates of the points on this grid.

5 a Copy the axes from Q4. Do not copy the points.
  b Plot these points.

  P(3, 3)   Q(−1, 3)   R(−1, 1)

  c Join P and Q, and Q and R.
  d Reasoning You have drawn two sides of a rectangle.
  Write down the coordinates of the 4th vertex.

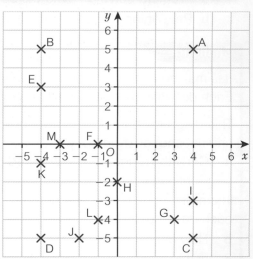

6 a Copy and complete the table of values for this function machine.

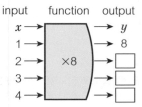

x	1	2	3	4	5
y	8				

  b Write the pairs of x and y values as coordinates.
  c The y-values make a sequence.
  What is the term-to-term rule of the sequence of y-values?

7 This function machine generates coordinates.
When you input an x-coordinate, it outputs the y-coordinate.

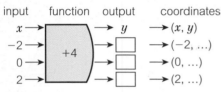

a Work out the missing y-coordinates.

b Copy the axes from Q4. Do not copy the points.
Plot the coordinates from part **a** on the grid.

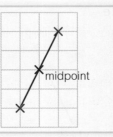
8 a Copy the coordinate grid.

b Plot these points on your grid.
A(1, 4)   B(5, 4)

c Join the points with a straight line.

d Put a cross at point C, the **midpoint** of the line joining A and B.

e Write down the coordinates of point C.

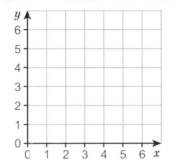

9 Copy and complete the table to work out the **midpoint** of each of these **line segments**.

Line segment	Beginning point	Endpoint	Midpoint
CD	(5, 6)	(5, 4)	
EF	(−1, −1)		

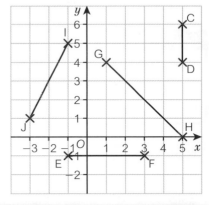

Worked example

Work out the midpoint of this line segment.

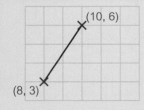

$(8 + 10) \div 2 = 9$ — Add the two x-coordinates together and divide by 2.

$(3 + 6) \div 2 = 4.5$ — Add the two y-coordinates together and divide by 2.

midpoint = (9, 4.5) — These are the x- and y-coordinates of the midpoint.

10 Check your answers in Q9 using the method in the Worked example.

11 Work out the midpoint of each of these line segments.

a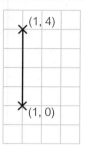

b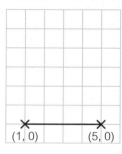

c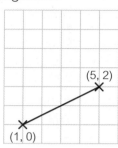

d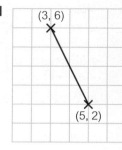

e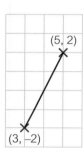

f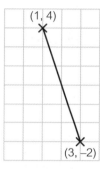

g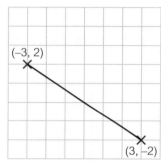

h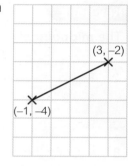

i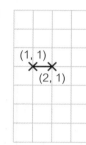

j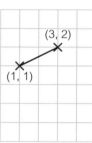

k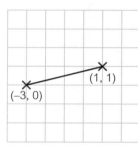

l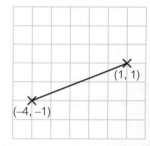

Challenge   A right-angled triangle has:
- a vertical side of 4 units with midpoint (−2, 1)
- a horizontal side of 4 units with midpoint (0, −1)

Find the midpoint of the third side of the triangle.

Reflect   Patty, Sally and Dave are talking about this line.

Patty says 'One end is at the point (2, −2).'

Sally says 'The midpoint is at (−3, 0).'

Dave says 'The other end is at the point (2, 2).'

a Who is right and who is wrong?

b Write a hint on reading coordinates, in your own words, to help the students who are wrong.

Check your hint. Will it stop them making their mistakes?

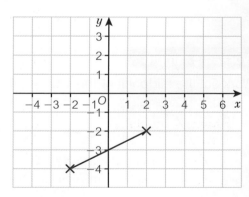

9.4 Extending sequences

Active Learn
Homework

- Describe and continue special sequences
- Use the term-to-term rule to work out more terms in a sequence
- Recognise an arithmetic sequence and a geometric sequence

Warm up

1 Fluency   Work out the outputs for each function machine.

a   6 → [÷2] → □
   3 →   → □
   8 →   → □

b   4 → [+0.5] → □
   1.2 →   → □
   2.8 →   → □

2 Fluency   Work out

   a $2 \times 3 + 4$ b $1 + 9 \times 5$ c $15 \div 3 - 1$ d $(10 + 6) \div 4$

3   What is the term-to-term rule for each sequence?

   a   2, 6, 10, 14, ...   b   4, −1, −6, −11, ...   c   1, 4, 16, 64, ...

Key point

An **arithmetic sequence** goes up or down in equal steps.
For example, the sequence 14, 11, 8, 5, 2, ... goes down in steps of 3.

4   Which of these sequences are arithmetic?

   a   3, 6, 9, 12, 15, ...   b   30, 20, 0, −30, −70 ...   c   The multiples of 12

   d   1, 4, 9, 16, 25, ...   e   2, 3, 5, 7, 11, 13, ..   f   The odd numbers

   g   −1, −4, −7, −10, −13, ...   h   0.2, 0.5, 0.8, 1.1, 1.4, ...   i   0, 10, 100, 1000, ...

5   Look at these patterns made from counters.

   a   Draw the next pattern in the sequence.

   b   The first three terms of the sequence are
     1, 4, 9.
     Write down the next three terms.

   c   What is the special name for the numbers in this sequence?

   d   **Reasoning**   Is this an arithmetic sequence? Explain your answer.

6   **Reasoning**   The table shows the first four triangle numbers.

Pattern	•	• • •	• • • • • •	• • • • • • • • • •		
Number of dots	1	3	6			

   a   Copy and complete the table.

   b   Work out the differences between consecutive terms.

   c   How many dots will there be in the 7th term? Draw the pattern to check your answer.

   d   Do the triangle numbers make an arithmetic sequence?

7   Which of these sequences are arithmetic?
For each arithmetic sequence, write down the first term and the common difference.
a   3, 5, 6, 8, 9, 11, ...   b   0.5, 1.5, 2.5, 3.5, 4.5, ...   c   1, 2, 3, 1, 2, 3, 1, ...
d   1, 2, 4, 8, 16, ...   e   25, 20, 15, 10, 5, ...   f   98, 89, 80, 71, 62, ...

8   The first 7 terms of the **Fibonacci sequence** are 0, 1, 1, 2, 3, 5, 8.
Write down the 8th, 9th, 10th and 11th terms.

9   **Problem-solving**   The population of rabbits in a field over 6 months is recorded.

Month	1	2	3	4	5	6
Number of rabbits	2	2	4	6	10	16

How long will it be before there are more than 200 rabbits in the field?

Worked example

A sequence has first term −4.
The term-to-term rule is 'add 5, then multiply by 2'.

Write the first three terms in the sequence.

1st term = −4

2nd term   −4 → +5 → ×2 → 2

3rd term   2 → +5 → ×2 → 14

−4, 2, 14

10   Write the first five terms of each sequence.
a   first term 5, term-to-term rule 'multiply by 2, then add 1'
b   first term 7, term-to-term rule 'multiply by 2, then subtract 3'
c   first term 127, term-to-term rule 'subtract 1, then divide by 2'
d   first term 2, term-to-term rule 'multiply by 3, then subtract 4'
e   first term 32, term-to-term rule 'divide by 2, then add 4'

Q10c hint

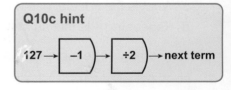

127 → −1 → ÷2 → next term

11   **Problem-solving / Reasoning**   Use first term 2. Choose three term-to-term rules with two steps, for example 'multiply by 2, then add 3'.
a   Write down the first five terms of each sequence.
b   Which sequence increases or decreases fastest?

12 Reasoning This pattern sequence is made from sticks.

Pattern 1   Pattern 2   Pattern 3

  a Describe how the sequence grows.
  b Cara says, 'The number of times you add 5 is one less than the pattern number.'
  Copy and complete Cara's working to find out if she is correct:
  Pattern 1   6 sticks
  Pattern 2   6 + 5 = ☐ sticks
  Pattern 3   6 + 5 + 5 = ☐ sticks
  Is she correct?
  c Use Cara's method to work out the number of sticks in pattern 10.

Key point   In a **geometric sequence** the term-to-term rule is 'multiply or divide by a number'.

For example:
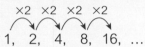
×2 ×2 ×2 ×2
1, 2, 4, 8, 16, ...

Each term is multiplied by 2.

---

13 Work out the next two terms of each geometric sequence.
  a 1, 10, 100, 1000, 10 000, ...   **b** 1, 5, 25 125, ...

14 Use the first term and the term-to-term rule to work out the first four terms of each geometric sequence.
  a First term 3, term-to-term rule '×10'   **b** First term 10, term-to-term rule '×2'
  c First term 500, term-to-term rule '÷5'   **d** First term 800, term-to-term rule '×0.5'

15 Is each sequence geometric, arithmetic or neither?
  a 3, 6, 12, 24, 48, ...   **b** 2, 4, 6, 3, 10, ...
  c 2, 2, 4, 6, 10, 16, ...   **d** 1, 4, 9, 16, 25, 36, ...
  e −10, −25, −40, −55, −70, ...   **f** First term 4, term-to-term rule '+6'

16 Problem-solving The second term of a geometric sequence is 6.
  The term-to-term rule is '×2'. What is the first term?

17 Problem-solving A sequence begins 1, 3, 9, 27, ...
  How many terms smaller than 100 are there in the whole sequence?

Challenge   Polly is told to exercise her leg after an injury.
She exercises for an hour each day for the first week, then decreases the daily time by 12 minutes each week.
  a In which week does Polly first exercise her leg for less than half an hour each day?
  b In which week doesn't she need to exercise her leg any more?
  c Do the daily exercise times each week form an arithmetic sequence?
  Explain your answer.

Reflect   Hassan says this riddle about a sequence:
'My first term is 3. My third term is 27. I am geometric. What s in between?'
Write down every step you take to work out Hassan's sequence.
You might begin, 'I write down all the numbers I know, leaving gaps for any missing numbers: 3, ___, 27, ...'

9.5 Straight-line graphs

- Recognise, name and plot graphs parallel to the axes
- Recognise, name and plot the graphs of $y = x$ and $y = -x$
- Plot straight-line graphs using a table of values
- Draw graphs to represent relationships

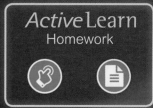
Active Learn
Homework

Warm up

1 Fluency Work out $3 \times 2 + 4$

2 $y = 5x$. Work out the value of y when x is equal to
  a 6   **b** 1   **c** 4

3 Work out the value of $3p + 4$ when p is equal to
  a 1   **b** 2   **c** 6

4 Reasoning

a Aisha's rule for generating coordinates is, 'Whatever the x-coordinate is, the y-coordinate is always 4.'
Which of these coordinate pairs satisfy Aisha's rule?
$(5, 5), (4, 4), (1, 4), (4, 3), (-1, 4), (5, -4), (0, 4), (4, 5)$

b Copy this grid and plot your points from part **a**.
Join the points with a straight line.
What do you notice?

c Elsie's rule for generating coordinates is, 'Whatever the x-coordinate is, the y-coordinate is always 2.'
Elsie generates these coordinates:
$(5, 2), (4, 2), (0, 2), (3, 2)$.
Where do you think these points will be on the grid?

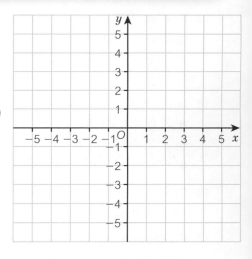

5 a Write down the coordinates of all the points on line A.
  b What do you notice about the coordinates?
  c Copy and complete these sentences.
    i The equation for line A is $y = \ldots$
    ii The equation for line B is $x = \ldots$
    iii The equation for line C is $y = \ldots$

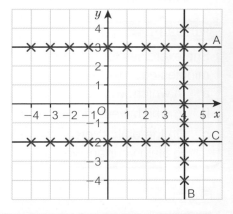

Key point

The equation $y = 2$ means that the y-coordinate is always 2, whatever the x-coordinate is. The line is **parallel** to the x-axis.
The equation $x = 3$ means that the x-coordinate is always 3, whatever the y-coordinate is. The line is parallel to the y-axis.

6 Write down the equations of the lines labelled A, B, C and D.

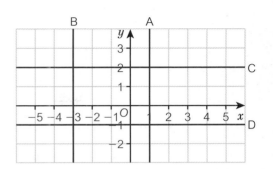

> **Key point** When you draw a graph, it should go to the edge of the grid.
> Label your graph by writing the equation next to the line.

7 Copy the grid from Q4.
Draw and label these graphs.
    a $y = 5$ **b** $x = 2$ **c** $y = 3$

Worked example

a Complete the table of values for $y = x - 2$.

x	0	1	2	3
y				

```
x                y
0 →   ┌─────┐  → -2
1 →   │     │  → -1        ← Draw a function machine.
2 →   │ - 2 │  →  0           Use the x-values in the table.
3 →   └─────┘  →  1
```

x	0	1	2	3
y	-2	-1	0	1

← Work out the y-values in the function machine. Write them in the table.

b Draw the graph of $y = x - 2$.

(0, −2), (1, −1), (2, 0), (3, 1)   ← Write down the coordinates.

```
y
3
2       y = x - 2
1
O
-1   1 2 3 4 5 6 7 8 9 10  x
-2
-3
```

← Draw a coordinate grid. Plot the points. Join them with a straight line to the edge of the grid. Label the line $y = x - 2$.

---

8 a Copy and complete this table of values for the equation $y = 3x$.

x	0	1	2	3	4
y					

b Write down the coordinates from the table.
c Draw a coordinate grid with x-axis from 0 to 5 and y-axis from 0 to 15. Plot the coordinates.
    Draw and label your graph.
d What is the value of y when $x = 6$?

> **Q8 hint**
>
>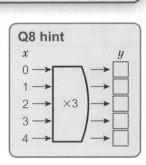

9 **Reasoning** Copy and complete this table of values from the graph.

x	0	1	2	3	4
y					

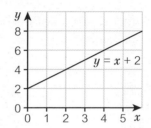

10 a Copy and complete this table of values for the equation $y = 3x + 4$.

x	0	1	2	3	4	5
y						

Q10a hint

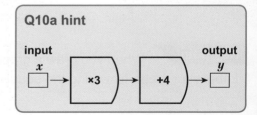

b From your table, write down the
  • smallest and largest x values
  • smallest and largest y values
  Draw a coordinate grid between these values.

c Draw the graph of $y = 3x + 4$.

d What is the value of y when $x = \frac{1}{2}$?

> **Q10c hint** Plot the coordinates from the table.

11 a Copy and complete the table from Q10a for the equation $y = 3x - 1$.

b Draw the graph of $y = 3x - 1$.

c **Reasoning** Look at the graphs you drew for Q8, Q10 and Q11. What do you notice?

12 a Complete this table of values for the equation $y = x$.

x	−3	−2	−1	0	1	2	3
y	−3		−1			2	

b Draw the graph of $y = x$.

13 **Problem-solving / Reasoning** This is the graph of $y = -x$.

a Write down three pairs of coordinates that lie on the line $y = -x$.

b Write down three pairs of coordinates that you know will *not* lie on the line $y = -x$.

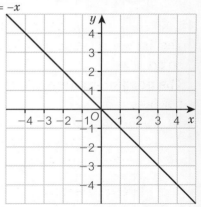

Challenge A spring is 10 cm long.
Jermaine puts different masses on the end and measures the length of the spring.
He records his results in this table.

a Draw the graph for Jermaine's results.

b Find the length of the spring when he adds 200 g.

c What mass gives a length of 11 cm?

Mass (g)	Length of spring (cm)
100	12
150	13
200	
250	15

Reflect Which of these did you find easiest? Which did you find hardest? Explain.

A Completing a table of coordinate pairs.

B Drawing a straight-line graph.

C Reading coordinates from a straight-line graph.

9.6 Position-to-term rules

- Generate terms of a sequence using a position-to-term rule
- Use linear expressions to describe the *n*th term of simple sequences

Warm up

1 Fluency Write down the missing rule for each function machine.

a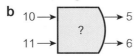

b

c
4 → ? → 20
5 → ? → 25

2 Fluency Work out the value of $3n$ when
    a $n = 1$ **b** $n = 2$ **c** $n = 3$ **d** $n = 8$

3 A sequence starts 10, 20, 30, 40.
What is
    a the 5th term   **b** the 10th term?

> **Key point**   Each term in a sequence has a position.
> The 1st term is in position 1, the 2nd term is in position 2, the 3rd term is in position 3, and so on.
> The **position-to-term** rule tells you how to work out a term in a sequence when you know its position.

4 Copy and complete this table to work out the first six terms of each sequence for these position-to-term rules.

Position number	1	2	3	4	5	6
Term	7	8	9	10	11	12

   a position number + 6   **b** 2 × position number
   c (2 × position number) + 1   **d** (3 × position number) − 2

> **Key point**   You use algebra to write the position-to-term rule.
> It is called the **nth term** because it tells you how to work out the term at position n (any position).

Worked example

The nth term of a sequence is $3n$.

Work out the first five terms.

> Substitute the position number into the expression $3n$.

Position (n)	1	2	3	4	5
Term ($3n$)	3 × 1 = 3	3 × 2 = 6	3 × 3 = 9	3 × 4 = 12	3 × 5 = 15

5   Copy and complete the tables to work out the first five terms of each sequence.
   a General term = $n + 2$ *TT +1*

Position (n)	1	2	3	4	5
Term ($n + 2$)	1 + 2 = 3	☐ + 2 = ☐			

   b General term = $2n$ *TT +2*

Position (n)	1	2	3	4	5
Term ($2n$)	2 × 1 = ☐	2 × ☐ = ☐			

   c General term = $n - 3$ *TT +1*

Position (n)	1	2	3	4	5
Term ($n - 3$)	− 2	− 1	0	1	2

   d General term = $5n$ *TT +5*

Position (n)	1	2	3	4	5
Term ($5n$)	5	10			

6   Work out the 10th term of the sequence with nth term
   a $3n$ b $n + 12$ c $5n$ d $n - 8$
   e $n - 15$ f $n + 0.5$ g $n - 0.7$ h $0.5n$

Q6 hint For the 10th term, $n = 10$.

7   **Problem-solving**   The nth term of a sequence is $4n$.
   How many terms are less than 25?

8   **Problem-solving**   The nth term of a sequence is $n - 5$.
   How many negative terms are there in the sequence?

Worked example

Work out what you do to the position number to get the term.

Work out the nth term of this sequence.
6, 12, 18, 24, 30, …

Position number: 1 ⟩×6 2 ⟩×6 3 ⟩×6 4 ⟩×6 5 ⟩×6 … n ⟩×6

Term: 6   12   18   24   30   … $6 × n$

nth term is $6n$ —— $6 × n = 6n$

9   Work out the nth term of each sequence.
   a 8, 16, 24, 32, 40, ….   b 11, 22, 33, 44, 55, …
   c 10, 20, 30, 40, 50, …   d 9, 18, 27, 36, 45, …
   e 1, 2, 3, 4, 5, …

10 **Reasoning**
   a What is the nth term of the multiples of 2?
   b What is the nth term of the multiples of 12?

11 Work out the nth term of each sequence.
The first one is started for you.

a 11, 12, 13, 14, 15, …

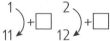

Position: 1) 2) 3) 4) 5) n)
)+□   )+□   )+□   )+□   )+□   )+□
Term: 11   12   13   14   15   □

term = position number + □ nth term = n − □

b 5, 6, 7, 8, 9, …

c 12, 13, 14, 15, 16, …

d 21, 22, 23, 24, 25, …

e 0, 1, 2, 3, 4, 5, …

f −3, −2, −1, 0, 1, 2, …

g −9, −8, −7, −6, −5, −4, …

Q11e hint

Position:)
)−□
Term:

12 For each sequence
  i Work out the nth term.
  ii Work out the common difference.

a 7, 14, 21, 28, 35, …

b 25, 50, 75, 100, 125, …

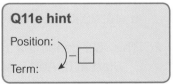

c **Reasoning** What do you notice about the nth term and the common difference?

13 Elena is training for a marathon.
She runs for 12 minutes on her first day,
then 12 minutes more each time she goes for a run.

a Continue this sequence up to the 6th term.

  12, 24, …

b What is the nth term for this sequence?

c For how many hours will Elena run on the 10th day?

d **Reasoning** Do you think that Elena will be able to stick to the model for 50 days? Explain your answer.

Q13c hint Convert the number of minutes to hours.

Challenge

a Work out the nth term of this arithmetic sequence.
  □, 18, □, 36, □, 54, □.

b Work out the 100th term.

Reflect Think about the term-to-term rule and the position-to-term rule for the 5 times table.
Which rule is '×5' and which is '+5'?
Make sure that you know the difference between the term-to-term rule and the position-to-term rule.
Write a hint, in your own words, to help you to remember which is which.

9 Check up

Sequences

1 What are the next three terms in each sequence?
   a 10, 20, 30, …   **b** 1, 3, 5, …   **c** 4, 7, 10, …

2 This rule generates a sequence:
   Start at 0. Add 2 each time.
   a Write down the first five terms.   **b** What is this sequence called?

3 Use the first term and the term-to-term rule to generate the first five terms of each sequence.
   a first term 9, term-to-term rule '−4'   **b** first term 0, term-to-term rule '+8'
   c first term 2, term-to-term rule '×3'   **d** first term 48, term-to-term rule '÷2'

4 **a** Draw the next shape in this pattern sequence.

   Pattern 1   Pattern 2   Pattern 3

   b Describe how the sequence grows.
   c Write down the number of dots in the first three patterns of this sequence.
   d Write down the next four terms of the number sequence.

5 Write down the next four terms of each sequence.

   a 6, 13, 20, 27, … **b** −8, −10, −12, −14, … **c** $\frac{1}{3}, \frac{1}{5}, \frac{1}{7}, \frac{1}{9}, …$ **d** 2, 4, 8, 16, 32, …
   e State whether each sequence is arithmetic, geometric or neither.

6 Work out the missing terms in each sequence.
   a 6, ___ , 0, −3, ___   **b** 1.5, ___ , 0.5, ___ , ___
   c ___ , 4, ___, −10, −17   **d** ___ , −3.2, −2.3, ___ , −0.5, ___

The nth term

7 Work out the first five terms of the sequence with nth term
   a $n + 7$ **b** $4n$

8 Work out the 10th term of the sequence with nth term
   a $n + 4$ **b** $6n$ **c** $n − 0.5$ **d** $0.25n$

9 **a** Find the position-to-term rule for this sequence.

Position in sequence	1	2	3	4	5
Term	6	12	18	24	30

   b Use your rule to find the 10th term and the 50th term.
   c What is the nth term of this sequence?

10 Find the nth term of this sequence: 3, 4, 5, 6, …

11 A sequence begins 3, 6, 9, 12, …
   Nawaz says, 'The nth term of this sequence is $n + 3$.'
   a What mistake has he made?   **b** Find the nth term of the sequence.

Graphs

12 Draw a grid with x- and y-axes from −6 to +6.
   a Plot the points (2, 1), (4, 3) and (5, 4).
   b Plot the points (0, −1) and (−2, −3).
      Join them with a straight line.
   c Find the midpoint of your line segment from part **b**.

13 Write down the equation of each line marked with a letter.

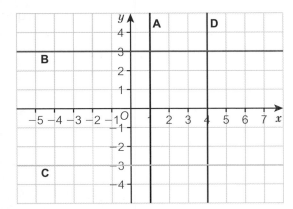

14 Draw a grid with x- and y-axes from −5 to +5.
   Draw and label the graphs of
   a $y = 2$ **b** $x = 4$ **c** $y = -1$ **d** $x = -2$

15 a Copy and complete this table of values for the equation $y = 4x + 2$.

x	0	1	2	3	4	5
y	2	6				

   b Use the values in the table to draw the graph of $y = 4x + 2$.
   c Use your graph to find the value of y when $x = \frac{1}{2}$.

16 Draw a grid with x- and y-axes from −6 to +6. Draw and label the graphs of $y = x$ and $y = -x$.

Challenge

a Draw these lines on a grid: $x = 5$, $x = 1$, $y = -2$, $y = 2$
b Copy and complete this sentence.
      The lines make the four sides of a _____ .
c Write down equations of four lines that make the sides of a rectangle.
d Write down equations of three lines that make the sides of a triangle.

Reflect   How sure are you of your answers? Were you mostly

   ☹ **Just guessing**   😐 **Feeling doubtful**   ☺ **Confident**

What next? Use your results to decide whether to strengthen or extend your learning.

9 Strengthen

Sequences

1 Write down the first five terms of each sequence.
   a First term 3, term-to-term rule 'add 2'

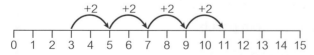

   b First term 10, term-to-term rule 'add 6'

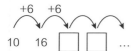

   c First term 20, term-to-term rule 'subtract 4'

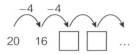

   d First term 15, term-to-term rule 'subtract 6'
   e First term 3, term-to-term rule 'multiply by 10'
   f First term 36, term-to-term rule 'halve'

> **Q1e hint** Multiply the first term by 10 to get the second term.

2 **a** Write the term-to-term rule for each sequence.

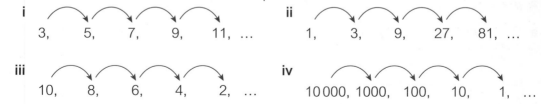

   b Which of the sequences in part **a** are arithmetic? Which are geometric?
   c Describe each sequence by giving the first term and the term-to-term rule.

> **Q2b hint** If you are Adding or subtracting a common difference, the sequence is Arithmetic.
> If you are multiplyinG or dividinG, the sequence is Geometric.

3 Here is a sequence of patterns made from yellow hexagons.

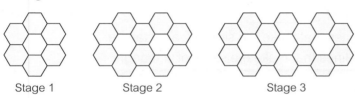

Stage 1   Stage 2   Stage 3

   a Write down the terms of the number sequence.
   This is an arithmetic sequence.

b How many yellow hexagons are added between

   i Stage 1 and Stage 2   **ii** Stage 2 and Stage 3?

c How many yellow hexagons will be in Stage 4?

d **Reasoning** What method did you use to work this out? What different method could you have used?

e The number sequence begins 6, 10, 14, …
Write down the next five terms.

4 Write down the first five terms of each sequence.

a First term 4, term-to-term rule 'multiply by 2, then add 3'

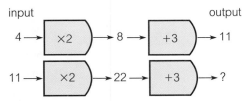

b First term 0, term-to-term rule 'multiply by 2, then add 5'

c First term 5, term-to-term rule 'subtract 1, then multiply by 2'

5 **a** What is the first term of this sequence?

b What is the term-to-term rule?

c Write down the next three terms.

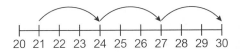

6 Write down the next three terms and the term-to-term rule for each sequence.

a 45, 50, 55, ...

b 12, 20, 28, ...

c 100, 96, 92, ...

d −15, −7, 1, ...

e −20, −13, −6, ...

Q6a hint

term-to-term rule

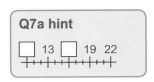

45   50   55

first term

7 Work out the missing terms in each sequence.

a ___ , 13, ___ , 19, 22   **b** 30, ___ , ___ , 42, 46

c ___ , 4, ___ , −10, ___   **d** ___ , −1, ___ , 0, ___

Q7a hint

☐ 13 ☐ 19 22

The nth term

1 Copy and complete the tables to show the first five terms of the sequences with these position-to-term rules.

a $n - 10$

Position number	1	2	3	4	5
Term					

b $100n$

Position number	1	2	3	4	5
Term					

Q1 hint Draw a function machine.

2 A sequence starts

4, 5, 6, 7, 8, ...

Gill draws a function machine.

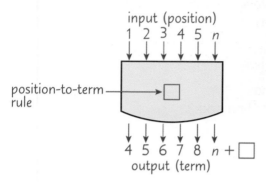

input (position)
1 2 3 4 5 n

position-to-term rule

4 5 6 7 8 $n +$ ☐
output (term)

a Find the position-to-term rule.

b Complete the general term of the sequence: $n +$ ☐

3 Use the method in Q2 to find the nth term of each sequence.

a 10, 11, 12, 13, 14, ...

b −2, −1, 0, 1, 2, ...

> **Q3b hint** nth term is $n -$ ☐

c 0, 1, 2, 3, 4, 5, ...

d 101, 102, 103, 104, ...

4 Find the position-to-term rule for each sequence.

a

Position in sequence	1	2	3	4	5
Term	5	6	7	8	9

b

Position in sequence	1	2	3	4	5
Term	11	12	13	14	15

c

Position in sequence	1	2	3	4	5
Term	5	10	15	20	25

d

Position in sequence	1	2	3	4	5
Term	7	14	21	28	35

5 Use the position-to-term rule to write the 10th term and the 50th term for each sequence in Q4. The first one has been done to help you below.

The position-to-term rule is +4.

So the 10th term is 10 + 4 = 14 and the 50th term is 50 + 4 = 54.

6 Write down the nth term for each sequence in Q4. Show that you have checked your answers. The first one has been done to help you below.

The position-to-term rule is +4. So the nth term is $n + 4$.

Check: 1st term is 1 + 4 = 5   2nd term is 2 + 4 = 6   3rd term is 3 + 4 = 7

7 Look at this sequence.

   a Copy and complete the sequence, filling in the numbers for the 3rd term.

   b Draw the next two terms, including the numbers.

   c What is the name of this number sequence?

   d Uzma says, 'The 4th term is found by adding a row of four dots on the bottom.'
   Is Uzma using the term-to-term rule or the position-to-term rule?

   e Craig says, 'The 4th term is found by adding all the numbers up to 4.'
   Is Craig using the term-to-term rule or the position-to-term rule?

   f Use the position-to-term rule to find the 10th term of the sequence.

Graphs

1 Write down the coordinates of each point marked with a letter.

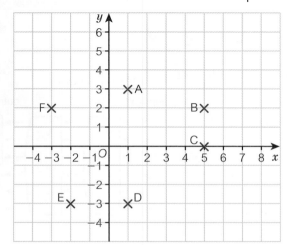

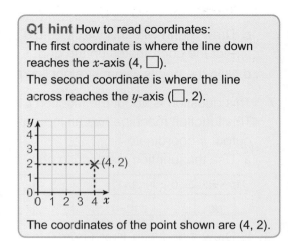

Q1 hint How to read coordinates:
The first coordinate is where the line down reaches the x-axis $(4, \square)$.
The second coordinate is where the line across reaches the y-axis $(\square, 2)$.

The coordinates of the point shown are (4, 2).

2 Copy the grid from Q1.

   a Plot the points A(3, 2), B(1, 5), C(6, 3), D(−1, −4), E(−2, 2) F(4, −3) and G(3, −4).

   b Join A and G.
   Join C and F.
   Use a cross to mark the midpoints of each line.
   Write down the coordinates of the midpoints.

3 **a** Draw a grid with x- and y-axes from −5 to +5.
   Plot four points with y-coordinates of 2.

   b Join all your points with a straight line. Where does the line cross the y-axis?

   c The equation of the line is $y = 2$.
   Which of these points will also lie on the line?
   (0, 2), (4, −2), (5, 2), (−2, −2), (−1, 2), (2, 2)

   d Write a sentence about the y-coordinates of the points that do *not* lie on the line.

4 Draw a grid with the x- and y-axes from −5 to +5.
   Draw the lines
   a $x = 4$ **b** $x = 1$ **c** $x = -3$

5 **a** Sophie tried to plot the line $x = 3$.
What mistake did she make?
b **Reasoning** Write Sophie a hint, explaining how to plot the
graph of a line like $x = 3$. Use these words to help:

x-coordinate   straight   axis

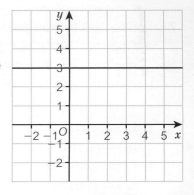

6 This question is about the function $y = 4x$.
a Copy and complete this table with the value of y for
each value of x.

x	0	1	2	3	4
y	0	4			

> **Q6a hint** $y = 4x$, so
> multiply each value of x
> by 4 to get the value of y.

b Write down the coordinates.
Coordinates   (0, __), (__, __), (__, __), (__, __), (__, __)
c Draw a grid with the x-axis from 0 to 4 and the y-axis from 0 to 16.
Plot the coordinates on the grid.
d Join the points with a straight line and label the line $y = 4x$.

7 This question is about the function $y = 3x + 1$.
This function machine has an input x-coordinate and an
output y-coordinate.
a Use the function machine to copy and complete this table.

input x output y

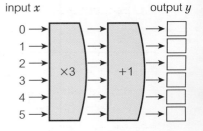

x	0	1	2	3	4	5
y		4				

b Draw a grid with x-axis from 0 to 5 and y-axis from 0 to 20.
Plot the points from your table of values.
c Draw a straight line through the points and label the line $y = 3x + 1$.
d Write the coordinates of the point where the graph crosses the y-axis.

Challenge   Look at this sequence of dominoes.
a How many dots will be in the top section of the
next domino?
b What is the term-to-term rule for the sequence of
dots in the bottom sections?
c Draw the next domino in the sequence.
d You can use the dominos to generate coordinates: (top, bottom).
The first domino gives the coordinates (0, 2).
Write down the next three pairs of coordinates.
e Plot the four pairs of coordinates on a grid.
f Complete this function machine.

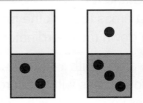

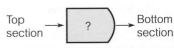

Top   ?   Bottom
section   section

Reflect   Lars says, 'The words *term* and *coordinates* are used a lot in these lessons.
These must be important words for understanding sequences and graphs.'

Write definitions, in your own words, for *term* and *coordinates*.

9 Extend

1 For each of these sequences, write down the first term and the term-to-term rule.
    a 16, 160, 1600, 16 000, …   **b** 0.25, 0.5, 1, 2, …   **c** 1, 1, 2, 3, 5, 8, …

2 **Reasoning** Here is a sequence of whole numbers: 88, 44, 22, …
Is the sequence infinite? Explain your answer.

3 **Problem-solving** Write a sequence containing these numbers, with at least one term in between them. Describe the term-to-term rule that you use.
    a 1 and 12   **b** 3 and 15   **c** 6 and 20
    d 1 and 100   **e** 4 and 10   **f** 35 and 30

4 **Reasoning** Look at this pattern made from squares.

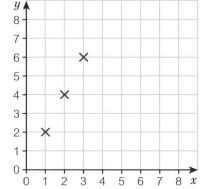

    a How many shaded squares will there be in the 4th pattern?
    b What is the special name for the number sequence generated by the shaded squares?
    c How many unshaded squares will there be in the 4th pattern?
    d How many unshaded squares will there be in the 100th pattern?
    e Think about the sequences of numbers generated by the pattern of shaded squares and the pattern of unshaded squares. Is either of them an arithmetic sequence? Explain your answer.
    f What calculation will you do to work out the total number of squares in the 6th pattern?
    g Write down the sequence generated by the total numbers of squares in the first four patterns.

5 Susannah uses this function machine to generate coordinates.

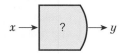

She plots them on the graph.

    a What is the rule for the function machine?
    b Write down the equation of the line through Susannah's points.
    c Copy the grid and plot the graph of Susannah's function.
    d Plot the graph of $y = x$ on the grid.
    e **Reasoning** Write down one difference and one similarity between the two graphs.

> **Q5a hint** Write down the coordinates of the points to help you.

> **Q5b hint** Write $y = …$
> Check that your equation works for two points on the line.

6 a Write down the coordinates of the points marked A, B and C.
   b **Reasoning** Points A, B and C are three vertices of a rectangle.
      Write down the coordinates of the 4th vertex.

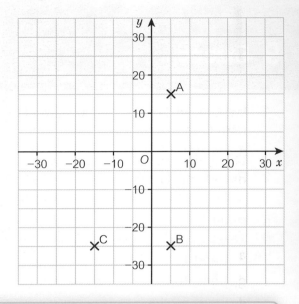

7 **Reasoning** Write the nth term for a sequence where the terms are
   a the multiples of 3
   b the multiples of 6
   c the even numbers.

> **Key point** The nth term is sometimes called the **general term**.

8 Find the first five terms and the 100th term of each sequence.
   a general term $n - 4$
   b general term $n^2 + 1$
   c general term $2n - 6$
   d general term $200 - n^2$

9 **Problem-solving / Reasoning** Look at these sequences.
   A 2, 5, 8, 11, 14, 17, 20, …
   B 1, 2, 4, 8, 16, 32, …
   C 4, 5, 6, 7, 8, 9, 10, …
   D 3, 12, 27, 48, 75, …
   a Which ones are arithmetic sequences?
   b Plot a graph for each sequence.
   c What do you notice about the graphs of the arithmetic sequences?
   d Here is another sequence.

      1, 5, 1, 5, 1, 5, …

      Sketch what you think its graph will look like.
      Now plot the sequence.
      Were you correct?

   > **Q9d hint** Sketching a graph means drawing it without a ruler or numbers on the axes. It helps you see the shape of a graph quickly.

10 a Copy this table of values.

x	0	1	2	3	4
y					

   Complete it for **i** $y = x$ **ii** $y = 2x$
   b Draw a grid with x- and y- axes from 0 to 6. Plot the graph of **i** $y = x$ **ii** $y = 2x$
   c Write down the coordinates of the point where the two lines meet.

> **Reflect** Sandra says, 'Sequences and straight-line graphs are all about following patterns.'
>
> Look back at the work you have done in this unit. Write three sentences that describe how what you have learned is all about 'following patterns'.

9 Unit test

1 Write down the next term in each sequence.
a 0, 50, 100, 150, … **b** 17, 15, 13, 11, … **c** 1, 10, 100, 1000, …

2 Write down the term-to-term rule and the next three terms for each sequence.
a 100, 90, 80, 70, … **b** 40, 52, 64, 76, …

c 8, 9.5, 11, 12.5, … **d** $\frac{1}{2}, \frac{1}{4}, \frac{1}{8}, \frac{1}{16}$, …

3 Write down the missing numbers in each sequence.
a ___ , 7, ___, 17, 22, ___   **b** ___, 3, 6, 12, 24, ___

4 Sushma makes a bracelet from beads.
It grows like this.
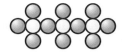
a Copy and complete this table.

Number of flowers	1	2	3	4	5
Number of beads	5	9			

b What is the term-to-term rule for the sequence?

5 **a** Write down the coordinates of points A, B, C, D and E.
b Write down the coordinates of the midpoint of the line segment BC.

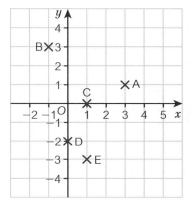

6 Jemima generates this table of coordinates from an experiment.

x	1	2	3	4	5
y	3	3	3	3	3

a Copy this grid and plot the coordinate pairs.
b Draw a line through the points.
What is the equation of the line?
c Copy and complete this sentence.
When $x = 50$, the y-coordinate is ___.

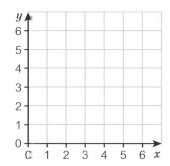

7 **a** Copy and complete this table of values for the graph of $y = 2x + 1$.
b What is the y-coordinate when the x-coordinate is 10?

x	0	1	2	3	4	5
y		3	5			

8 **a** Copy and complete this table of values for the graph of $y = 3x - 3$.

x	0	1	2	3	4
y					

   b Draw a grid with x-axis from 0 to 10 and y-axis from −3 to 10. and plot the graph of $y = 3x - 3$.

9 **a** Find the position-to-term rule for this sequence.

Position in sequence	1	2	3	4	5
Term	0.5	1	1.5	2	2.5

   b Use your rule to find the 10th term and the 50th term.

   c What is the nth term of this sequence?

10 Write the first five terms of the sequence whose nth term is
   a $3n + 4$ **b** $6 - n$

11 Look at your sequences from Q10.
   Write down whether each sequence is ascending or descending.

12 Is each sequence arithmetic or geometric?
   a 96, 48, 24, 12, ... **b** 96, 88, 80, 72, ...

13 Look at this sequence: 4, 8, 12, 16, …
   a Write down the next three terms of the sequence.
   b Write a formula for the nth term of the sequence.

Challenge   Investigate the number of squares in each of these shapes. Use the steps below to begin.

a Write a sequence for the *total* number of squares you can see in each pattern. Don't forget to count squares made of smaller squares.

Hint Use a systematic method for counting the total number of squares.

b Use your sequence to predict the *total* number of squares in the 5th pattern.

c Can you find a way of working out the *total* number of squares in the 6th pattern using your answer to part **b**?

d Describe any other number patterns in these shapes.

Reflect   Write a heading 'Five important things about sequences and graphs'. Now look back at the work you have done in this unit and list the five most important things you think you have learned. For example, you might include:
   • words (with their definitions)
   • methods for working things out
   • mistakes you made (with tips on how to avoid them in future).

10 Transformations

Master   Check up p278   Strengthen p280   Extend p286   Unit test p288

10.1 Congruency and enlargements

- Identify congruent shapes
- Use the language of enlargement
- Enlarge shapes using given scale factors
- Work out the scale factor given an object and its image

Active Learn
Homework

Warm up

1 Fluency What is the missing number?

a $6 \times 3 = \square$ **b** $5 \times 2 = \square$ **c** $2 \times \square = 8$ **d** $\square \times 3 = 12$

2 Which one of each set of shapes is not the same size and shape as the other two?

a A   B   C

b A   B   C

Key point

Shapes are **congruent** if they are the same shape and size. For example, these shapes are all congruent.

3 Which of these pairs of shapes are congruent?

A   B   C   D

   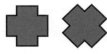

4 Trace each shape. Shade in the congruent parts of each shape in the same colour. The first one is done for you.

a

b   **c**   **d**

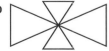 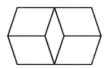 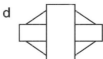

5 A computer programmer is working on a game where shapes fit together.
She uses 5 pairs of congruent shapes.
Which pairs of shapes are congruent?

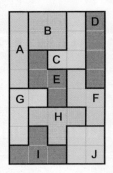

Q5 hint Congruent shapes can be different colours.

6 Reasoning A graphic designer draws two congruent trapeziums.

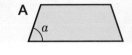
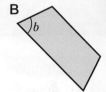

a Trace the longest side of trapezium A.
Lay this over the longest side of trapezium B.
Are both sides the same length?

b Repeat for the other sides. Are they all the same length?

c Copy and complete this sentence.
If two shapes are congruent, the lengths of matching sides are _____.

d Measure angle a in trapezium A. Measure angle b in trapezium B.
Are they the same size?

e Repeat for the other angles. Are they the same size?

f Copy and complete this sentence.
If two shapes are congruent, the matching angles are _____.

7 Copy each shape. Then split them into the number of congruent shapes shown.
The first one is done for you.

a four congruent triangles   **b** two congruent triangles   **c** two congruent triangles and two congruent rectangles

> **Key point**   In congruent shapes, **corresponding sides** (matching sides) and **corresponding angles** (matching angles) are equal.

8 These two triangles are congruent.
Copy and complete these sentences.

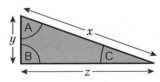
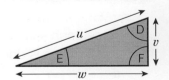

a Side x and side ☐ are corresponding sides.

b Side ☐ and side v are corresponding sides.

c Angle A and angle ☐ are corresponding angles.

d Angle ☐ and angle E are corresponding angles.

9 Reasoning Which two of these triangles are congruent? Explain how you know.

A   **B**   **C**

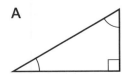
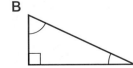
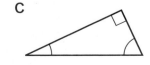

10 a Copy this triangle onto squared paper.

b Copy and complete this enlargement
of the triangle by scale factor 2.

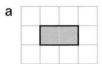

Q10b hint To enlarge by
scale factor 2, the height
has been multiplied by 2.
Multiply the base by 2. Now
join to make the third side.

11 Copy each shape on to squared paper. Now **enlarge** each shape by
  i scale factor 3   **ii** scale factor 5

a 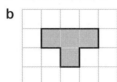   **b** 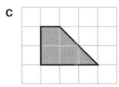   **c**   **d**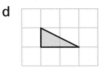

12 A photograph measuring 15 cm by 10 cm is enlarged by scale factor 3.
What are the new length and width?

13 In a school play, all the props need to be 10 times their real size.
  a A real DVD case is 15 mm thick. How thick is a DVD case in the play?
  Give your answer in centimetres.
  b A real calculator is 14 cm long. How long is a calculator in the play?
  Give your answer in metres.

Worked example

a Write the ratio of the length of the object to the corresponding length in the image.
Give the ratio in its simplest form.

object   image

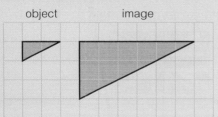

Ratio = 2 : 6

= 1 : 3

Length of object = 2 squares
Length of image = 6 squares

Write the ratio in
its simplest form.

Each side length of the image
is 3 times the corresponding
side length of the object.

b Write down the scale factor of the enlargement.

Scale factor = 3

14 For each of these enlargements:

    i Write the ratio of the length of a side of the object to the corresponding length in the image. Give the ratio in its simplest form.

    ii Write the scale factor of the enlargement.

a

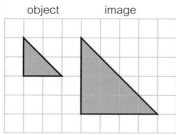

b

15 Reasoning Look at the shapes B to F. Which shapes are

    a congruent to shape A

    b an enlargement of shape A

    c neither congruent to, nor an enlargement of shape A?

16 Problem-solving A photo printing service offers the following size prints.

5″ × 5″,   7″ × 5″,   8″ × 6″,   8″ × 8″,   16″ × 12″

Sam wants two prints, one to be an enlargement of the other.

    a Would a 160 × 120 print be an enlargement of an 80 × 60 print? Explain your answer.

    b Would an 80 × 60 print be an enlargement of a 70 × 50 print? Explain your answer.

    c Would an 80 × 80 print be an enlargement of a 50 × 50 print? Explain your answer.

> **Q16 hint** ″ is the symbol for inches. You might want to draw a sketch to help you.

17 Reasoning Amanda says, 'If I enlarge a shape by scale factor 3, the perimeter of the image will be 3 times the perimeter of the object.'

Is she correct? Explain your answer.

Challenge   What congruent shapes are there in the flags of Great Britain?

Reflect   Look back at this lesson.

List all the new mathematical words you learned. Be careful to spell them correctly.

Write a short definition for each of them. Where possible, use your own words.

Draw your own shape or shapes with each definition, to show what you mean.

10.2 Symmetry

- Recognise reflection and rotational symmetry in 2D shapes
- Solve problems using line symmetry
- Identify all the symmetries of 2D shapes
- Identify reflection symmetry in 3D shapes

Active Learn
Homework

Warm up

1 **Fluency**   What are the names of these 3D shapes?

a 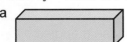   b   c   d

2 For each triangle, write down if it is scalene, isosceles or equilateral.
   Give a reason for your answers.

a    b   c

Key point

A shape has **reflection symmetry** if one half folds exactly on top of the other half.

fold

The dashed line is called a **line of symmetry** or mirror line.

3 **a** Trace each triangle in Q2 and cut it out.
        Try folding it in half in different ways.
   b Copy and complete these sentences.
        An equilateral triangle has ☐ lines of symmetry.
        An isosceles triangle has ☐ lines of symmetry.
        A scalene triangle has ☐ lines of symmetry.

4 Copy these shapes. Draw all the lines of symmetry on the shapes.
   The first one is done for you.

a 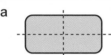   b   c   d

5 How many lines of symmetry does each shape have?

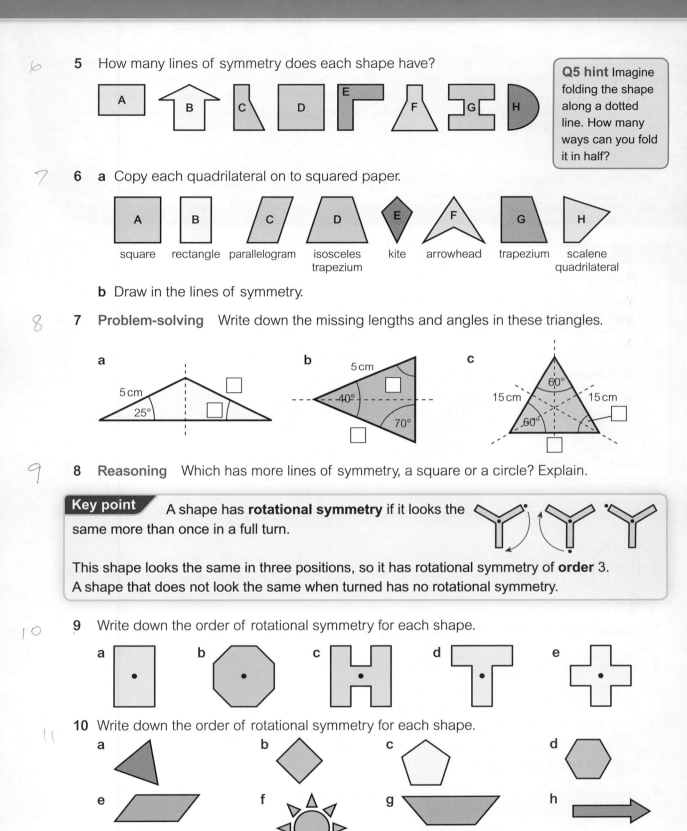

Q5 hint Imagine folding the shape along a dotted line. How many ways can you fold it in half?

6 a Copy each quadrilateral on to squared paper.

square   rectangle   parallelogram   isosceles trapezium   kite   arrowhead   trapezium   scalene quadrilateral

b Draw in the lines of symmetry.

7 Problem-solving Write down the missing lengths and angles in these triangles.

a

5 cm
25°

b

5 cm
40°
70°

c

60°
15 cm   15 cm
60°

8 Reasoning Which has more lines of symmetry, a square or a circle? Explain.

Key point A shape has **rotational symmetry** if it looks the same more than once in a full turn.

This shape looks the same in three positions, so it has rotational symmetry of **order** 3. A shape that does not look the same when turned has no rotational symmetry.

9 Write down the order of rotational symmetry for each shape.

a   **b**   **c**   **d**   **e**

10 Write down the order of rotational symmetry for each shape.

a   **b**   **c**   **d**

e   **f**   **g**   **h**

11 For each shape, write down
   i the number of lines of symmetry   ii the order of rotational symmetry
   a square   b rectangle   c parallelogram
   d isosceles trapezium   e kite   f rhombus
   g isosceles triangle   h equilateral triangle

> **Key point** If a 3D shape has reflection symmetry, the mirror line is called a
> **plane of symmetry**.

12 This cuboid has reflection symmetry.
   The three **planes of symmetry** are
   shaded red in the diagrams.

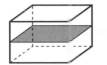

   Which of these 3D shapes have reflection
   symmetry?

13 **Reasoning**   a   Does a cylinder have reflection symmetry?
   b Does a dice have reflection symmetry?

14 **Problem-solving**   Scientists look for symmetry in molecules because it
   can help predict chemical properties.
   a How many planes of symmetry does the
      3D model of a water molecule have?
   b How many planes of symmetry does the
      3D model of an ammonia molecule have?

> ### Challenge
>
> 1   a   How many lines of symmetry does the letter I have?
>   b   What is the order of rotational symmetry of the letter I?
>
> 2   Investigate the reflection and rotational symmetry of
>   a   your name written using capital letters
>   b   the digits 0 to 9
>
> 3   The word MUM has 1 line of symmetry.   **MUM**
>   Write some more words that have line symmetry.
>
> 4   The number **916** has rotational symmetry.
>
>   Find two more numbers that have rotational symmetry.
>
> 5   Find a word that has rotational symmetry.

> **Reflect**   After this lesson, Evan said, 'A trapezium is *always* symmetrical.'
> Robyn said, 'A trapezium is *never* symmetrical.'
> Claire said, 'A trapezium is *sometimes* symmetrical.'
> Who is correct, Evan, Robyn or Claire? You could draw diagrams to help you.
> Use what you have learned in this lesson, and what you know about trapeziums, to explain.

10.3 Reflection

Active Learn
Homework

- Recognise and carry out reflections in a mirror line
- Reflect a shape on a coordinate grid
- Describe a reflection on a coordinate grid

Warm up

1 Fluency Here is a coordinate grid.
  a Which is the x-axis and which is the y-axis?
  b What are the coordinates of point A?

2 Make a copy of the coordinate grid from Q1.
  Label the x and y axes.
  Plot and label these points.
  B (3, 2)   **C** (−2, 3)   **D** (2, −1)   **E** (−1, −2)

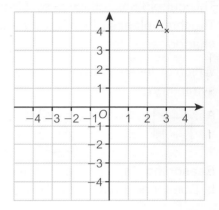

3 The diagram shows some straight lines on a
  coordinate grid.
  Match each line with the correct equation.

  $x = -2$ $y = 1$ $x = 3$ $y = -2$ $y = x$

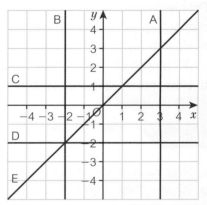

4 Place a mirror on the dotted line on shape X.
  Choose the correct **reflection** of shape X.

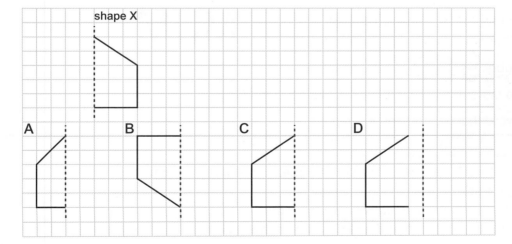

Key point

A **reflection** is a type of transformation.
You reflect shapes in a mirror line.
All points on the image are the same distance from the mirror line as the points on the object, but on the opposite side.

Worked example

Is the bottom shape a correct reflection of the top shape in the mirror line?
Give a reason for your answer.

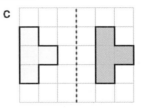

This is a correct reflection.

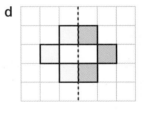

No. The bottom shape should be one square from the mirror line, not on the mirror line.

5 **Reasoning**   Decide whether each diagram shows a correct reflection in the mirror line.
If the reflection is not correct, give a reason why.
Then copy the shape and draw a correct reflection.

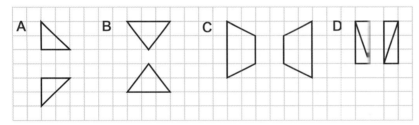

a   b   c   d

6 **a** Copy each pair of shapes. Draw in the mirror line for each pair.

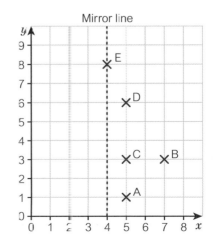

b **Problem-solving**   Are the shapes in each pair congruent? Explain how you know.

7 The points A to E are reflected in the mirror line shown.
Write down the coordinates of the image of each point.

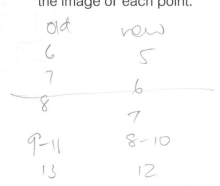

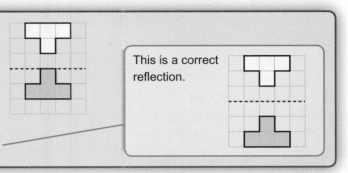

Mirror line

8 Copy this diagram.
Draw the image of the triangle after a reflection in the line
a $x = 2$ **b** $y = 3$

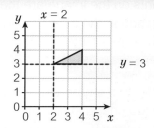

9 Copy this diagram.
Draw the image of the shape
after a reflection in the line
a $x = -3$
b $y = -1$
c $x = 0$

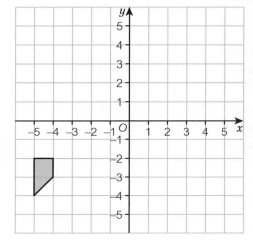

10 Draw a grid with x- and y-axes from −5 to +5.
A triangle has vertices at (1, 2), (5, 2) and (1, 4).
  a Draw the triangle on the coordinate grid. Label the triangle A.
  b Reflect the triangle in the line $y = 2$. Label the reflected shape B.
  c Reflect the triangle and its image from part **b** in the line $x = 1$.
    Label the reflected shape C.
  d What shape have you made?

11 **Problem-solving** The diagram shows five
congruent shapes on a coordinate grid.
Copy and complete these statements.
The first one is done for you.
  a A is a reflection of B in the line $x = -1$.
  b A is a reflection of D in the line _____.
  c D is a reflection of E in the line _____.
  d E is a reflection of F in the line _____.
  e C is a reflection of F in the line _____.

Q11 hint Count the empty squares between the original
shape and its reflection. The mirror line must be half way
between the two for all points on the object to be the same
distance from the mirror line as all points on the image.

12 Copy this diagram.
Draw the images of these shapes after reflection in the lines given.
The first one is done for you.

   a Shape A in the line $y = x$.

   b Shape B in the line $y = -x$.

   c Shape C in the line $y = x$.

   d Shape D in the line $y = -x$.

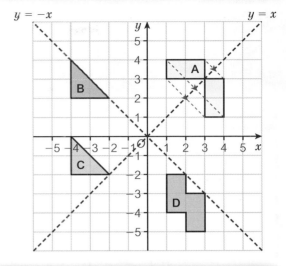

Challenge   In the diagram, triangle ABC has been reflected in the line $y = x$ to become triangle DEF.

1 a Write down the coordinates of the vertices of the triangles ABC and DEF.
Put your answers in a table like this.

	Coordinates of vertices					
Object ABC	A	(1, 4)	B		C	
Image DEF	D		E		F	

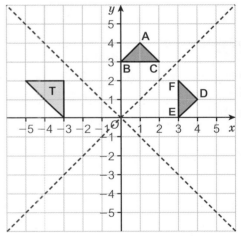

   b What do you notice about the coordinates of the object and its image?

   c Draw other shapes on a coordinate grid and reflect them in the line $y = x$.

   d What can you say about the coordinates of each object and its image?

2 On a new grid, label the vertices of triangle **T** and reflect it in the line $y = -x$.

   a What do you notice about the coordinates of the object and its image?

   b Draw other shapes on the grid to check this is always true.

Reflect

a Look back at Q5.
Write down the steps you took to draw the reflected images.
You might begin with:
Step 1: I found the mirror line.

b Look back at Q12.
Write down the steps you took to draw the reflected images.

c Which steps were the same or different for Q5 and Q12?

10.4 Rotation

- Describe and carry out rotations on a coordinate grid

*Active*Learn
Homework

Warm up

1 Fluency

a Does each diagram show a quarter turn or a half turn?

i   **ii**   **iii**   **iv**

b How many degrees in a quarter turn?

c How many degrees in a half turn?

2 Which of these arrows is **clockwise**
and which is **anticlockwise**?

 A   B

Key point   A **rotation** is a type of transformation.

You rotate a shape by turning it around a point, called the **centre of rotation**.
To describe a rotation you also need to give the **angle** and direction (**clockwise** or
anticlockwise).

---

3 Each grey shape has been rotated about the centre of rotation ✗.
Describe each rotation. The first one is done for you.

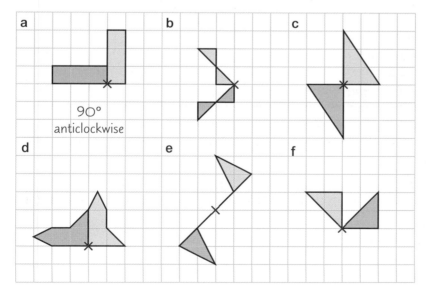

Q3 hint
Turn the page ↰ or ↱
until the grey shape
looks like the pink
shape. Have you
turned through 90° or
180°? In a clockwise or
anticlockwise direction?

4 Reasoning There is no need to state the direction (clockwise or anticlockwise) for a 180° rotation. Explain why.

5 The shapes marked A have been rotated about the centre of rotation ×.
Describe the rotations.

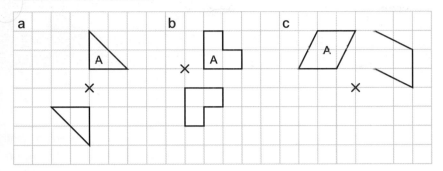

6 The grid shows four flags.
Describe these rotations. The first one is done for you.
  a A on to B
  90° rotation clockwise about the point (1, 0)
  b B on to A
  c D on to B
  d C on to D
  e A on to C
  f C on to A

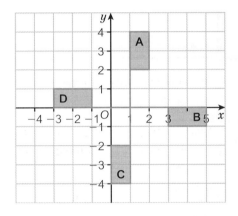

7 Write down the coordinates of the triangle after a rotation of
  a 90° clockwise about (0, 1)
  b 90° anticlockwise about (0, 0)
  c 180° about (0, 0).

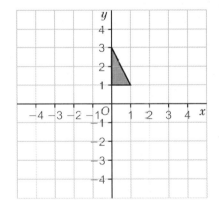

> **Q7 hint** Copy the grid and use tracing paper to draw the rotations.

8 a Draw a coordinate grid with x and y-axes from −5 to +5.
  Plot and join the points (0, 0), (2, 2), (4, 0) and (2, −2) on the coordinate grid.
  Label the shape A.
  b Name the shape you have drawn.
  c Rotate your shape by 90°, 180° and 270° clockwise about (0, 0), showing all of these on the same diagram.
  d Describe the shape you now have.

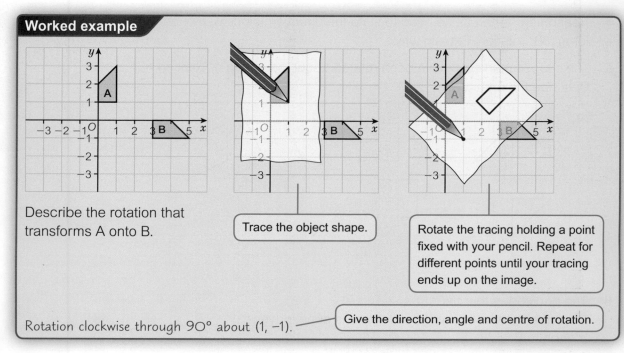

Describe the rotation that transforms A onto B.

Trace the object shape.

Rotate the tracing holding a point fixed with your pencil. Repeat for different points until your tracing ends up on the image.

Rotation clockwise through 90° about (1, −1).

Give the direction, angle and centre of rotation.

9 Describe the rotation that transforms
   a A onto B
   b A onto C
   c B onto D
   d B onto E

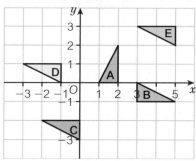

Q9 hint Use tracing paper to help you.

Challenge This is the logo for a company.
The logo designer draws shape A,
then rotates the shape three times.
a Describe the three rotations.
b Design your own company logo involving at least two rotations.
   Describe the rotations.

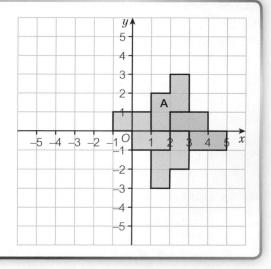

Reflect In the last lesson you learned about reflection. In this lesson you learned about rotation. Look carefully at some of the shapes you reflected and rotated in these lessons. Can a reflection of a shape and a rotation of a shape give the same result?

10.5 Translations and combined transformations

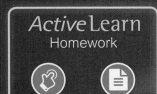
Active Learn
Homework

- Translate 2D shapes
- Transform 2D shapes by combinations of rotations, reflections and translations

1 Fluency Which arrow points left and which points right?

A   B

2 The diagram shows a green triangle on a coordinate grid. Copy the diagram.

a Draw the image of the triangle after a reflection in

  i $x = 2$, label this image A

  ii $y = 1$, label this image B

  iii $x = -1$, label this image C

b Draw the image of the triangle after a rotation

  i 90° clockwise about (0, 2), label this image D

  ii 180° about (1, 3), label this image E

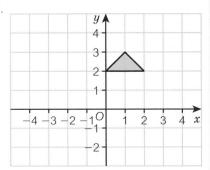

Key point A **translation** is a type of transformation. A translation of a 2D shape is a slide across a flat surface. To describe a translation you need to give the movement left or right, followed by the movement up or down.
A translation does not change the size or shape of an object.

3 Describe each translation.
The first two are started for you.

  a A to B

    □ squares right

  b A to C

    □ squares right, □ squares down

  c A to D

  d A to E

  e A to F

  f A to G

  g A to H

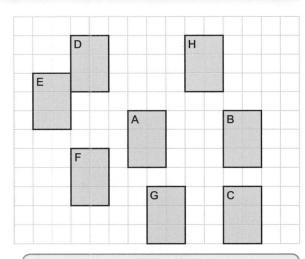

Q3 hint Choose a vertex of shape A. Count how many squares left or right and up or down you need to move that vertex to its new position.

4 The diagram shows four triangles on a coordinate grid.

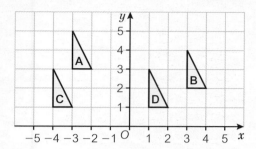

Triangle A to triangle B is a translation 6 squares right and 1 square down.
Describe each of these translations.

a triangle A to triangle C

b triangle A to triangle D

c triangle D to triangle B

d triangle C to triangle B

5 Copy the orange shape onto squared paper.
Draw the image of the shape after these translations.
Part **a** is done for you.

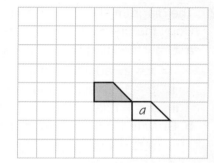

a 2 squares right, 1 square down

b 3 squares right, 2 squares up

c 1 square left, 3 squares up

d 2 squares left, 2 squares down

e 4 squares left

6 Copy this grid and shape X.
Translate shape X using these translations.

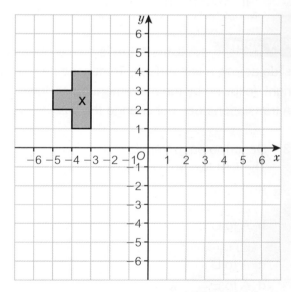

a 5 squares down.
Label it A.

b 3 squares right.
Label it B.

c 1 square right, 2 squares down.
Label it C.

d 1 square left, 7 squares down.
Label it D.

e 8 squares right, 2 squares up.
Label it E.

7 **Reasoning / Problem-solving** In the game of chess, in one turn,
a knight can move either 2 squares left or right followed by 1 square
up or down, or 1 square left or right followed by 2 squares up or down.
Can this knight move to a white square?
Draw a grid and show the knight's possible positions to support
your answer.

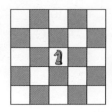

Transform the shape using these transformations:

translation 3 squares left and 1 square up, followed by a reflection in the line $x = -2$.

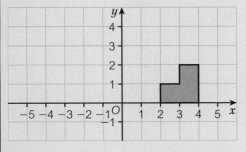

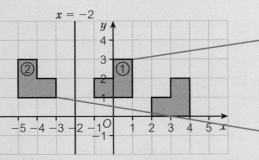

$x = -2$

1 First step is the translation ①.

2 Draw in the line $x = -2$.
3 Reflect ① in the line to make ②.

8 Copy the diagram four times.
   On separate copies, transform the shape
   using these transformations.
   a Reflection in the x-axis followed by
      a reflection in the y-axis.
   b Translation 3 left and 4 down followed by
      a translation 1 left and 2 down.
   c 180° rotation about the point (0, 0)
      followed by a reflection in the y-axis.
   d Reflection in the line $y = 2$,
      followed by a reflection in the x-axis.

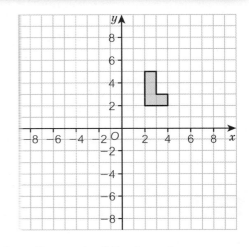

9 **Reasoning / Problem-solving** Larry says, 'You can describe each of the two-step
   transformations in Q8 as a single transformation.'
   Is he correct? If so, write the single transformations.

10 **Reasoning / Problem-solving** Charlie transforms shape A using
   two translations: 4 squares left and 2 squares down followed by
   3 squares right and 4 squares up.
   He labels the image B.
   Charlie says, 'If I translate shape A 1 square left and 2 squares up,
   I'll also get shape B.'
   Is he correct? Explain your answer.

   Q10 hint Draw a
   shape on a grid,
   label it A, then follow
   Charlie's instructions.

11 Copy this diagram twice.

On separate copies, transform the triangle using these two-step transformations.

a Rotation 90° anticlockwise about (−1, 4) followed by a translation 2 squares right and 4 squares down.

b Reflection in the line $x = -1$ followed by a rotation 90° clockwise about (−1, 1).

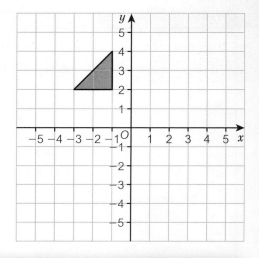

> **Key point**   When a shape is transformed by a translation, rotation or reflection, the image has exactly the same side lengths and angles as the object.
> This means that the original shape and the transformed shape are congruent.

12 The diagram shows shapes A to J.
Write true (T) or false (F) for each of these statements.
If your answer is false, explain why.

a G is a translation of B.   **b** F is a reflection of B.

c E is a rotation of F.   **d** H is a translation of C.

e J is a reflection of H.   **f** G is a rotation of D.

g C is a reflection of A.   **h** A is a translation of B.

i B is a rotation of D.   **j** D is a translation of I.

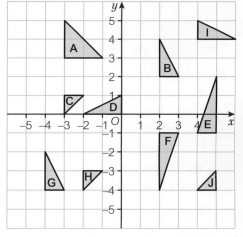

> **Challenge**   Sara says, 'There is a two-step transformation that will transform shape D to shape B in Q12.
>
> **a** Copy and complete Sara's description of the two-step transformation.
> Translation 2 squares right and 2 squares up followed by a rotation _____
>
> **b** Describe a two-step transformation that transforms shape D to shape I.
>
> **c** Write another two-step transformation that transforms shape D to shape I.

> **Reflect**   Write these transformations in order, from the one you find easiest to the one you find hardest. Look back at some of the transformation questions in this lesson to help you.
>
> **A** Reflection in a horizontal line
>
> **B** Reflection in a vertical line
>
> **C** Rotation of 90° (either clockwise or anticlockwise)
>
> **D** Rotation of 180°
>
> **E** Translation
>
> Write a hint, in your own words, for the transformation you found the hardest.

10 Check up

Shapes and symmetry

1   Which pairs of arrows are congruent?

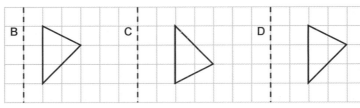

2   Write down the number of lines of symmetry for each of these shapes.

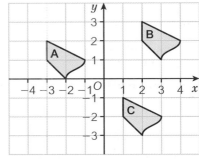

3   Sketch this equilateral triangle.
Draw in the lines of symmetry.

4   Write down the order of rotational symmetry of each of these shapes.

5   a   How many lines of symmetry does this regular octagon have?
    b   What is the order of rotational symmetry?

Translations, reflections and enlargements

6   Copy this shape onto squared paper.
Draw the image of the shape after these translations.
    a   3 squares right, 2 squares down. Label this shape A.
    b   4 squares left, 1 square up. Label this shape B.

7   Describe each translation.
    a   Shape A to shape B
    b   Shape B to shape C

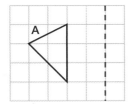

8   Which triangle, B, C or D, is the correct reflection of shape A?

9 a Draw the original shape from Q6 enlarged by scale factor 2.
  b Write down the ratio of the length of the sides of the object to the image.

10 Copy this diagram.
   Draw the image of the shape after a reflection
   in these lines.
   a $y = 1$. Label your reflected shape A.
   b $x = -1$. Label your reflected shape B.

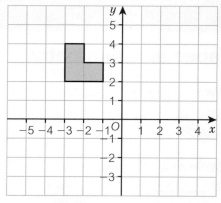

Rotations and combined transformations

11 The diagram shows two shapes, A and B.
   Shape A has been rotated to give shape B.
   Arthur describes the rotation as rotation, 90°, about (2, 1).
   What is missing from Arthur's description?

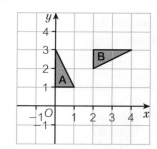

12 Use your diagram from Q10.
   Draw the image of the shape after these rotations.
   a 90° clockwise about (−2, 1). Label this rotated shape A.
   b 180° about (−3, 1). Label this rotated shape B.

13 Copy this diagram and transform the shape using
   these transformations.
   a A translation 2 squares left and 1 square down
     followed by a reflection in the line $x = -1$.
     Label the image A.
   b A rotation 90° anticlockwise about (2, 2) followed
     by a translation 2 squares right and 1 square up.
     Label the images B.

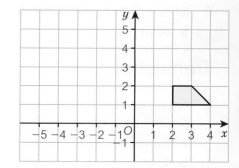

Challenge

a Nigel says, 'Every shape that has 2 lines of symmetry also has order of rotational
  symmetry 2.'
  Is Nigel correct? Explain your answer.
b Paul says, 'Every shape that has order of rotational symmetry 2 also has 2 lines of
  symmetry.'
  Is Paul correct? Explain your answer.

Reflect   How sure are you of your answers? Were you mostly

   😞 Just guessing   😐 Feeling doubtful   🙂 Confident

   What next? Use your results to decide whether to strengthen or extend your learning.

10 Strengthen

Shapes and symmetry

1 Trace shape A.
Use your tracing to help you find out which of these shapes are congruent to shape A.

> **Q1 hint** You are looking for shapes that are exactly the same size and shape as A.

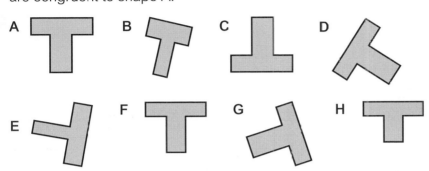

2 In each of these shapes, which other parts are congruent to the part labelled 1?

a
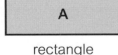

b

c

> **Q2 hint** You could trace part 1 of each shape to help you.

3 Which shapes have line symmetry?

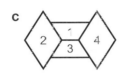

> **Q3 hint** Check using a mirror. Are the two halves the same?

4 Trace and cut out each quadrilateral.

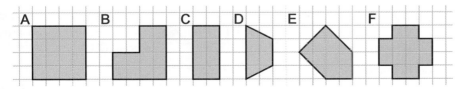

a Fold the shape so that the two halves fit on top of each other.
b Draw a dotted line along the crease.
c Do this as many ways as possible. How many lines of symmetry does each shape have?

5 a Copy each shape on centimetre squared paper.

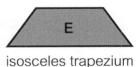

b Draw a line of symmetry. Check your line using a mirror.
c Draw any other lines of symmetry for each shape.
d Write the number of lines of symmetry beneath each shape.

6 a Trace each shape, including the dot and arrow. For E and F draw your own dot in the middle of the shape.

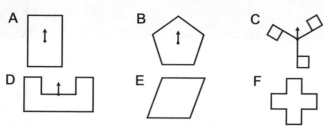

A   B   C

D   E   F

Q6 hint When rotating, count the number of times the traced shape fits the original. Only count the starting position once.

b Hold your pencil on the dot. Rotate the traced shape a full turn. How many times does it look the same?

c Write down the order of rotational symmetry.

7 Match the correct order of rotational symmetry card to each of these shapes.

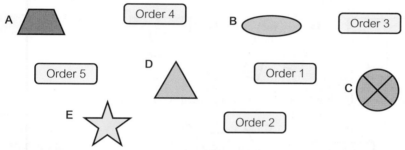

A   Order 4   B   Order 3

Order 5   D   Order 1   C

E   Order 2

Translations, reflections and enlargements

1 Copy and complete to describe how point A moves to point B in each diagram.

☐ squares _____   ☐ squares _____

Q1 hint left or right   Q1 hint up or down

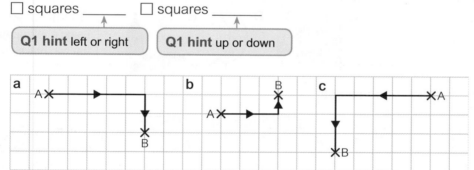

a   b   c

2 The grid shows triangles A to E.
Describe the translation that takes
a A to B   **b** D to E
c C to D   **d** B to C
e E to A   **f** E to C

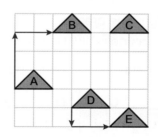

Q2 hint Choose one vertex (corner) of the shape. Give the number of squares right or left followed by the number of squares up or down.

3 Copy this shape onto squared paper.
Draw the image of the shape after these translations.
a 2 squares left, 1 square down. ⌐☐ Label your translated shape A.
b 3 squares right, 2 squares up. Label your translated shape B.

4   Which reflections have the mirror line in the correct place?
    Count the squares from one vertex (corner) of the shape to the mirror line.
    Now count the squares from the reflected vertex to the mirror line. It should be the same
    number of squares.
    Repeat for all the vertices (corners).

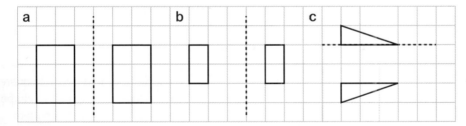

5   Copy the shapes and mirror lines onto squared paper.
    Reflect the shapes in the mirror lines.
    The first two are started for you.

Q5 hint Count the squares from each vertex to help you.

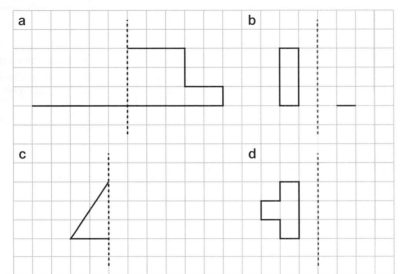

6   Copy this diagram.

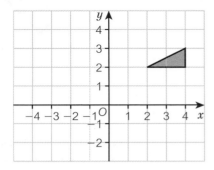

Q6a i hint Draw $y = 1$.
Check: put your finger at different points along the line. Do they all have a y-coordinate of 1?

Q6a ii hint Check that your original shape and your reflection are the same distance from your line $y = 1$.

    a   i   Draw the line $y = 1$.
        ii Draw the reflection of the triangle in the line $y = 1$. Label the image A.
    b   i   Draw the line $x = 2$.
        ii Draw the reflection of the triangle in the line $x = 2$. Label the image B.
    c   i   Draw the line $y = 3$.
        ii Draw the reflection of the triangle in the line $y = 3$. Label the image C.

7 Look at the diagram.

a How many squares is length
  i AB   **ii** BC   **iii** CD   **iv** DA?

b The shape is enlarged by scale factor 3.
This means that every length on the original shape is multiplied by 3.
How many squares is the new length
  i AB   **ii** BC   **iii** CD   **iv** DA?

c **i** On squared paper, draw the shape enlarged by scale factor 3.
  ii Write down the ratio of the length of the sides of the original shape to the enlarged shape.

d **i** Draw the shape enlarged by scale factor 2.
  ii Write down the ratio of the length of the sides of the object to the image.

> **Q7c ii hint** Every 1 square length on the original shape (object) is worth 3 square lengths on the enlarged shape (image).
> object : image
>   1 : ☐

8 **a** On centimetre squared paper, copy this shape, then draw it enlarged by scale factor 2.
b Use a ruler to measure the length AC on the object and on the image.
What do you notice?

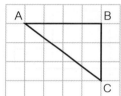

> **Q8a hint** Work out the length of the new AB and BC. Join AC.

Rotations and combined transformations

1 The shapes marked A have been rotated about a centre of rotation marked **✗**.
Copy and complete to describe each rotation.
  ☐° _____ ← clockwise or anticlockwise

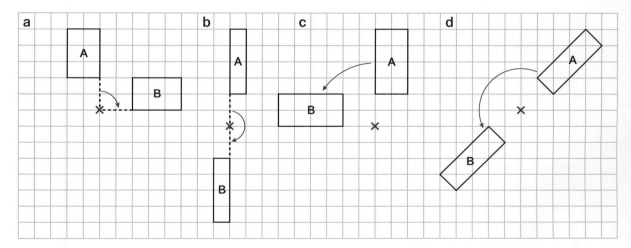

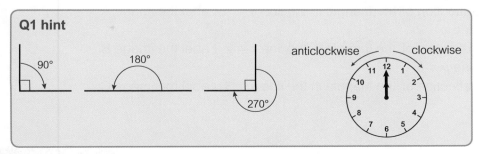

> **Q1 hint**

283

2 Copy the diagram and draw the image of the rectangle after these rotations.

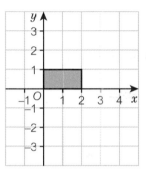

a 90° anticlockwise about (0, 0).
Label your rotated shape A.

b 90° clockwise about (2, 0).
Label your rotated shape B.

c 180° about (0, 0).
Label your rotated shape C.

> **Q2a hint** Trace the rectangle on tracing paper. Put your pencil on the point (0, 0) and turn your tracing paper 90° (quarter turn) anticlockwise.

3 Repeat Q2 for the shapes on these diagrams.

a

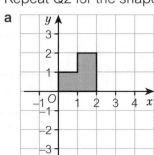

b

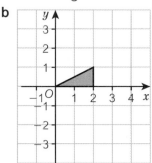

4 Copy this diagram.

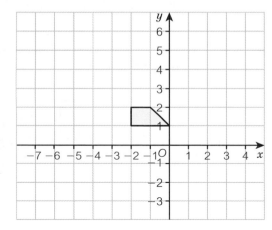

Do you have enough information to carry out these rotations? If the answer is no, explain what extra information you need. If the answer is yes, draw the image of the shape.

a Rotation 90° about (0, 1).

b Rotation 90° clockwise about (1, 1).

c Rotation 90° anticlockwise.

d Rotation clockwise about (0, 1).

e Rotation 180° about (1, 0).

f Rotation 90° anticlockwise about (−4, 2).

> **Q4 hint** For a rotation about a point, you must know
> • angle of rotation
> • direction of rotation (clockwise or anticlockwise)
> • centre of rotation.

5 Copy this diagram. Transform the shape using these transformations.
The first one is started for you.

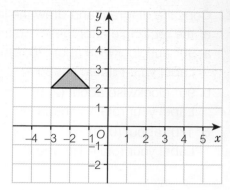

a A translation 3 squares right and 1 square up.
Follow this with a reflection in the line $y = 2$.

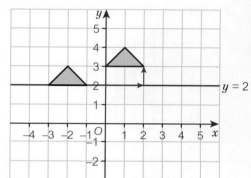

b A rotation 90° clockwise about (−1, 2).
Follow this with a translation 2 squares right and 3 squares down.
c A reflection in the y-axis followed by a translation 2 squares left and 2 squares up.
d A reflection in the line $y = 1$ followed by a rotation 180° about (0, 0).

Challenge

a Copy this rectangle onto a squared grid.
Draw the image of the rectangle after a rotation of 90° clockwise about the red dot.
Repeat the rotation of the image, 90° clockwise about the red dot.
Finally repeat the rotation of the new image, 90° clockwise about the red dot.

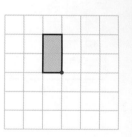

b What is the order of rotational symmetry of your combined final shape?
c Does your new shape have any lines of symmetry?
Explain your answer.

Reflect   In these lessons you have answered questions about these transformations:

reflection   rotation   translation

Write a definition for each of them, using one of these descriptions:

flips over   changes position   turns around

For each definition, draw a sketch to show what the definition means.
How did your definition help you choose the shapes?

10 Extend

1   The diagram shows four triangles. Triangle A has
    been rotated 180° about the point (−1.5, 3) to give
    triangle B.
    Describe the rotation that transforms
    a   B onto C
    b   C onto D
    c   D onto A

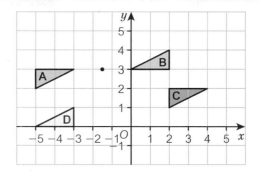

2   **Reasoning**   The diagram shows three
    rectangles A, B and C on a centimetre
    square grid.
    B and C are both enlargements of A.
    a   What is the scale factor of
        enlargement of
            i   A to B   ii   A to C   iii   B to C?
    b   Work out the perimeter of each rectangle.
    c   Work out the area of each rectangle.
    d   Copy and complete this table. Write each ratio in its simplest form.

Rectangles	Ratio of lengths	Ratio of perimeters	Ratio of areas
A : B	1 : 2		
A : C			

    e   What do you notice about the ratios you found in part **d**?
        Explain what these ratios tell you.

3   **Problem-solving**   This shape is enlarged by scale factor 5.
    a   Without drawing the enlargement, work out
            i   the perimeter of the enlargement
            ii   the area of the enlargement.
    b   Draw the enlargement to check that your
        answers to parts **a i** and **ii** are correct.

4   Copy this diagram.
    a   Reflect the shape in the line $y = x$.
    b   Reflect the combined shape in the x-axis, then
        reflect the newly combined shape in the y-axis.
    c   How many lines of symmetry does your
        final shape have?
    d   What is the order of rotational symmetry of
        your final shape?

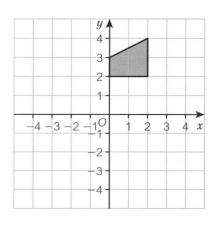

5 The diagram shows triangles ABC and DEF.
Triangle ABC has been reflected to make triangle DEF.
  a Write down the equation of the mirror line.
Triangle DEF is reflected in the line $x = 6$ to become triangle GHI.
  b Copy and complete this table showing the coordinates of the vertices of the triangles.

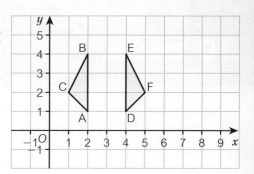

Triangle ABC	A (2, 1)	B (2, 4)	C (1, 2)
Triangle DEF	D (☐, ☐)	E (☐, ☐)	F (☐, ☐)
Triangle GHI	G (☐, ☐)	H (☐, ☐)	I (☐, ☐)

Triangle GHI is reflected in the line $x = 9$ to become triangle JKL.
  c Reasoning Without drawing triangle JKL, work out the coordinates of the vertices of triangle JKL.
Explain how you worked out your answer.

6 Reasoning Are these statements true or false?
  a When you rotate a shape the image is congruent.
  b When you enlarge a shape the image is congruent.
  c The perimeters of two congruent shapes are *not* equal.
  d The areas of two congruent shapes are equal.
  e When you reflect a quadrilateral in a mirror line that sits on one of its sides, the image is an octagon.

7 Greenhouse gases slow down or prevent the loss of heat from Earth to space. These diagrams show the representations of molecules of three different greenhouse gases.
How many planes of symmetry does each molecule have?
  a Methane (CH_4) **b** Carbon dioxide (CO_2) **c** Chlorofluorocarbon (CFC)

> **Q7a hint** Imagine a mirror cutting through the shape. Will the shape look the same?

> **Q7c hint** CFC is a powerful greenhouse gas, thousands of times worse than CO_2.

Challenge

A pentomino is a shape made from five congruent squares that touch side-to-side. The diagram shows four congruent pentominoes.
Design your own pentominoes.
How many *different* pentominoes can you draw?

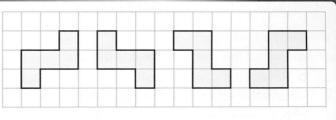

Reflect   Look back at the questions you answered in this Extend lesson.
Beside each one, write the type of transformation you used.
List all the other mathematics topics you used to answer these questions.

10 Unit test

1   Write the letters representing the congruent parts of each shape.

    a 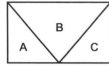   b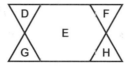

2   How many lines of symmetry does each shape have?

    a   b   c   d   e

3   Katie says, 'This triangle has these 3 lines of symmetry.'
    Is she correct? Explain your answer.

    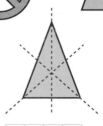

4   Copy this shape onto squared paper.
    Draw the image of the shape after the translation
    2 squares right, 4 squares down.

5   Copy this shape onto squared paper.
    Reflect it in the mirror line.

6   For each shape in Q2, write the order of rotational symmetry.

7   In the diagram, shape A has been rotated to give shape B.
    Dianne describes the rotation as, 'Rotation 90°, clockwise'.
    What is missing from Dianne's description?

    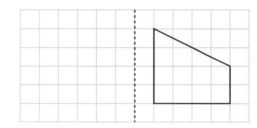

8   Copy the diagram and draw the image of the triangle after
    these transformations.
    a Reflection in $x = 1$.
       Label this shape A.
    b Rotation 90° clockwise about $(3, -1)$.
       Label this shape B.
    c Rotation 180° about $(1, 0)$.
       Label this shape C.

    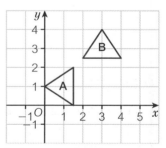

9 On this grid, Brian reflects the shape on the right in the line $y = 2$ and gets the shape on the left. Explain the mistake that Brian has made.

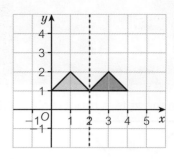

10 Copy the diagram in Q8.
  a Transform the shape using a rotation 90° anticlockwise about $(1, -1)$ followed by a translation 3 squares right and 1 square down. Label the image A.
  b Transform the original shape using a translation 2 squares left and 1 square down, followed by a reflection in the x-axis. Label the image B.

11 a Draw the shape from Q4 enlarged by scale factor 3.
  b Write the ratio of the length of the sides of the original shape to the enlarged shape.

12 The diagram shows two shapes, A and B. Describe a two-step transformation that transforms shape A to shape B.

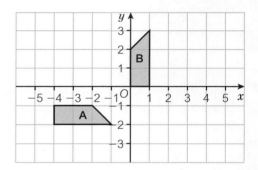

13 How many planes of symmetry do these 3D solids have?

  a 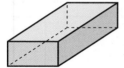 **b** 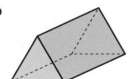 **c**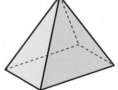

Challenge Split each diagram into four congruent shapes.
Each shape must contain a triangle, a square, a hexagon and a circle.

  a 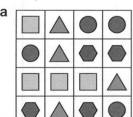 **b**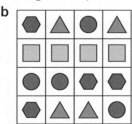

Hint The triangle, square, hexagon and circle can be in different places within the four congruent shapes, i.e. the outlines of the shapes must be congruent, but the patterns within them do not need to be.

Reflect List the four transformations that you have learned about in this unit.
Draw a sketch for each of them to remind you what each transformation does.
Now read this sentence carefully.
After _____ the shape and its image are congruent.
Complete this sentence with one or more transformations.
Explain your choice of word(s).

Answers

UNIT 1

1.1 Mode, median and range

1 a 4   b 34
2 10
3 a 7   b 1, 1, 2, 2, 4, 7, 7, 7, 8   c 4
4 a 9   b 13   c 80
   d 33   e 250   f 0.8
5 £3.50
6 a phone   b 2   c 7
   d 0.4   e 40   f 0
7 a impossible, 5, 5, 0.5, 50, 20
   b There is no range for a because the values
      are not numbers.
8 7 visits
9 a 5, 6   b 0.6, 0.8   c 1, 2
10 a Students' own answers, e.g. 2, 2, 3, 4, 4
   b Students' own answers, e.g. 2, 3, 4
11 a 10   b 5
12 a range = 4 cm, mode = 8 cm
   b range = 7 g, mode = 20 g
   c range = 400 mm, mode = 300 mm
   d range = 20°, mode = 50°
13 a 2 cm, 4 cm, 4 cm, 5 cm, 7 cm. Median = 4 cm
   b 11, 12, 12, 15, 15, 16, 16, 17, 18. Median = 15
   c 15 g, 20 g, 30 g, 50 g, 55 g, 70 g, 75 g. Median = 50 g
14 a 7.5   b 8.5   c 6   d 5
   e 5.5   f 7
15 a 4   b 9.5   c 13.5
   d 14   e $\frac{1}{2}$
16 a 2.5 children   b no
17 3 or any number greater than 3
18 a £18   b £13
19 a Mode = 5, range = 6, median = 5
   b Mode = 30, range = 90, median = 35
   c Mode = 200, range = 500, median = 350
   d Mode = 0.3, range = 0.7, median = 0.3
   e Mode = 1, range = 9, median = 1
   f Mode = 50, range = 40, median = 50

Challenge

1 yes, yes, yes
2 median = mode

1.2 Displaying data

1 a 0, 2, 4, 6, 8, …   b 0, 10, 20, 30, 40, …
2 A = 9, B = 16
3 a 10   b 14   c 44   d 1
4 a i 2   ii 1
   b

Day	Sat	Sun	Mon	Tues	Wed
Texts	8	11	2	5	3

   c Sunday   d 29
5 a Wednesday 40, Thursday 50, Friday 45
   b

   Saturday

   Sunday

   c 260 messages
6 a Fizzy   b 3 students   c 8 students
   d 28 students   e the highest bar

7 a 17 students
   b

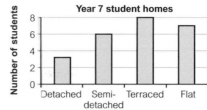

Year 7 student homes

   c Terraced house
8 a 10 girls   b size 6   c 5 sizes
9 a size 8   b 5 sizes
   c

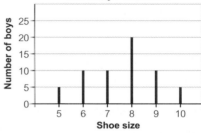

Year 9 boys' shoe sizes

   d 60
10 a 21 students   b ⦀⦀ ⦀⦀ ⦀⦀ | c Walk
   d

Method of travel	Frequency
walk	21
car	16
bicycle	3
bus	15
other	2

11

Website	Tally	Frequency
Facebook	⦀⦀ ⦀⦀ ||	12
Instagram	||||	4
Snapchat	||	2
YouTube	⦀⦀		6
Twitter	⦀⦀ ||	7

12 a

Number of food items	4	5	6	7	8	9
Frequency	3	6	7	3	0	1

   b

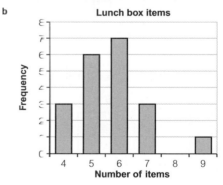

Lunch box items

   c 6   d 5
13 Ashley is correct. The mode is the category with the
    greatest frequency.

Challenge

   Bar chart is missing title, axis, labels, scale.

1.3 Grouping data

1 20, 21, 22, 23, 24, 25, 26, 27, 28, 29
2 a 15   b 8   c 22

3 a

Pulse rate	Tally	Frequency
70–79	I	1
80–89	ℍℍ IIII	9
90–99	ℍℍ II	7
100–109	III	3

b 80–89 beats per minute **c** 10

4 a 25–29 jumps **b** 29 students

  c Yes. The 30–34 class now has a frequency of 12, and is the modal class.

5 a

Number of coins	Frequency
0–2	6
3–5	12
6–8	6
9–11	5
12–14	2

b 3–5 coins **c** 7 students **d** no

6 a 0–4 books

b

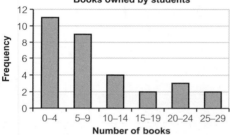

Books owned by students

7 a

Drill size, mm	Tally	Frequency
1–5	ℍℍ	5
6–10	IIII	4
11–15	II	2
16–20	I	1

b

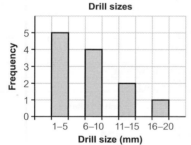

Drill sizes

c 1–5

8 a

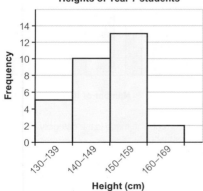

Heights of Year 7 students

b 150–159 cm

c No, only the category that the student's height is in.

9 a Yes; most of these students drink more than 1.2 litres a day – the three bars on the right represent 65 students drinking more than 1.2 litres a day.

  b Students' own answers, e.g. use smaller class intervals

Challenge

Students' own answers

1.4 Averages and comparing data

1 15

2 6

3 a 25 **b** 5

4 a 8 **b** 6 **c** 7

  d 5.6 **e** No

5 a 3.5 **b** 8 **c** 3 **d** 30p

  e 5.5 m **f** No

6 196.4 cm

7 a Manjit 11, Tony 20 and 60, Sebastian no mode

  b Manjit 8.5, Tony 40, Sebastian 12

  c Manjit 8.75, Tony 40, Sebastian 13.5

  d Manjit 10, Tony 40, Sebastian 16

8 a 9 min **b** Mean 17 min; Median 17 min

9 a i Team 1 range: 4; Team 2 range: 6

    ii Team 2

  b i Team 1 mode: 13; Team 2 mode: 11

    ii Team 1

10 a Kieran's friends like gym least; Robin's friends like swimming least (frequency is 0).

  b Kieran's friends like football most; Robin's friends like tennis most. Comparison uses the mode.

11 a Men 10, women 9

  b Men 74.5, women 78

  c The women performed better than the men, because their median score is higher.

    The men's scores were more variable than the women's, because their range is greater.

12 a Steel: mean = 9.7 minutes, range = 1 minute

    Aluminium: mean = 9.66 minutes (2 d.p.), range = 1.6 minutes

  b The aluminium bikes are slightly faster than the steel ones, because the mean is slightly smaller.

    The aluminium bikes vary more than the steel ones, because the range is greater.

Challenge

  a Shotput athletes (mean 191.5, median 190) are taller on average than javelin athletes (mean 187.8, median 188.5)

  b-c Students' own answers

1.5 Line graphs and more bar charts

1 Title, axis labels

2 a 3 **b** 10 **c** 21.5 **d** 25

3 a day; week

  b i 6 °C ii 4 °C iii 7 °C

  c Wednesday and Friday

  d 4 °C **e** 7 °C

4 a 39 °C **b** 10 am **c** 4 readings

  d Between 8 am and 10 am and between 11 am and noon

5 a During March and August **b** Decreasing

  c During January, April and June **d** During June

6

Michelle's time on Facebook

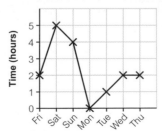

7 **a** 4 junk emails   **b** Saturday
    c She did not receive any genuine emails on Sunday.
    d 35 junk, 60 genuine emails; no, she is not correct, because she has 95 emails in total and 35 is less than half of 95

8 **a**

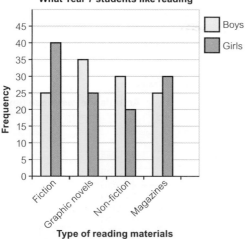

What Year 7 students like reading

Type of reading materials

    b Yes. More girls than boys like reading fiction and magazines. More boys than girls like reading graphic novels and non-fiction.

9 **a** **i** 80   **ii** 30
    b **i** 30   **ii** 50
    c **i** Chemistry and biology
       ii Biology
    d 55

10 **a**

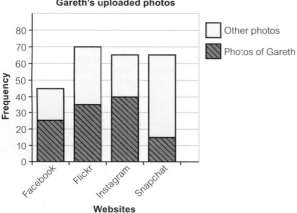

Gareth's uploaded photos

Websites

    b 115 photos of Gareth, 130 others; no, because 130 is greater than 115

Challenge

Students' own answers

1 Check up

1 3.2 m
2 **a** 12   **b** 2   **c** 5   **d** 5
3 **a**

Big cat	Tally	Frequency
lion	|||||		6
tiger	||||	4
cheetah	||	2
leopard	||	2
jaguar			1

    b lion
4 **a** 10 families   **b** 2 children
    c No families had 5 children. **d** 90 families
5 **a** 23 °C   **b** 7 am
    c The temperature rose from 14 °C to 24 °C.

    d 24 °C
6 **a** Statistics   **b** Algebra   **c** Number
    d 5 more   **e** 50 students
7 **a** 5 gold medals   **b** Denmark
    c Denmark   **d** Denmark and Hungary
8 **a** Oakbridge 1.7 km, St John's 4 km
    b Oakbridge 2.2 km, St John's 9.4 km
    c On average, the students at Oakbridge travel shorter distances to school (smaller mean). The distances for St John's vary more than the distances for Oakbridge (larger range).

Challenge

    a 2   **b** 5   **c** 90 (or −10)
    d Any number less than or equal to 4

1 Strengthen: Averages and range

1 **a** crisps   **b** 50p
2 5, 8
3 **a** 7 cm   **b** 8 cm
4 **a** **i** 12 cm, 14 cm   **ii** 13 cm   **b** 14 cm
5 4
6 23
7 **a** 2   **b** 13   **c** 11
8 **a** 13 press-ups   **b** 1.5 cm
9 **a** 12   **b** 12, 3
10 **a** 30   **b** 5   **c** 6
11 **a** 3   **b** 5   **c** 5   **d** 17

1 Strengthen: Charts and tables

1 **a** 15 bowling balls   **b** 4 kg   **c** 4 kg
2 **a–c**

Prize	Tally	Frequency
$20	||||	5
$50	|||	3
$100	||	2

    b $20
3 **a** 10–19 minutes
    b

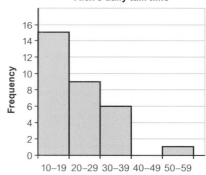

Alex's daily talk time

Time (minutes)

4 **a-c**

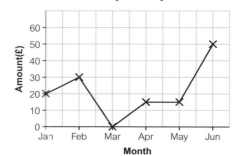

Money in charity box

Month

    d Increase

1 Strengthen: Comparing data

1 a 25

b

Visitors to a sports centre

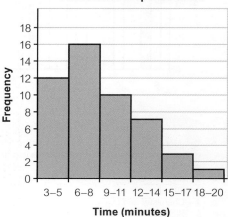

Key: Adults, Children

c August

2 a i 19 minutes **ii** 17 minutes
iii Travelling by car is quicker than by bus on average, because the median time by bus is more than by car.
b i 7 minutes **ii** 22 minutes
iii Car travel times vary more than by bus, because the range of car times is greater than the range of bus times.

3 a 30 mins **b** 20 mins **c** Atifa **d** 100 mins

Challenge

a

Letters	Tally	Frequency
1–2	IIII	5
3–4	IIII III	8
5–6	IIII	4
7–8	II	2

b Students' own answers
c Look for comparison of modes and ranges.

1 Extend

1 a i 152.5 km/h **ii** 55 km/h
b i 185 km/h **ii** 145 km/h
c On average, the wind speed of hurricanes in 2017 was the higher than 2012, because the median is the higher (185 km/h vs 152.5 km/h). The wind speeds in 2017 also varied more than in 2012, because the range is greater (145 vs 55).

2 a Dave's friends might all be tall.
b If Dave's friends are tall, their mean will be higher than the mean for all Year 7 students.
c Measure the heights of more students.

3 a Fitness

b

Challenge award badges

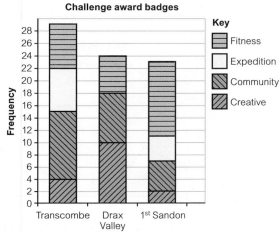

Key: Fitness, Expedition, Community, Creative

c Trascombe

---

4 a i Wigan 40–49 points, Bradford 10–19 points
ii Wigan, because the modal class has a bigger point value than Bradford's.
b No, because we do not know any of the values in a class. The maximum possible range for Wigan is 49 points, minimum range is 31 points. The maximum possible range for Bradford is 49 points, minimum range is 31 points.

5 a 6
b Multiply the mean by the number of values.

6 a 6–8 minutes
b i 28 ambulances
ii 21 ambulances

c

Ambulance response times

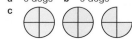

d i 95 minutes is an unusually long response time. It could be due to an ambulance breakdown, difficult location, human error, etc.
ii mode = 7 minutes, median = 7 minutes, mean = 17.75 minutes
iii The median or the mode, because most values are 10 minutes or under, and both the mean and the mode are under 10. The mean does not represent the data well because it is much higher than most of the values.

Challenge

a 2, 2, 5
b Students' own answers, e.g. 3, 5, 6, 7 or 3, 4, 7, 7.
c Students' own answers, e.g. 1, 1, 2, 5, 6
d 12
e 10
f Students' own answers, e.g. −3, 0

1 Unit test

1 a 6 dogs **b** 9 dogs
c

d 36 dogs

2 a i 5 months **ii** 6 months **iii** 1 accident
iv 6 accidents
b i 15 months **ii** 0 accidents **iii** 5 accidents
c Students' own answers, e.g. The modal number of accidents after speed cameras were fitted is less than the modal number before (0 vs 1). The number of accidents after speed cameras were fitted varies less than before (5 vs 6).

3 a 4 **b** 16–18 **c** 13–15 **d** 13
4 a 5 **b** 8 **c** 5 **d** 4
5 a 7 cl **b** Wednesday **c** 13 cl

6 **a**

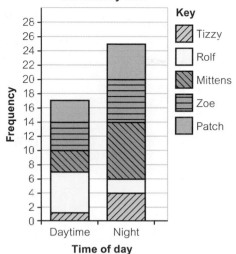

Weekly distance travelled by cats

Key:
- Tizzy
- Rolf
- Mittens
- Zoe
- Patch

Time of day

b The total distance travelled by the cats is further at night (25 km) than during the day (17 km). Looking at individual cats, all the cats except Rolf travel further at night time.

7 **a** 30 kg

b

Mass (kg)	Tally	Frequency
15–24	IIII	5
25–34	IIII IIII	9
35–44	IIII	5
45–54	I	1

c 62 kg is greater than any of the values.

d 30.4 kg

Challenge

a **i** 7 km/h   **ii** 6.8 km/h (1 d.p.)

    iii 10 km/h

b The median is unchanged. The mean has increased to 7.3 km/h. The range has increased to 13 km/h.

UNIT 2 Number skills

2.1 Mental maths

1 **a** Same (8)   **b** Different (4, −4)

    c Same (12)   **d** Different (3 and $\frac{1}{3}$)

2 **a** 70   **b** 900   **c** 5000   **d** 250

    e 4000   **f** 18000   **g** 3   **h** 54

    i 72   **j** 8

3 **a** 7   **b** 8   **c** 8   **d** 12

4 **a** **i** 24   **ii** 24   **iii** 24

    iv 24   **v** 24

    b No. The answers are all the same.

5 **a** 80   **b** 70   **c** 54   **d** 72

6 **a** 80   **b** 120   **c** 1200   **d** 12000

    e 300   **f** 3000   **g** 30000   **h** 300000

    i 360   **j** 4800   **k** 660   **l** 660

7 **a** 192   **b** 256   **c** 252   **d** 315

    e 504   **f** 288

8 **a** 14   **b** 16   **c** 14   **d** 26

    e 2   **f** 2   **g** 38   **h** 16

9 **a** e.g. $2 \times 5 + 10 = 20$; $5 \times 2 + 10 = 20$; $10 + 2 \times 5 = 20$

    b $10 \div 2 - 5 = 0$; $2 \times 5 - 10 = 0$; $5 \times 2 - 10 = 0$

10 **a** 10   **b** 11   **c** 10   **d** 11

    e 2   **f** 18   **g** 1   **h** 29

11 **a** **i** 84   **ii** 84

    b **i** 37   **ii** −9

    c **i** 16   **ii** 16

12 **a** 6   **b** 24   **c** 1   **d** 12

    e 75   **f** 60

13 **a** 11   **b** 21   **c** 40   **d** 28

    e 2   **f** 48   **g** 2

14 e.g.

$45 = (4 + 2) \times 8 - 6 \div 2$

$6 = 4 + 2 \times (8 - 6) \div 2$

$7 = (4 + 2 \times 8 - 6) \div 2$

$9 = 4 + (2 \times 8 - 6) \div 2$

$14 = 4 + 2 \times (8 - 6 \div 2)$

$17 = 4 + (2 \times 8) - 6 \div 2$ or $4 + 2 \times 8 - (6 \div 2)$

    or $(4 + 2 \times 8) - 6 \div 2$ or $4 + (2 \times 8 - 6 \div 2)$

15 No; they only have $(4 \times 2) + (3 \times 4) = 20$ batteries.

Challenge

a **i** 13   **ii** 13   **iii** 45   **iv** 30

    v 7

b Students' own answers

c Students' own answers

2.2 Addition and subtraction

1 **a** 61   **b** 65

2 **a** 90, 100   **b** 540, 500   **c** 150, 100

    d 500, 500   **e** 1550, 1500   **f** 1550, 1600

3 **a** 5000   **b** 3000   **c** 6000

    d 1000   **e** 16000   **f** 17000

4 **a** 110   **b** 120   **c** 190

    d 170   **e** 105, 171

5 Calculations A and C are wrong.

In A, the 1 ten 'carry over' has not been added.

Correct answer is 576.

In C, the 1 hundred 'carry over' has been written as part of the answer. Correct answer is 753.

6 444 hits

7 **a** 435   **b** 1931   **c** 2255   **d** 1634

8 2182 cars

9 **a** 91   **b** 125   **c** 188   **d** 179

    e 117   **f** 145   **g** 2465   **h** 1666

10 No, $7 - 4 \neq 4 - 7$

11 **a** 4 thousands   **b** 4 ten thousands

    c 4 hundred thousands   **d** 4 hundred thousands

    e 4 million   **f** 4 ten thousands

    g 4 thousands   **h** 4 hundred thousands

12 a 90 000   **b** 760 000   **c** 1 760 000
     d 14 000   **e** 19 000   **f** 21 000

13 a 300 000   **b** 600 000   **c** 300 000
     d 4 700 000   **e** 5 400 000   **f** 100 000

14 a 2 000 000
     b 5 000 000
     c 5 000 000

15 £50 460

Challenge

a Concert A 39 500, concert B 376 245, concert C 896 023

b 1 310 000

c 489 878

2.3 Multiplication

1 a 72   **b** 7   **c** 10   **d** 3, 7

2 a 500   **b** 500   **c** 600

3 a 200   **b** 700   **c** 100   **d** 1000

4 a 350   **b** 800   **c** 54 000

5 a 484   **b** 6393   **c** 372   **d** 8324
     e 3120   **f** 1434   **g** 29 712   **h** 41 607

6 Calculations A and C are wrong.
In A, the 1 hundred 'carry over' has been written as part of the answer. Correct answer is 726.
In C, the 'carry overs' have been ignored. Correct answer is 2256.

7 £4096

8 a 920 books   **b** 578 non-fiction books

9 £16 878

10 a 480   **b** 2484   **c** 2418   **d** 1980
     e 6318   **f** 13 407

11 £1105

12 8745 people

13 Loss of £800

Challenge

a 2380

b $36 \times 58 = 2088$, $83 \times 65 = 5395$

2.4 Division

1 a 32   **b** 4   **c** 8   **d** 63
     e 7   **f** 9

2 a 3   **b** 2 r2   **c** 2 r6

3 a 22   **b** 24   **c** 16   **d** 13

4 a 31   **b** 312   **c** 21   **d** 211
     e 14   **f** 141   **g** 142   **h** 1421

5 a Calculations A and C are wrong.
     b A 209, C 346

6 a 123   **b** 63   **c** 63   **d** 64
     e 133   **f** 83   **g** 469   **h** 207

7 £15 per hour

8 64

9 a 31 r1   **b** 25 r3   **c** 61 r3   **d** 73 r2
     e 488 r5   **f** 1169 r2

10 a 18   **b** 29   **c** 114   **d** 290
     e 150   **f** 21

11 a 22 r2   **b** 24 r5   **c** 36 r1   **d** 49 r3

12 a 21 rows   **b** 7 left

13 13

14 £72

15 £94

Challenge

a 288

b $1234 \div 65 = 18$ r64

c $6543 \div 12 = 545$ r3

2.5 Money and time

1 100

2 a 0.25 and $\frac{1}{4}$; 0.5 and $\frac{1}{2}$; 0.75 and $\frac{3}{4}$

b **i** $\frac{1}{4}$ hour = 15 minutes

     ii $\frac{1}{2}$ hour = 30 minutes

     iii $\frac{3}{4}$ hour = 45 minutes

3 500p and £5.00; 50p and £0.50; 5p and £0.05

4 a £0.80   **b** £0.75   **c** £0.07
     d £1.25   **e** £5.20   **f** £5.02

5 a £5   **b** £10   **c** £39
     d £1   **e** £1041   **f** £0

6 yes

7 £6

8 a £2.70   **b** £7.50   **c** £7.05
     d £3.20   **e** £3.02   **f** £0.06

9 £12, 1200 pence

10 £79.90

11 a £15   **b** £4.80

12 £4.30

13 a £13.23   **b** £27.26   **c** £23.90
     d £89.09   **e** £73.00   **f** £100.00

14 a £334.88   **b** £0.44 or 44p   **c** £1.20
     d £694.89   **e** £293.13   **f** £11.33

15 a £15.25
     b The full amount of the bill will not be covered if rounding to the nearest pound or nearest penny has rounded each person's share down.

16 a C   **b** B   **c** A

17 15 minutes or $\frac{1}{4}$ hour

18 a 5 hours   **b** 7 hours 30 minutes
     c 2 hours 30 minutes   **d** 45 minutes
     e 15 hours 30 minutes   **f** 6 hours 15 minutes
     g 3 hours 45 minutes   **h** 2 hours 6 minutes

19 5 hours 15 minutes

Challenge

Channel 3 is best value for money.

2.6 Negative numbers

1 5, 4, 3, 2, 1, 0, −1, −2, −3, −4, −5

2 a i 3 °C   **ii** 1 °C   **iii** −1 °C   **iv** −4 °C
     b Colder

3 9 °C

4 a 7 °C   **b** 1 °C   **c** 4 °C

5 a

```
     F        E     B                    C       A          G     D
  +--+--+--+--+--+--+--+--+--+--+--+--+--+--+--+--+--+--+--+--+--+
 −10 −9 −8 −7 −6 −5 −4 −3 −2 −1  0  1  2  3  4  5  6  7  8  9 10 °C
```

     b −9 °C   **c** 10 °C

6 a <   **b** >   **c** <   **d** >
     e <   **f** >

7 a <   **b** >   **c** >   **d** <
     e >   **f** <

8 a −14, −9, −5, −2, 7, 8, 19
     b −13, −11, −6, −6, −4, −2, 7
     c −41, −30, −21, −15, −12, 23, 24, 63

9 a −4   **b** −5   **c** −6   **d** −7
     e −7   **f** −8   **g** −9   **h** −9
     i −10   **j** −11

10 a −1   **b** −2   **c** −3   **d** −1
     e 0   **f** 1   **g** 2   **h** 1
     i 2   **j** −2

11 a 6   **b** −6   **c** 4   **d** −4
     e 10   **f** −10   **g** 6   **h** −6
     i 48   **j** −48

12 a −12   **b** −14   **c** −40   **d** −27
     e −6

Challenge

a Students' own answers

b Students' own answers

2.7 Factors, multiples and powers

1
a 10, 12, 20, 22, 24
b 10, 15, 20
c 10, 20

2
a 54   b 20   c 28   d 48
e 72   f 84

3
a 3   b 5   c 8   d 4
e 7   f 7

4 7, 14, 21, 28, 35

5
a 30   b 25, 30, 45
c 18, 30, 32   d 18, 27, 45
e i 5, 9   ii 2, 9   iii 2, 5, 10
f 37 because it is a prime number

6
a 2, 3, 4 or 6   b 2, 3, 5, 6, 10 or 15
c 2, 4 or 8   d–f Students' own answers

7
a 25: not multiple of 10   b 15: not multiple of 4
c 10: not multiple of 25   d 42: not multiple of 9
e 21: not multiple of 5   f 25: not multiple of 2

8
a i 4, 8, 12, 16, 20, 24, 28, 32
   ii 6, 12, 18, 24, 30, 36, 42, 48
b 12, 24

9
a 2, 4, 6, 8, 10, 12, 14, 16, 18
b 3, 6, 9, 12, 15, 18
c 6, 12, 18
d

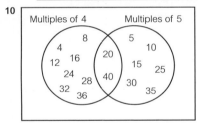

10

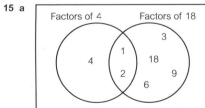

11
a i 3, 6, 9, 12, 15, 18, 21, 24, 27, 30
   ii 8, 16, 24,32,40, 48, 56, 64, 72, 80
b 24

12
a 56   b 60   c 75

13 1 and 24, 2 and 12, 3 and 8, 4 and 6,
6 and 4, 8 and 3, 12 and 2, 24 and 1

14
a i 1, 2, 3, 6   ii 1, 3, 5, 15
b 3

15 a

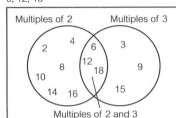

b 1, 2   c 2

16 a

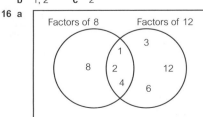

b 1, 2, 4   c 4

17
a 3   b 9   c 8   d 6
e 5   f 8

18
a 2, 3, 5, 7, 11, 13, 17, 19
b No, because 2 is a prime number and it is even.

19
a i 1, 23   ii 1, 5, 7, 35   iii 1, 29
b 23 and 29

20
a e.g. 30 and 60   b e.g. 6 and 15

Challenge

a 3   b 18, 36, 54 etc.   c 53 or 59
d 1 and 30 or 3 and 10 or 5 and 6 or 2 and 15

2.8 Square numbers

1
a 16   b 81   c 49   d 8

2
a 9   b 18

3
a 38   b 2   c 10

4
a
• • • •
• • • •
• • • •
• • • •
b 4 × 4 = 16   c 5 rows, 5 columns
d 100   e 100

5
a 36   b 64   c 81   d 121

6
a 400   b 225   c 10000   d 169

7
a 6   b 5   c 8   d 10

8
a 15   b 19   c 20   d 100

9 4, 36, 49, 81, 100

10 No; 41 is not a square number.

11
a 16   b 8   c 24   d 8
e 14   f 25   g 0

12
a 2   b 16   c 128   d 16
e 62   f 0   g 254

13
a $4 \times 4 = 4^2$ b $2 \times 2 \times 2 = 2^3$
c $4 \times 4 \times 4 = 4^3$ d $5^3 = 5 \times 5 \times 5$
e $10 \times 10 = 10^2$ f $10^3 = 10 \times 10 \times 10$

Challenge

a

b 1, 3, 6, 10, 15, 21
c +2, +3, +4, +5, +6; add on 1 extra each time
d 28, 36
e No; 36 + 9 = 45
f The number of dots along each side of the triangles increases
by one each time.
The pattern is: 1, 1 + 2, 1 + 2 + 3,
1 + 2 + 3 + 4, etc., which gives the triangle numbers.
g square numbers

2 Check up

1
a 789   b 720   c 2184   d 4055

2 101

3
a 10052   b 952   c 22 r1

4
a 30000   b 560000

5
a 240   b 4200   c 100000

6 207

7
a 2°C   b 9 degrees

8
a >   b <   c >

9
a 2, 4, 9, 12, 18
b 16   c 2, 19, 23   d 4, 9, 16

10
a 3   b −3   c −4   d −7

11
a i 12   ii 2
b 35

12
a 22   b 54   c 24   d 8
e 36   f 1   g 21   h 21
i 2

13 £474

14 Layla has enough money to buy all three items.

15 158 followers

16 £85.80

17 45 minutes

Challenge

1 e.g. $(2 + 3 + 5) \times 6 = 60$

2 Any two consecutive numbers, e.g. $4 - 5 = -1$

3 The other pairs of 2-digit primes with this property are 17 and 71, 37 and 73, 79 and 97 (and 11 and 11).

2 Strengthen: Written methods

1 a 672   b 964   c 839   d 808
   e 2630   f 216   g 214   h 5031

2 a 419   b 448   c 164   d 254
   e 282   f 271   g 1972   h 6471

3 a 187   b 576   c 159   d 54
   e 117   f 147   g 1674   h 789

4 a 462   b 948   c 12436   d 28882

5 a 714   b 4114   c 806   d 9906
   e 15687

6 a 62   b 82   c 37   d 47
   e 37 r1   f 47 r2

7 a 34   b 21   c 34   d 52
   e 52 r3

2 Strengthen: Mental work

1 a 40000   b 37000

2 a 2000000 b 2500000   c 2460000

3 a 28   b 280   c 2800   d 18
   e 180   f 1800   g 10   h 100
   i 1000   j 48   k 480   l 480
   m 4800

4 a 294   b 150   c 272   d 212
   e 360

5 a 14   b 32   c -2 d 18
   e 6

6 a 9   b 6   c 42   d 8

7 a 9   b 11   c 18   d 35

8 a 5°C   b 5°C   c 1°C   d 5°C

9 a 5°C   b 1°C   c -3°C   d -2°C

10 a >   b <   c >   d >

11 a $-4, -3, 0, 5, 7, 8, 9$
   b $-8, -7, -4, -3, 2, 6$
   c $-5, -4, -3, 1, 2, 3, 5$

12 $-1 + 1 = 0$

13 a 0   b -1
   c i $-5 + 2 = -3$
     ii $-3 + 5 = 2$
     iii $-7 + 3 = -4$
     iv $-4 + 2 + 5 = 3$ or $-4 + 7 = 3$

14 a no   b yes   c yes   d no

15 a $2^2 = 4, \sqrt{4} = 2$ b $3^2 = 9, \sqrt{9} = 3$
   c $4^2 = 16, \sqrt{16} = 4$ d $5^2 = 25, \sqrt{25} = 5$
   e $6^2 = 36, \sqrt{36} = 6$ f $10^2 = 100, \sqrt{100} = 10$

16 a 11   b 21   c 16   d 3

17 a 16   b 4   c 81   d 9

18 a i 1 and 8, 2 and 4, 4 and 2, 8 and 1
     ii 1 and 20, 2 and 10, 4 and 5,
       5 and 4, 10 and 2, 20 and 1
   b Factors of 8: 1,2, 4, 8
     Factors of 20: 1, 2, 4, 5, 10, 20
     Common factors: 1, 2, 4
   c 4
   d 15

19 a Multiples of 5: 5, 10, 15, 20, 25, 30, 35, 40, 45, 50
     Multiples of 7: 7, 14, 21, 28, 35, 42, 49, 56, 63, 70
   b 35
   c 42

2 Strengthen: Problem solving

1 a John £3, Richard £13, Lucy £9
   b about £25

2 984 people

3 3 hours 15 minutes

4 £224

5 £78

Challenge

14 should be in 'Multiples of 7'.

16 should be in 'Square numbers'.

25 should be in 'Square numbers'.

1 should be in 'Factors of 24' or 'Square numbers'.

12 should be in 'Factors of 24'.

21 should be in 'Multiples of 7'.

2 Extend

1 a i 1   ii 2
     iii 3   iv 4
     v 5   vi 6
   b The answers are the numbers 1 to 6 in order.
   c e.g. $7 = 4 + 4 - 4 \div 4$,
     $8 = 4 + 4 - (4 - 4) = 4 + 4 \times (4 \div 4)$

2 a 144, 169   b yes

3 2021

4 a 7 girls and 6 boys (6 tables)
   b Yes; 14 girls and 12 boys (3 tables)
   c 6

5 Calculations A and C are wrong.

6 4 hours 45 minutes

7 No; there are only 29827 vaccines.

8 2539

9 a 1   b 1   c 6   d 6
   e 18   f 18

10 a 2959   b 948   c 2908

11 a Box 2
   b 177 chocolates (11 Box 2 and one Box 1)

12 Rufus is paid £300.80 for working 32 hours and Mamadou is paid £288.20. This means Rufus is paid £12.60 more than Mamadou.

13 two 12 × 12 trays and one 15 × 15 tray

14 a 3 and 4   b 8 and 9   c 5 and 6
   d 9 and 10 e 6 and 7   f 7 and 8

15 a 36   b 60

16 a 6   b 8

2 Unit test

1 2500000

2 766

3 a 492   b 23002

4 a 21   b 651 r10

5 a 5°C   b warmer   c -4°C or -5°C

6 £130

7 £4.50

8 -6°C

9 a 360   b 2100   c 60000

10 60

11 1176 books

12 23 bunches

13 a -4 b -16 c -12

14 a 0   b 25   c 35   d 55
   e 8

15 7670 wristbands

16 Steven, £2.88 more

Challenge

a 1, 4, 9, 16, 25, 36, 49, 64, 81, 100, 121, 144, 169, 196, 225, 256, 289, 324, 361, 400

b 1, 4, 9, 6, 5, 6, 9, 4, 1, 0, ...

c 2, 3, 7, 8

d Students' own answers

UNIT 3 Expressions, functions and formulae

3.1 Functions

1 a 22   b 6   c 5
2 a 35   b 0   c 20
3 a ×   b −   c ÷
4 a 4, 7, 15   b 3, 6, 9.5   c 10, 13, 17   d 3, 6, 9

5

Input	0	1	2	3
Output	0	8	16	24

6 a 'add 4' or '+ 4'   b 'multiply by 7' or '× 7'
   c 'subtract 1' or '− 1'   d 'divide by 3' or '÷ 3'

7 14

8

Input (hours)	4	6	8
Output (£)	28	42	56

9 a ÷ 0.12, 40
   b i 16   ii 34
10 a 3, 5   b 19, 25
11 a 16, 20   b 3, 18, 33
12 a

Input	5	13	19
Output	4	8	11

   b No; the output number will not be a whole number if the input number is even.

13 29

Challenge

Any two-step function machines with input 6 and output 20.
Examples: × 2, + 8; + 4, × 2; − 4, × 10

3.2 Simplifying expressions 1

1 a −3   b 2   c −3   d −7
2 $4 + 4 + 4 = \underline{3} \times 4 = \underline{12}$
3 a 6   b 2   c 4   d −2
4 a $4x$ b $2y$ c $3z$ d $3x$
   e $5y$ f $2z$
5 6 + 6 + 6 and 3 × 6
   $z + z + z + z + z$ and $5 \times z$
   $x + x$ and $2 \times x$
   $y + y + y + y$ and $4 \times y$
6 a $7m$ b $9n$ c $12q$ d $11x$
7 a $2y$ b $4y$ c $8b$ d $-2r$
8 a $7m$ b $8x$ c $3x$ d $8p$
   e $-5z$ f $4y$ g $4n$ h $-t$
9 Any three calculations that give an answer of $8x$.
   Examples: $x + 7x$, $2x + 6x$, $3x + 5x$, $10x - 2x$
10 a $10r + 5b$ b $13a + 3c$
    c $10t + 11$ d $9x + 4y$
11

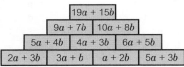

12 a $5g + 5h$ b $3x - 2y$ c $4d + 2k$
    d $3x - 3y$ e $-4a + 5b$ f $2n - 3p$
    g $w - 3u$ h $6 + 2b$ i $8b + 18$
13 a top row $3a + 5$; middle row $2a + 3$, $a + 2$
    b top row $8b - 2$; middle row $4b + 2$, $4b - 4$
    c middle row $2c + 6$; bottom row $c - 4$, $c + 2$
14 A and C
15 $2w + 3z + 8w + 7z + 6w + 5z = 16w + 15z$
16 a $4a - 2b$ b $5a + 2b + 3$
    c $8a - 2b - 8$ d $12a + 4b + 9$

Challenge

Any divisions of a square into identical parts.
Examples:

 = 2p   = 4h

3.3 Simplifying expressions 2

1 area = length × width
2 a 5^2 b 11^2
3 a 6   b 4   c 3
4 a $4n$ b $8x$ c y d $-2w$
5 a $3m$ b $7n$ c $5p$ d de
   e eg f ek g k^2 h ak
6 $4a$ and $4 \times a$
   $a + 4$ and $4 + a$
   $3a$, $3 \times a$ and $a \times 3$
   $5b$ and $5 \times b$
   ab and $a \times b$
   $b \times 3$ and $3 \times b$
7 a $10a$ b $32b$ c $6y$ d $4y$
   e $2a$ f $45c$ g $6b$ h $10z$
8 a e.g. $6 \times a$, $3 \times 2a$, $12a \div 2$, $18a \div 3$
   b e.g. $2c \times 1$, $c \times 2$, $4c \div 2$, $6c \div 3$
   c e.g. $4 \times t$, $2t \times 2$, $8t \div 2$, $20t \div 5$
   d e.g. $10 \times 1s$, $2s \times 5$, $20s \div 2$, $100s \div 10$
9 a top row $200t$; middle row $5t$, 40
   b top row $72y$; middle row 12, $6y$
10 a 72   b 160   c 38
11 $7 \times 15 = 7(5 + 10)$
    $4 \times 13 = 4(3 + 10)$
    $4 \times 17 = 4(20 - 3)$
    $3 \times 39 = 3(40 - 1)$
    $5 \times 17 = 5(7 + 10)$
12 a 236   b 87   c 196   d 441
13 a $p + 3$ b area $= (p + 3)$ c $4p$, 12
    d $4p + 12$
14 a $2x + 14$ b $7c - 14$ c $7c + 14$ d $2x - 14$
    e $12x - 6$ f $12x - 2$ g $15x + 5$ h $15x + 3$
15 a $6x + 2$ b $11x + 24$ c $6x + 28$ d $13x + 9$

Challenge

Any six of these:

$2(2a + 4) = 4a + 8$	$3(2a + 4) = 6a + 12$
$2(a + 4) = 2a + 8$	$3(a + 4) = 3a + 12$
$2(2a - 3) = 4a - 6$	$3(2a - 3) = 6a - 9$
$2(6a - 9) = 12a - 18$	$3(6a - 9) = 18a - 27$
$2(a + 2) = 2a + 4$	$3(a + 2) = 3a + 6$
$4(2a - 4) = 8a + 16$	$5(2a + 4) = 10a + 20$
$4(a + 4) = 4a + 16$	$5(a + 4) = 5a + 20$
$4(2a - 3) = 8a - 12$	$5(2a - 3) = 10a - 15$
$4(6a - 9) = 24a - 36$	$5(6a - 9) = 30a - 45$
$4(a + 2) = 4a + 8$	$5(a + 2) = 5a + 10$
$6(2a - 4) = 12a + 24$	$8(2a + 4) = 16a + 32$
$6(a + 4) = 6a + 24$	$8(a + 4) = 8a + 32$
$6(2a - 3) = 12a - 18$	$8(2a - 3) = 16a - 24$
$6(6a - 9) = 36a - 54$	$8(6a - 9) = 48a - 72$
$6(a + 2) = 6a + 12$	$8(a + 2) = 8a + 16$

c Any of these:
   $3(2a + 4) = 6(a + 2) = 6a + 12$
   $2(2a + 4) = 4(a + 2) = 4a + 8$
   $4(2a + 4) = 8(a + 2) = 8a + 16$
   $2(6a - 9) = 6(2a - 3) = 12a - 18$

3.4 Writing expressions

1 a 30   b 18   c 22   d 6
   e 8   f 14   g 45   h 12
2 a $3x$ b $2y$ c $-2t$ d $11p + 4$
   e $6a + 2b$
3 a $d + 7$ b $d - 9$
4 $x - 3$
5 a $y + 2$ b $y + 6$ c $y - 3$ d $y - 9$
6 a $w + 3$ b $r + 10$ c $y - 7$ d $w + 8$
   e $w - 3$
7 a $a + 4$ b $b - 3$ c $x - 20$ d $y + 7$
   e $m - 4$ f $\frac{l}{2}$

8 **a** $n + 3$ **b** $n - 21$ **c** $50 - n$ **d** $n + 8$

9 £2x

10 **a** £3y **b** £$\frac{y}{2}$

11 6x

12 **a** 9y **b** £45

13 **a** $2x$ **b** $5x$ **c** $x + n$ **d** nx
    e $\frac{n}{2}$ **f** $\frac{n}{3}$

14 x add $9 = x + 9$
    x subtract $9 = x - 9$
    multiply x by $9 = 9x$
    subtract x from $9 = 9 - x$
    multiply x by 9 then add $9 = 9x + 9$
    multiply x by 9 then subtract $9 = 9x - 9$

15 **a** $3g$ **b** $5b$ **c** $7b$ **d** $b + g$
    e $y + g + 3$

Challenge

1 $10x + 50$

2 Greg's total travelling time that week was $10x + 30$, so he is not correct.

3.5 Substituting into formulae

1 **a** 18 **b** 28 **c** 6 **d** 3

2 **a** 180 **b** 280 **c** 195 **d** 2
    e 4

3 **a** $2t$ **b** xy

4 **a** 2 **b** 3 **c** 5

5 **a** 3mg **b** 4mg

6 **a** 5 **b** 8

7 **a** £57.75 **b** £330.00

8 **a** 6 **b** 8 **c** 10

9 **a** d = distance, s = speed and t = time
    b **i** $d = s \times t = 50 \times 2 = 100$ km
       ii $d = s \times t = 60 \times 3 = 180$ km

10 **a** 100 miles **b** 280 miles **c** 195 miles

11 **a** 2 m/s **b** 3 m/s **c** 4 m/s

12 **a** 80 **b** 100 **c** 72

13 **a** 2 **b** 2 **c** 8

14 **a** £8.00 **b** £11.60

15 **a** 20 **b** 48 **c** 24 **d** 35

16 £1120

17 daughter 167.5 cm; son 180.5 cm

Challenge

Hand span (cm)	15	13	16	14
Height (cm)	145	128	155	138
Sadie's values (H)	135	117	144	126
Zosha's values (H)	150	130	160	140

Zosha's values are closer to the real heights.

3.6 Writing formulae

1 **a** $3k$ **b** $4m$ **c** $2n$ **d** $5y$

2 **a** $y + 5$ **b** $y - 2$ **c** $2y$

3 **a** $x = 8$ **b** $m = 10$ **c** $F = 20$

4 **a** **i** 3 **ii** 4 **iii** 5
    b Add 2
    c number of floats = number of students + 2
    d $f = s + 2$

5 **a** **i** £5 **ii** £10 **iii** £15
    b sale price = original price − 10
    c $x = y - 10$

6 $M = R + 12$

7 $M = y + z$

8 **a** × 4 **b** $C = 4s$ or $C = s \times 4$

9 $F = 12b$

10 $P = 8h$

11 **a** $F = M \div 40$
    b Aurora 13 kg, Bluegrass 12.5 kg, Flanagan 12 kg, Phantom 15 kg, Summer 9 kg, Tonto 11.5 kg

12 **a** $6p + 12$ **b** 42 cm²

Challenge

152 faces

3 Check up

1 **a** 13, 15, 19 **b** 1, 3, 7

2 **a** 'multiply by 2' or '× 2' **b** 'subtract 5' or '− 5'

3 **a** 5, 11, 20, 29 **b** 3, 9, 18, 27

4 **a** $11k$ **b** $12h$ **c** $6b$ **d** $7y$
    e $2x$ **f** $9z$ **g** fg or gf **h** pq or qp
    i $24c$ **j** $27t$ **k** $3y$ **l** m^2

5 **a** $4a$ **b** $9b$ **c** $a + b$ **d** ab

6 **a** 56 **b** 114

7 **a** $6c + 12$ **b** $8d + 7$

8 **a** $4x + 16$ **b** $6x - 12$ **c** $16x + 40$

9 $17x + 4$

10 **a** 250 g **b** 1250 g or 1.25 kg

11 **a** 45 **b** 150

12 $w = d + 30$

13 $M = E - S$

Challenge

a Yes, because multiplying any odd or even number by 5 gives a number that ends in 0 or 5. Adding 10 will not change this.

b e.g. × 2 + 8, × 3 + 2, × 4 − 4

3 Strengthen: Functions

1 **a** 11 **b** 2

2 **a** 4, 9 **b** 16, 24, 40

3 **a** × 2, + 5 **b** × 2 **c** Yes **d** × 2

4 **a** 'add 4' or '+ 4' **b** 'subtract 2' or '− 2'
    c 'divide by 6' or '÷ 6'

5 **a** 7, 15 **b** 7, 9

3 Strengthen: Expressions

1 **a** 36 **b** 33 **c** 112 **d** 63

2 **a** $2 \times (20 - 1) = 2 \times 20 - 2 \times 1 = 40 - 2 = 38$
    b $5 \times (30 - 1) = 5 \times 30 - 5 \times 1 = 150 - 5 = 145$
    c $3 \times 29 = 3 \times 30 - 3 \times 1 = 90 - 3 = 87$

3 **a** $2b$ **b** $3b$ **c** $7b$ **d** $6b$
    e $6b$

4 **a** $4b$ **b** $5b$ **c** $3b$ **d** b
    e $6b$ **f** b

5 **a** $3y - 3x$ **b** $4b - 3a$ **c** $6z - 4w$ **d** $6x - 5y$
    e $8a - 4b$ **f** $2q - 12p$ **g** $6z + 10$ **h** $2d + 5e + 5$

6 **a** $4x$ **b** $2z$ **c** $6n$ **d** $4p$
    e $3y$ **f** $11w$ **g** ab or ba **h** ac or ca
    i at or ta

7 **a** $12x$ **b** $16y$ **c** $30b$ **d** $5c$
    e $4y$ **f** $2f$ **g** $2z$ **h** g

8 **a** $3 + 2$ **b** $3 - 1$ **c** $x + 3$ **d** $x - 2$

9 **a** $5 + 4$ **b** $x + 4$

10 **a** $9 - 5$ **b** $y - 3$

11 **a** $x + 6$ **b** $x - 5$ **c** $x + 4$ **d** $10 - x$

12 **a** 8×2 **b** 4×5 **c** $2x$ **d** $5x$

13 **a** 24 minutes **b** $4r$

14 **a** £40 **b** £3h

15 **a** $2y$ **b** $6y$ **c** $x + y$ **d** xy

16 **a** $3x + 6$ **b** $4y + 20$ **c** $6c + 8$ **d** $16 + 12p$

17 **a** $2w - 10$ **b** $8u - 12$ **c** $15 - 3a$ **d** $12 - 20b$

3 Strengthen: Formulae

1 **a** 28 **b** 56 **c** 12, 84

2 **a** 12 **b** 5, 20 **c** 9, 36

3 **a** 10 **b** 16 **c** 7, 30

4 **a** 12 **b** 15, 27 **c** 9, 0, 9

5 **a** 18 **b** 5, 30 **c** 10, 60

6 **a** 6 **b** 50 **c** 9

7 a

number of runners ⟶ +5 ⟶ number of medals

number of medals ⟶ −5 ⟶ number of runners

b 25

c Number of medals = number of runners + 5

8 a $\frac{P}{2}$ or $P \div 2$

b $P, \div 2$

c $M = \frac{P}{2}$

9 $d = c + 6$

Challenge

$2(3x + 6), 6(6 − 2y)$

3 Extend

1 a 1. 1st and 2nd outputs 5, 8; 3rd input 72

2. multiply by 4; 2nd output 24; 3rd input 15

b The 4th input and output numbers could be any number in each case.

1. e.g. input 60, output 10

2. e.g. input 23, output 80

2 a One of the following:

$5a + 3a − a = 7a$

$5a + 6a − 2a = 9a$

$5a + 10a = 15a$

$6a − 2a + 3a − a = 6a$

$10a + 3a − a = 12a$

b One of the answers from part **a**.

c $5a + 10a = 15a$

3 a i 1.4 cm   **ii** 1.3 cm

b range = largest height − smallest height

4 $12x \div 2, 2x \times 3$ and $30x \div 5$ (all equivalent to $6x$)

$x \times 8, 24x \div 3, 4 \times 2x, 16x \div 2$ (all equivalent to $8x$)

$6x \times 2, 72x \div 6, 3 \times 4x, 0.5 \times 24x, 36x \div 3$

(all equivalent to $12x$)

5 a $2n$ **b** $2n + 5$

6 a $3m + 10$ **b** $\frac{m}{2}$ **c** $\frac{m}{5} + 3$

7 a 202   **b** 185   **c** 170

8 a $t = d \div 4$ or $t = \frac{d}{4}$

b Monday 5 hours, Tuesday 6 hours, Wednesday 4 hours, Thursday 4.5 hours, Friday 6.5 hours

Challenge

a 29

b Students' own answers

c i Each L-value is 11 greater than the number at the top of the L shape.

ii L-value = top number + 11

d i

n	
$n + 5$	$n + 6$

ii $n + 11$

e '$n + 11$' is the same as 'top number + 11' but uses the letter n to represent the top number.

f 53

Unit test

1 a 6, 9, 13   **b** 1, 7

2 'multiply by 4' or '× 4'

3 a 8, 14, 16   **b** 1, 3, 4, 6

4 a $3x$ **b** $7m$ **c** q **d** $−2a$

5 128

6 a £6   **b** £15

7 a 15   **b** 4   **c** 3

8 a $4y$ **b** $8z$ **c** mn **d** bc

e x^2

9 a $8a + 7b + 4$ **b** $14a + 6b − 4$

10 a $b + 2$ **b** $b − 5$

11 £2d

12 24

13 $S = R + 30$

14 a $r = s − 4$ **b** 16

15 $T = £20h$

16 $M = 60$

17 a $18a$ **b** $36b$ **c** $4c$

18 a $4x + 8$ **b** $15y + 20$ **c** $2z − 12$

19 a $5m$ **b** $m + n$ **c** mn

20 $T = b + y$

Challenge

a 30

b $D = A + 3$ so $D = 5 + 3 = 8$

$G = D − A$ so $G = 8 − 5 = 3$

$F = 2G + 4$ so $F = 2 \times 3 + 4 = 10$

$E = AD$ so $E = 5 \times 8 = 40$

$C = E \div 10$ so $C = 40 \div 10 = 4$

$H = 3C + 7$ so $H = 5 \times 4 + 10 = 30$

C, E, F and G could have been calculated in a different order so long as G was calculated before F and C was calculated before E.

c $B = 4G$

d $H = \frac{A^2 + 3A}{2} + 10$

UNIT 4

4.1 Decimals and rounding

1 10 mm

2 **a** **i** £48 **ii** £265 **iii** £497
   b **i** £4.18 **ii** £31.67 **iii** £47.10

3 9, 25, 36, 64, 81

4 **a** **i** 4.5 cm **ii** 45 mm
   b **i** 6.5 cm **ii** 65 mm
   c **i** 8.4 cm **ii** 84 mm
   d **i** 2.8 cm **ii** 28 mm
   e **i** 11.7 cm **ii** 117 mm
   f **i** 9.9 cm **ii** 99 mm
   g **i** 0.6 cm **ii** 6 mm

5 **a** **i** 6.4 cm **ii** 64 mm
   b **i** 3.2 cm **ii** 32 mm
   c **i** 5.8 cm **ii** 58 mm

6 Accurate lines drawn:
   a 5.4 cm **b** 7.5 cm **c** 2.6 cm
   d 9.1 cm **e** 42 mm **f** 18 mm

7 **a** one tenth **b** one ten
   c one hundredth **d** one thousandth
   e one hundredth

8 **a** 6.7 < 6.9 **b** 6.6 > 6.2
   c 1.7 > 1.4 **d** 3.7 < 3.86
   e 10.09 < 10.9 **f** 3.9 > 3.107
   g 21.299 < 21.92 **h** 0.400 < 0.6

9 No, because 0.35 < 0.5

10 **a** 6.08 > 6.03 **b** 6.01 < 6.06
   c 12.371 < 12.38 **d** 0.42 > 0.419
   e 7.624 > 7.621 **f** 9.909 > 9.099

11 **a** 5.21, 5.23, 5.25, 5.28
   b 4.19, 4.3, 4.67, 4.7

12 **a** 16.28, 16.3, 16.39, 16.4
   b 14.23, 14.2, 14.1, 14.08

13 **a** 0.111, 0.11, 0.10, 0.01
   b 9.98, 9.9, 9.89, 9.809

14 **a** 8 **b** 9 **c** 5 **d** 7
   e 15 **f** 18 **g** 3 **h** 8

15 13.9, 1.8, 2.0, 5.1

16 **a** 3.2 **b** 8.2 **c** 4.8 **d** 3.1
   e 15.6 **f** 0.6 **g** 3.0 **h** 8.0

17 **a** 40 **b** 30 **c** 21 **d** 32
   e 3 **f** 20 **g** 2 **h** 3

18 Students' own answer, e.g.
   Round 8.45 up to 9; 9 ÷ 3 = 3

Challenge

a 8 hundreds, 8 hundredths

b 8.8 means 8 ones and 8 tenths but 8.08 means 8 ones and 8 hundredths.

c 8.8 and 8.80 both mean 8 ones and 8 tenths.

d Yes, because 0.88 < 0.9

4.2 Length, mass and capacity

1 **a** 70 **b** 700 **c** 8 **d** 80

2 **a** 2300 **b** 2300 **c** 23 000 **d** 18 000
   e 64

3 millimetre, centimetre, metre, kilometre

4 **a** 45.21 **b** 236.7 **c** 452.1 **d** 2367
   e 4521 **f** 23 670 **g** 130.8 **h** 8.7
   i 13.08 **j** 8.70 **k** 1.308 **l** 0.870

5 **a** × 10 1 place to the left
   b × 100 2 places to the left
   c × 1000 3 places to the left
   d ÷ 10 1 place to the right
   e ÷ 100 2 places to the right
   f ÷ 1000 3 places to the right

6 **a** 56 **b** 455 **c** 12 020 **d** 40
   e 5.43 **f** 2.75 **g** 2.04 **h** 0.36

7 0.31 × 100

8 4200 ÷ 1000

9 **a** 6 cm = 6 × 10 = 60 mm
   b 8 m = 8 × 100 = 800 cm
   c 9 km = 9 × 1000 = 9000 m
   d 400 cm = 400 ÷ 100 = 4 m
   e 80 mm = 80 ÷ 10 = 8 cm
   f 25 000 m = 25 000 ÷ 1000 = 25 km

10 **a** multiply **b** multiply **c** multiply
   d divide **e** divide **f** divide

11 **a** 400 cm **b** 15 cm **c** 6 km

12 **a** 2.5 m = 2.5 × 100 = 250 cm
   b 12.5 km = 12.5 × 1000 = 12 500 m
   c 88 mm = 88 ÷ 10 = 8.8 cm
   d 160 cm = 160 ÷ 100 = 1.6 m
   e 7.3 m = 730 cm
   f 67 mm = 6.7 cm

13 **a** 103 mm, 13 cm, 301 mm, 31 cm
   b 345 cm, 3.5 m, 400 cm, 4.2 m
   c 6.7 m, 675 cm, 678 cm, 6.87 m
   d 2.12 m, 234 cm, 3 m, 303 cm

14 No, because 135 cm = 1350 mm

15 **a** 5 kg = 5 × 1000 = 5000 g
   b 7 litres = 7 × 1000 = 7000 ml
   c 15 000 ml = 15 000 ÷ 1000 = 15 litres
   d 6000 g = 6000 ÷ 1000 = 6 kg

16 **a** 4.2 kg = 4.2 × 1000 = 4200 g
   b 0.75 litres = 0.75 × 1000 = 750 ml
   c 4250 ml = 4250 ÷ 1000 = 4.25 litres
   d 875 g = 875 ÷ 1000 = 0.875 kg
   e 9.5 kg = 9500 g
   f 1260 ml = 1.26 litres

17 **a** 475 g, 0.54 kg, 1.45 kg, 1475 g
   b 0.03 kg, 30.3 g, 300 g, 0.303 kg

18 **a** 80 000 g **b** 80 kg

19 Yes, because 2050 ml < 2.5 litres

Challenge

a Empty the 500 ml jug, then fill the 500 ml jug from the 3-litre jug. The 3-litre jug now has 2500 ml in it.

b Empty the 500 ml jug and the 750 ml jug, then fill the 500 ml jug and the 750 ml jug from the 3-litre jug. The 3-litre jug now has 1750 ml in it.

4.3 Scales and measures

1 **a** 10 mm **b** 100 cm **c** 1000 m
   d 1000 g **e** 1000 ml

2 **a** 48 **b** 370 **c** 9100

3 **a** 4 m **b** 2 m

4 **a** length 4 m, width 1 m
   b length 4.5 m, width 1.5 m
   c length 8 m, width 2 m

5 Accurate lines drawn:
   a 8 cm **b** 6.5 cm

6
```
+---+---+---+---+---+
0  0.2 0.4 0.6 0.8  1
```

7 **a** **i** 7 **ii** 7.5
   b 10.5
   c **i** 250 **ii** 125 **iii** 475
   d 55 **e** 15.5 **f** 5.8 **g** −2

8 **a** 350 ml **b** 0.5 pints **c** −5°C **d** 280 g

9 **a** 66 mph **b** 400 ml **c** 0.8 m

10 No; the temperature is −18°C.

11 **a** 40 cm **b** 2250 g **c** 2750 m **d** 58 mm

12 200 ml

13 **a** 8.6 cm = 86 mm = 80 mm + 6 mm = 8 cm 6 mm
   b 1.5 m = 150 cm = 100 cm + 50 cm = 1 m 50 cm
   c 3.5 km = 3500 m = 3000 m + 500 m = 3 km 500 m
   d 2.41 m = 241 cm = 200 cm + 41 cm = 2 m 41 cm
   e 4.7 litres = 4700 ml = 4000 ml + 700 ml = 4 litres 700 ml
   f 9.3 cm = 9 cm 3 mm
   g 10.8 kg = 10 kg 800 g
   h 20.6 m = 20 m 60 cm
   i 7.3 litres = 7 litres 300 ml

14 B (12.4 m) and D (12 m 40 cm)

15 a 82 cm 5 mm   **b** 24 m 32 cm
    c 5 kg 100 g   **d** 10 litres 440 ml

16 5 litres 700 ml

17 31 m 50 cm

Challenge

a

b

c

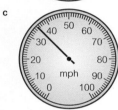

4.4 Working with decimals mentally

1 a 0.2   **b** 0.22   **c** 0.02
    d 0.022   **e** 0.002

2 a 80   **b** 1800   **c** 320

3 a $3 \times 24 = 3 \times 20 + 3 \times 4 = 60 + 12 = 72$
    b $8 \times 26 = 8 \times 20 + 8 \times 6 = 160 + 48 = 208$
    c $42 \times 7 = 40 \times 7 + 2 \times 7 = 280 + 14 = 294$
    d $56 \times 3 = 50 \times 3 + 6 \times 3 = 150 + 18 = 168$

4 a 0.6   **b** 1.8   **c** 2.8   **d** 3.2
    e 5.4   **f** 1.6   **g** 5.6   **h** 2.0
    i 4.8

5 a 0.35   **b** 0.12   **c** 0.18   **d** 0.09

6 4.2 GBP (or £4.20)

7 a 0.12   **b** 0.27   **c** 0.35   **d** 0.24
    e 0.72   **f** 0.044   **g** 0.024

8 a Yes, because $0.8 \times 0.5 = 0.40 = 0.4$
    b Students' own answer

9 a 33   **b** 64   **c** 84   **d** 84

10 a $22 \times 3.4 = 20 \times 3.4 + 2 \times 3.4 = 68 + 6.8 = 74.8$
    b $32 \times 1.2 = 30 \times 1.2 + 2 \times 1.2 = 36 + 2.4 = 38.4$
    c $21 \times 4.1 = 20 \times 4.1 + 1 \times 4.1 = 82 + 4.1 = 86.1$
    d $43 \times 3.1 = 40 \times 3.1 + 3 \times 3.1 = 124 + 9.3 = 133.3$

11 $42 \times 1.6 = 40 \times 1.6 + 2 \times 1.6 = 64 + 3.2 = \$67.20$

12 $64 \times 6.5 = 60 \times 6.5 + 4 \times 6.5 = 390 + 26 = 416$ g

13 a 232.5   **b** 2.325   **c** 2325   **d** 2.325
    e 0.2325

14 Yes; 0.93 is multiplied by 10 to give 9.3, but 25 is divided by 10 to give 2.5, so the ÷ 10 and × 10 cancel each other out and the answer stays the same.

15 a Estimate: $7 \times 0.3 = 2.1$, not 21
    b 8.4 is divided by 10 to give 0.84, but 0.17 is multiplied by 10 to give 1.7, so the × 10 and ÷ 10 cancel each other out and the answer stays the same.

Challenge

a 3666.63

b $0.3 \times 0.11 = 0.033$
    $3.3 \times 0.11 = 0.363$
    $33.3 \times 0.11 = 3.663$
    $333.3 \times 0.11 = 36.663$

c 3.6663; 333.3 is divided by 100 to give 3.333, so the answer is $366.63 \div 100 = 3.6663$

4.5 Working with decimals

1 a 13   **b** 5   **c** 64   **d** 7

2 a 125   **b** 284   **c** 206   **d** 537

3 a 288   **b** 1872   **c** 13   **d** 48

4 a 7.3   **b** 7.9   **c** 8.0   **d** 8.3
    e 9.7   **f** 10.9

5 a 1.3   **b** 1.0   **c** 0.9   **d** 1.9
    e 0.4   **f** 0.4

6 4.1 m

7 2.7 km

8 a 64.31   **b** 4.5   **c** 10.85   **d** 130.96

9 3.7 m

10 a 13.12   **b** 22.25   **c** 44.32   **d** 2.14
    e 17.51   **f** 27.879   **g** 79.9   **h** 21.59

11 a 0.4   **b** 0.7   **c** 0.87   **d** 0.38

12 Students' own answer, e.g.
    $10 - 3 = 7$ so $1 - 0.3 = 0.7$
    $100 - 38 = 62$ so $1 - 0.38 = 0.62$

13 a 18.3   **b** 26.4   **c** 59.2   **d** 26.1
    e 19.6   **f** 6.93   **g** 64.96   **h** 70.35
    i 398.15

14 a 17.1   **b** 32.8   **c** 17.3   **d** 12.4
    e 5.5   **f** 12.54   **g** 100.31   **h** 35.47

15 £30.72

16 18.5 cm

17 1.3 g

18 a 5.95   **b** 5.75

19 a 6.2   **b** 9.3   **c** 32.3   **d** 11.8

Challenge

1 a 2.48, 2.84, 4.28, 4.82, 8.24, 8.42
    b 10.9
    c 5.94
    d **i** 16.84   **ii** 33.68   **iii** 67.36
    e **i** 1.24   **ii** 0.62   **iii** 0.31

2 a 0.24, 0.42, 2.04, 2.40, 4.02, 4.20
    b 4.44
    c 3.96
    d **i** 8.40   **ii** 16.80   **iii** 33.60
    e **i** 0.12   **ii** 0.06   **iii** 0.03

4.6 Perimeter

1 a equilateral triangle, 3   **b** square, 4
    c regular pentagon, 5   **d** regular hexagon, 6
    e regular octagon, 8

2 a 16 cm   **b** 24 cm   **c** 8.4 m

3 a 14 cm   **b** 14 cm   **c** 12 cm   **d** 22 cm

4 13 cm

5 a 6 cm   **b** 30 cm   **c** 30 cm   **d** 32 cm

6 8.4 cm

7 a 3 cm   **b** 6 cm   **c** 5.5 cm

8 **Shape A:**   **a** $a = 10.5$ cm $b = 9$ cm   **b** 42 cm
   Shape B: **a** $a = 3$ cm $b = 9$ cm   **b** 52 cm
   Shape C: **a** $a = 2$ cm $b = 2$ cm   **b** 24 cm

9 a $x = 0.5$ m, $y = 0.8$ m
    b 12.6 m

10 a Both have perimeter 22 cm.
    b The sum of the vertical lines in the rectangle is the same as in the compound shape, and the sum of the horizontal lines in the rectangle is the same as in the compound shape.

11 a

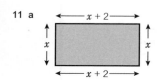

b $4x + 4$ cm

12 Shape D is the odd one out as it has perimeter $6x + 11y$. (The others have perimeter $4x + 12y$.)

13

Number of sides	3	4	6	8	12	24
Side length (cm)	8	6	4	3	2	1

14 a 5.4 cm **b** 9.2 cm

Challenge

a Students' own shapes

b Students' own answers

c Students' own answers

d Eight squares arranged as a 2 × 4 rectangle (perimeter 12) or any shape that fits into a 3 × 3 square and has perimeter 12.

e Eight squares arranged as an 8 × 1 rectangle (perimeter 18)

4.7 Area

1 a 16 **b** 4 **c** 25 **d** 4
e 9 **f** 8

2 Area 16 cm²: shapes B and D
Perimeter 16 cm: shapes A and B

3 a $5y$ **b** $8x$ **c** $9y$ **d** $9x$

4 a 6 cm² **b** 7 cm² **c** 9 cm²

5 a $12y$ cm² **b** $24w$ cm² **c** $5x$ cm²

6 a 1 cm² **b** 4 cm² **c** 9 cm²
d 16 cm² **e** 25 cm² **f** l^2 cm²

7 square numbers

8 area A = 5 × 3 = 15 cm²
area B = 8 × 7 = 56 cm²
total area = area A + area B = 15 + 56 = 71 cm²

9 Shape A: **a** 3 cm **b** 4 cm **c** 23 cm²
Shape B: **a** 4 cm **b** 3 cm **c** 18 cm²
Shape C: **a** 9 cm **b** 5 cm **c** 145 cm²

10 a 2 m, 1 m
b 40 m²
c Students' own answers, e.g.:
Area = 7 × 5 + 1 × 5 = 35 + 5 = 40 m²
Area = 5 × 6 + 2 × 5 = 30 + 10 = 40 m²
Area = 7 × 6 − 2 × 1 = 42 − 2 = 40 m²

11 a 63 cm² **b** 6300 mm²

12 a 9 cm **b** 36 cm

13 6 cm²

14 32 cm

15 a 1600 cm² **b** 4 booklets

Challenge

Students' own answers

4.8 More units of measure

1 a × 100 **b** × 10 **c** × 1000 **d** ÷ 1000
e × 1000

2 a 180 **b** 60 000 **c** 6100 **d** 610
e 610 **f** 3.2 **g** 6.4

3 a 230 mm > 9.5 cm **b** 0.05 kg < 400 g
c 250 ml < 1.5 litres

4 a centimetres or millimetres
b litres
c grams

5 a km² **b** cm² **c** m²

6 a 3t = 3 × 1000 = 3000 kg
b 9000 kg = 9000 ÷ 1000 = 9t
c 4.6t = 4.6 × 1000 = 4600 kg
d 5 ha = 5 × 10 000 = 50 000 m²
e 120 000 m² = 120 000 ÷ 10 000 = 12 ha
f 2 litres = 2000 ml = 2000 cm³
g 75 cm³ = 75 ml

h 3500 cm³ = 3500 ml = 3.5 litres

7 a multiply **b** divide

8 2.3 tonnes

9 a 6800 kg **b** 8600 ml

10 The land for grazing animals is greater.

11 a i 45 000 m² **ii** 4.5 ha
b 3750 spaces
c Students' own answers; e.g. This model does not consider the space needed between rows of cars.

12 1.04 tonnes

13 1 mile

14 a 6 ft = 6 × 30 = 180 cm
b 150 cm = 150 ÷ 30 = 5 ft
c 4 miles = 4 × 1.6 = 6.4 km
d 160 km = 160 ÷ 1.6 = 100 miles

15 5 miles

Challenge

Students' own answers

4 Check up

1 Accurate line drawn: 5 cm

2 a 15 °C **b** 0.4 pints

3 a 8 cm = 80 mm **b** 2 kg = 2000 g
c 600 cm = 6 m

4 a 3.2 m = 320 cm **b** 8.7 litres = 8700 ml
c 2400 m = 2.4 km

5 15 000 kg is heavier;
1.5 tonnes = 1.5 × 1000 = 1500 kg

6 23.8

7 8.165, 8.3, 8.35, 8.47, 8.9, 8.95

8 a 4.6 **b** 14.0

9 a 16.5 **b** 15.3

10 2.65 m

11 a 0.6 **b** 0.32 **c** 0.18

12 a 20.8 **b** 13.2

13 11.25 kg

14 a 32.4 **b** 32.4 **c** 0.324

15 42 cm

16 5 cm²

17 12 m

18 a 44 m **b** 114 m²

Challenge

1 a, b Students' own answers
c 532.0 and 0.235

2 Students' own answers

4 Strengthen: Scales and measures

1 a Students' own answers
b Accurate lines drawn:
i 2 cm **ii** 3 cm **iii** 5 cm

2 a 320
b i 550 **ii** 525 **iii** 575

3 83 cm

4 a 350 ml **b** 160 ml **c** 125 ml

5 a i 79 **ii** 790 **iii** 7900
b i 21.8 **ii** 218 **iii** 2180

6 a I 8.2 **ii** 0.82 **iii** 0.082
b i 32.9 **ii** 3.29 **iii** 0.329

7 a 7.3 cm, 46.2 cm, 5.91 cm
b 40 mm, 32 mm, 57.1 mm

8 a 8 m, 3.5 m, 0.17 m
b 500 cm, 450 cm, 620 cm

9 a 6 kg, 0.826 kg, 0.091 kg
b 4000 g, 6300 g, 7620 g

10 a 0.006 tonnes **b** 0.0092 tonnes
c 0.00385 tonnes **d** 4 000 000 kg
e 372 000 kg **f** 73 000 kg

4 Strengthen: Decimals

1 a 0.3 b 0.5
2 a 0.03 b 0.37 c 0.04
   d 0.56 e 0.74 f 0.19
3 a 0.5 b 0.3 c 0.7 d 4.6
   e 4.6
4 4.6, 12.8, 0.8, 156.0
5 a

4.5 4.51 4.52 4.53 4.54 4.55 4.56 4.57 4.58 4.59 4.6

   b 4.5 c 4.6 d 4.6
6 a 6.6 b 6.5 c 6.6 d 6.2
   e 6.1 f 6.2 g 14.3 h 141.4
   i 14.4
7 a 63.2 b 23.25 c 29.94
8 £33.50
9 a 13.2 b 2.33 c 18.16
10 1.25 m
11 a $2 \times 30 = 60$ $2 \times 3 = 6$
       $2 \times 0.3 = 0.6$ $2 \times 0.03 = 0.06$
    b $5 \times 80 = 400$ $5 \times 8 = 40$
       $5 \times 0.8 = 4$ $5 \times 0.08 = 0.4$
12 a 0.8 b 3.6 c 0.06 d 0.32
13 a 36.8 b 19.2 c 15.75 d 12.96
14 a 31.4 b 12.4 c 5.2 d 4.3

4 Strengthen: Perimeter and area

1 a 20 cm b 35 cm c 72 cm d 33 cm
2 a 1 cm² b $\frac{1}{2}$ cm² c $1\frac{1}{2}$ cm² d $\frac{1}{2}$ cm²
   e 1 cm² f $1\frac{1}{2}$ cm² g $\frac{1}{2}$ cm² h $1\frac{1}{2}$ cm²
   i $\frac{1}{2}$ cm²
3 a i 3 cm ii 2 cm iii 10 cm iv 6 cm²
   b i 6 cm ii 4 cm iii 20 cm iv 24 cm²
4 a 24 cm, 26 cm² b 18 cm, 14 cm²
   c 14 cm, 10 cm² d 15 cm, 11 cm²
   e 40 cm, 87 cm²
5 a 2 cm b 8 cm

Challenge

a hedgehogs A, B, C, E and I
b hedgehog G
c hedgehog H

4 Extend

1 a $2.08 \div 32 = 0.065$
       $0.208 \div 32 = 0.0065$
    b $714 \div 0.084 = 8500$
       $714 \div 0.0084 = 85\,000$
2 a 6 b 0.6 c 600 d 600
   e 6000 f 6
3 a £1.69 b Diamond
4 a 44 cm b 2.65 m
   c No; the increases are getting smaller and each is a long
      way from the mean.
5 1.136 m
6 3.5 cm
7 a 162 m² b 420 m
8 200 cm
9 a 46 cm b 54 cm²
10 53 084 m²
11 a 16 600 m² b £498 000
12 a height = 136.8 m, length = 935.4 m
    b Steel Dragon 2000 2.4 km (1 d.p.), The Ultimate
       2.2 km (1 d.p.)
    c 104.7 m

13 a 6.6 tonnes per second
    b 6600 litres per second
    c 506 000 litres per second

Challenge

1 1 cm²   100 mm²   1 cm² = 100 mm²
2 a 500 mm² b 900 mm² c 420 mm²
   d 3 cm² e 8 cm² f 3.6 cm²

4 Unit test

1 18 cm
2 a 12.39 < 12.55 b 8.6 > 8.07
   c 29.8 > 29.37
3 3.09, 3.3, 3.35, 3.41, 3.8, 3.85
4 a 0.8 b 0.24 c 0.36
5 2.4 litres
6 a mm² b cm
7 a 64 mph b −15 °C
8 55.3
9 No it should be 27.0.
10 a 800 cm = 8 m
    b 2 litres = 2000 ml
    c 9000 g = 9 kg
11 a 550 cm = 5.5 m
    b 400 g = 0.4 kg
    c 4.5 km = 4500 m
12 532 m 70 cm
13 a 158.25 kg b 0.25 m or 25 cm
14 a 34.8 b 21.15 c 107.1
15 17.28 kg
16 25 cm²
17 a 22 cm b 14 cm²
18 a 9.9 b 28.05
19 a 8.46 b 846 c 0.846
20 No; 8 miles ≈ 8 × 1.6 = 12.8 km
21 a i 2704 mm² ii 0.002 704 m²
    b 58 mm

Challenge

1 a
```
   2 3 . 7 2
+  1 8 . 9 6
-----------
   4 2 . 6 8
```
   b
```
   4 2 . 8 1
-  2 5 . 7 3
-----------
   1 7 . 0 8
```
2 3 cm and 8 cm

UNIT 5 Fractions and percentages

5.1 Comparing fractions

1 Shapes A and D

2 a $A = \frac{3}{5}$, $B = \frac{2}{5}$, $C = \frac{4}{5}$

 b $\frac{2}{5}, \frac{3}{5}, \frac{4}{5}$

3 a A **b** D **c** C **d** B

4 a $\frac{5}{8}$ **b** $\frac{2}{3}$ **c** $\frac{2}{5}$ **d** $\frac{7}{10}$

5 a i $\frac{3}{4}$ **ii** $\frac{1}{4}$

 b i $\frac{4}{5}$ **ii** $\frac{1}{5}$

 c i $\frac{5}{8}$ **ii** $\frac{3}{8}$

 d i $\frac{5}{9}$ **ii** $\frac{4}{9}$

6 a 1 square shaded, e.g.

 b 3 squares shaded, e.g.

 c 2 squares shaded, e.g.

7 $\frac{24}{48} = \frac{1}{2}$

8 a $\frac{1}{7}, \frac{3}{7}, \frac{6}{7}$ **b** $\frac{2}{9}, \frac{4}{9}, \frac{7}{9}$

 c $\frac{1}{6}, \frac{3}{6}, \frac{4}{6}$ **d** $\frac{3}{11}, \frac{4}{11}, \frac{5}{11}, \frac{10}{11}$

 e $\frac{2}{15}, \frac{3}{15}, \frac{6}{15}, \frac{7}{15}$

9 a > **b** > **c** < **d** <

10 a 1 section shaded, e.g.

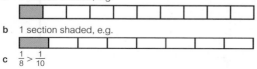

 b 1 section shaded, e.g.

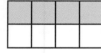

 c $\frac{1}{8} > \frac{1}{10}$

11 a > **b** < **c** < **d** >

12 Beach Buoys; $\frac{1}{3}$ is the larger fraction so it is the larger discount.

13 a $\frac{3}{4}$ **b** $\frac{3}{7}$ **c** $\frac{2}{7}$ **d** $\frac{7}{8}$

14 $\frac{5}{7}, \frac{2}{3}, \frac{3}{5}$

15 $\frac{3}{4}$

Challenge

a 15 ways

b 4 squares shaded, e.g.

70 ways

5.2 Simplifying fractions

1 a 3 **b** 4 **c** 5

2 a 6 **b** 6

3 a 2 **b** 5 **c** 4

4 a yes **b** yes **c** no

5 a 1 **b** 1 **c** 1 **d** 1

6 a $1\frac{2}{3}$ **b** $2\frac{2}{3}$ **c** $2\frac{2}{5}$ **d** $1\frac{2}{5}$

 e $1\frac{5}{6}$ **f** $2\frac{1}{6}$ **g** $1\frac{3}{4}$ **h** $2\frac{3}{4}$

 i $1\frac{7}{8}$ **j** $2\frac{5}{8}$

7 $1\frac{7}{10}$

8 $4\frac{1}{4}$ litres

9 a $\frac{5}{20} = \frac{10}{40}$ **b** $\frac{1}{5} = \frac{5}{25}$ **c** $\frac{5}{7} = \frac{15}{21}$

 d $\frac{2}{5} = \frac{16}{40}$ **e** $\frac{8}{10} = \frac{4}{5}$ **f** $\frac{4}{12} = \frac{1}{3}$

 g $\frac{30}{36} = \frac{5}{6}$ **h** $\frac{12}{21} = \frac{4}{7}$

10 a $\frac{10}{20} = \frac{1}{2}$ **b** $\frac{50}{60} = \frac{5}{6}$ **c** $\frac{90}{110} = \frac{9}{11}$

11 a $\frac{5}{15} = \frac{1}{3}$ **b** $\frac{25}{35} = \frac{5}{7}$ **c** $\frac{15}{55} = \frac{3}{11}$

12 a $\frac{2}{4} = \frac{1}{2}$ **b** $\frac{6}{10} = \frac{3}{5}$ **c** $\frac{14}{20} = \frac{7}{10}$

13 a $\frac{6}{7}$ **b** $\frac{3}{4}$ **c** $\frac{2}{3}$ **d** $\frac{2}{3}$

 e $\frac{1}{2}$ **f** $\frac{2}{3}$

14 a $\frac{3}{10}$ **b** $\frac{7}{10}$ **c** $\frac{1}{2}$ **d** $\frac{2}{5}$

 e $\frac{1}{5}$ **f** $\frac{1}{4}$ **g** $\frac{3}{4}$ **h** $\frac{11}{50}$

 i $\frac{8}{25}$ **j** $\frac{21}{25}$

15 ÷ 2 then ÷ 2 is the same as ÷ 4

16 a $1\frac{1}{3}$ **b** $1\frac{2}{3}$ **c** $1\frac{1}{4}$ **d** $1\frac{3}{4}$

 e $1\frac{2}{5}$ **f** $2\frac{1}{2}$

Challenge

a Students' own answers, e.g. $\frac{1}{2}, \frac{2}{4}, \frac{3}{6}, \frac{5}{10}$

b Unlimited/infinite number of possible equivalent fractions

5.3 Working with fractions

1 a 5 **b** 2 **c** 9 **d** 9

2 a £6 **b** 5 m

3 a $1\frac{4}{5}$ **b** $2\frac{2}{3}$ **c** $1\frac{1}{2}$ **d** $1\frac{2}{7}$

 e $1\frac{3}{4}$ **f** $1\frac{5}{6}$

4 a $\frac{2}{5}$ **b** $\frac{5}{7}$ **c** $\frac{7}{9}$ **d** $\frac{7}{11}$

5 a $\frac{1}{5}$ **b** $\frac{2}{7}$ **c** $\frac{1}{5}$ **d** $\frac{7}{9}$

6 a i $\frac{3}{5} + \frac{2}{5} = \frac{5}{5} = 1$

 ii $\frac{4}{7} + \frac{3}{7} = \frac{7}{7} = 1$

 iii $\frac{5}{11} + \frac{6}{11} = 1$

 b Students' own answers

7 a $1 - \frac{11}{12} = \frac{12}{12} - \frac{11}{12} = \frac{1}{12}$

 b $1 - \frac{3}{7} = \frac{7}{7} - \frac{3}{7} = \frac{4}{7}$

 c $\frac{3}{5}$ **d** $\frac{5}{9}$ **e** $\frac{1}{3}$ **f** $\frac{1}{8}$

8 a $\frac{2}{5}$ **b** $\frac{1}{2}$ **c** $\frac{3}{4}$ **d** $\frac{2}{3}$

e $\frac{1}{2}$ f $\frac{1}{4}$ g $\frac{3}{5}$ h $\frac{1}{3}$

i $\frac{1}{3}$

9 a $\frac{7}{9} + \frac{3}{9} = \frac{10}{9} = 1\frac{1}{9}$ b $1\frac{2}{9}$ c $1\frac{2}{7}$

   d $1\frac{2}{5}$ e $1\frac{2}{9}$ f $1\frac{3}{11}$ g $1\frac{1}{3}$

   h $1\frac{4}{5}$

10 Students' own answers, e.g. $\frac{1}{16} + \frac{3}{16}, \frac{2}{20} + \frac{3}{20}$

11 a $\frac{5}{9}$ b $\frac{4}{9}$

12 a 5 kg   b 6 kg   c £2   d 4 cm
   e 3 t   f 25 ml

13 a 4   b 70   c 5.5   d 9.9
   e 12.3   f 15   g 27.4   h 125

14 a $18   b 15 m   c 10 km   d 24 kg
   e £35   f £7.50

15 a 18 g   b 6 g

16 14 litres

17 37.5 miles

18 No; although Hannah's statement fits the first two sets of values fairly well, but it doesn't fit the third and fourth sets of values as well, so it is not always true.

19 £13.50

Challenge

Dominoes in order:

5.4 Fractions and decimals

1 a $\frac{1}{2}$ b $\frac{1}{4}$ c $\frac{3}{4}$

2 a 7 tens or 70   b 7 tenths or 0.7
   c 7 hundredths or 0.07

3 a $\frac{1}{5}$ b $\frac{9}{50}$ c $\frac{9}{20}$

4 a

$$0 \quad \frac{1}{10} \quad \frac{2}{10} \quad \frac{3}{10} \quad \frac{4}{10} \quad \frac{5}{10} \quad \frac{6}{10} \quad \frac{7}{10} \quad \frac{8}{10} \quad \frac{9}{10} \quad 1$$
$$0 \quad 0.1 \quad 0.2 \quad 0.3 \quad 0.4 \quad 0.5 \quad 0.6 \quad 0.7 \quad 0.8 \quad 0.9 \quad 1$$

   b

$$0 \quad \frac{1}{5} \quad \frac{2}{5} \quad \frac{3}{5} \quad \frac{4}{5} \quad 1$$
$$0 \quad 0.2 \quad 0.4 \quad 0.6 \quad 0.8 \quad 1$$

5 a $\frac{9}{10}$ b $\frac{3}{5}$ c $\frac{1}{2}$ d $\frac{13}{100}$

   e $\frac{17}{50}$ f $\frac{31}{50}$ g $\frac{1}{4}$ h $\frac{3}{4}$

   i $\frac{7}{20}$ j $\frac{3}{50}$ k $\frac{1}{50}$ l $\frac{1}{20}$

6 a 0.3   b 0.33   c 0.03   d 0.07

7 a $\frac{16}{25}$ b $\frac{18}{25}$

8 a 0.2   b 0.8   c 0.15   d 0.45
   e 0.08   f 0.48   g 0.14   h 0.34

9 a 0.6

b Students' own answers, e.g.
$\frac{12}{20} = \frac{6}{10} = 0.6$
$\frac{12}{20} = \frac{120}{200} = \frac{60}{100} = 0.6$

10 a 0.8   b 0.4   c 0.2   d 0.3
   e 0.2   f 0.35   g 0.28   h 0.8

11 $\frac{7}{17}$

12 $\frac{1}{6}$

13 $\frac{1}{5}$

14 $\frac{3}{8}$

15 $\frac{3}{5}$

16 a $\frac{4}{5}$ b 0.8

Challenge

a $\frac{1}{9} = 0.111..., \frac{2}{9} = 0.222..., \frac{3}{9} = 0.333...$

b 0.777...

c Students' calculator check

d $\frac{9}{9} = 1, \frac{10}{9} = 1\frac{1}{9} = 1.111..., \frac{11}{9} = 1\frac{2}{9} = 1.222...$ etc.

5.5 Understanding percentages

1 $\frac{9}{100}$

2 a $\frac{2}{5}$ b $\frac{1}{2}$ c $\frac{2}{25}$ d $\frac{1}{4}$

3 a 0.4   b 0.5   c 0.08   d 0.25

4 a i 30%   ii 70%
   b i 77%   ii 23%
   c i 50%   ii 50%
   d 100%

5 a i 14% ii 12%   iii 10%
   b Used floor space = 14% + 12% + 10% = 36%
   74% + 36% = 110%
   But 'empty floor space + used floor space' must be 100%'.

6 20%

7 2%

8 a $\frac{27}{100}$ b $\frac{99}{100}$ c $\frac{3}{10}$ d $\frac{1}{2}$

   e $\frac{1}{10}$ f $\frac{3}{20}$ g $\frac{1}{4}$ h $\frac{3}{4}$

9 a $\frac{37}{100}$ b $\frac{79}{100}$ c $\frac{61}{100}$ d $\frac{119}{100}$

10 a $\frac{1}{5}$ b $\frac{3}{5}$ c $\frac{2}{5}$ d $\frac{9}{10}$

11 No, because 150% > 100%, and a discount cannot be more than the original price.

12 a 0.35   b 0.81   c 0.09   d 0.01
   e 0.4   f 1.1

13 a 45%   b 72%   c 3%   d 80%
   e 120%

14 a 27%   b 90%   c 22%   d 16%
   e 65%   f 60%

15 70% = $\frac{7}{10}$, 51% = $\frac{51}{100}$, 80% = $\frac{4}{5}$, 21% = $\frac{21}{100}$

   10% = $\frac{1}{10}$, 39% = $\frac{39}{100}$

16 a 4/5 = 80%   b 43/50 = 86%   c 7/10 = 70%

17 45%

Challenge

a Students' diagrams

b 13 squares shaded, e.g.

c 6 squares shaded, e.g.

d Students' own answers

e Students' own explanations

5.6 Percentages of amounts

1 a $\frac{1}{2}$ b $\frac{1}{4}$ c $\frac{3}{4}$

2 a £9 b 3 cm c 9 cm

3 a 4% b 70% c 22% d 45%

4 a 2 b 3.2 c 0.6 d 6
  e 60 f 90

5 a 15 b 45 c 80 d 27
  e 9 f 9.5

6 a £10 b £30 c 75 kg d 225 kg
  e 50 mm f 150 mm

7 £68.38

8 a 8 kg b 15 ml c 150 m d £4.50

9 a £300 b £120 c £18 d £44.50
  e £18.30 f £96.30

10 a £5 b £10 c £40 d 140 g
   e 7.5 km f £7.20 g £9 h £13.50

11 a 450 tickets b £5400

12 a 7 cm b 3.5 cm c 0.7 cm d £2.50
   e £1.25 f £0.25 or 25p

13 a 6 b 9.2 c 30 d 0.5
   e 0.75 f 1.99

14 a £6 b 31.5 kg c 10.5 m d 102 km

15 a £1.60 b 10.5 kg
   c 12.8 litres d 4.2 t

16 £12

17 £170 or £171 or £170.97

Challenge

a i £100 ii £10 100
b i £101 ii £10 201
c £10 303.01

5 Check up

1 a $\frac{1}{6}$ b $\frac{3}{5}$

2 $\frac{1}{6}$ or $\frac{1}{7}$

3 a $\frac{4}{9} < \frac{8}{9}$ b $\frac{3}{8} > \frac{1}{4}$ c $\frac{1}{3} > \frac{1}{5}$

4 a $\frac{3}{7} = \frac{12}{28}$ b $\frac{30}{40} = \frac{6}{8}$

5 a $1\frac{1}{4}$ b $3\frac{5}{6}$

6 a $\frac{3}{5}$ b $\frac{5}{9}$ c $\frac{3}{5}$

7 a $\frac{1}{2}$ b $\frac{3}{10}$

8 a $\frac{3}{5}$ b $\frac{3}{4}$

9 a $\frac{3}{4}$ b $\frac{9}{10}$

10 a £3 b 9 km c 15 kg d 24 mm

11 a $\frac{13}{100}$ b $\frac{7}{10}$ c $\frac{1}{5}$ d $\frac{21}{50}$

12 a 0.9 b 0.49 c 0.35 d 0.4

13 a $\frac{23}{100}$ b $\frac{3}{5}$ c $\frac{2}{25}$ d $\frac{3}{4}$

14 a 42% b 30% c 82% d 44%

15

Fraction	Decimal	Percentage
$\frac{1}{2}$	0.5	50%
$\frac{7}{10}$	0.7	70%
$\frac{1}{4}$	0.25	25%
$\frac{3}{50}$	0.06	6%

16 $\frac{5}{12}$

17 $\frac{3}{7}$

18 a £4 b 9 cm c 18 km d 50 kg

19 70%

20 £4.80

21 a 0.1 b 0.2 c 0.7

Challenge

a–c Students' own answers

5 Strengthen: Fractions

1 a 3 squares shaded, e.g.

  b 7 squares shaded, e.g.

  c 5 squares shaded, e.g.

2 a 2 b 7 c $\frac{2}{7}$ d $\frac{5}{7}$

3 a $\frac{4}{5}$ b $\frac{4}{9}$ c $\frac{3}{5}$ d $\frac{1}{5}$

4 a $\frac{6}{7}$ b $\frac{7}{12}$

5 a three quarters b three fifths
  c $\frac{7}{9}$ d $\frac{11}{12}$ e $\frac{11}{15}$

6 $\frac{1}{5}$

7 a $\frac{1}{2}$ b $\frac{1}{3}$ c $\frac{1}{7}$

8 a 6 b 4 c 10

9 $2\frac{1}{4}$

10 a $1\frac{1}{5}$ b $2\frac{2}{5}$ c $2\frac{1}{6}$ d $3\frac{1}{2}$

11 a $\frac{3}{5}$ b $\frac{5}{8}$ c $\frac{5}{9}$ d $\frac{7}{10}$
   e $\frac{3}{5}$ f $\frac{2}{7}$ g $\frac{5}{9}$

12 a $\frac{4}{5}$ b $\frac{6}{7}$ c $\frac{2}{3}$ d $\frac{3}{4}$

13 a $\frac{2}{5}$ b $\frac{1}{6}$ c $\frac{1}{3}$ d $\frac{2}{9}$

14 a $\frac{3}{4}$ b $\frac{4}{5}$ c $\frac{9}{10}$ d $\frac{3}{5}$
   e $\frac{2}{3}$ f $\frac{5}{6}$ g $\frac{2}{3}$

15 a Missing value 5   b Missing value 1
  c Missing value 3   d Missing value 2
  e Missing value 2   f Missing value 5
16 a 5   b 4   c 9   d 6
17 a 5   b 2   c 2   d 3
18 a 4   b 5   c 7   d 9
19 a 4   b 8   c 12   d 16

5 Strengthen: Fractions, decimals and percentages

1

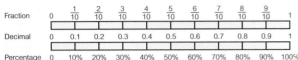

Fraction	0	$\frac{1}{10}$	$\frac{2}{10}$	$\frac{3}{10}$	$\frac{4}{10}$	$\frac{5}{10}$	$\frac{6}{10}$	$\frac{7}{10}$	$\frac{8}{10}$	$\frac{9}{10}$	1
Decimal	0	0.1	0.2	0.3	0.4	0.5	0.6	0.7	0.8	0.9	1
Percentage	0	10%	20%	30%	40%	50%	60%	70%	80%	90%	100%

2 a $0.5 = \frac{1}{2} = 50\%$ b $\frac{1}{4} = 0.25 = 25\%$

  c $75\% = \frac{3}{4} = 0.75$ d $0.25 = \frac{1}{4} = 25\%$

3 a $80\% = \frac{4}{5} = 0.8$ b $0.6 = \frac{3}{5} = 60\%$

  c $20\% = \frac{1}{5} = 0.2$ d $\frac{4}{5} = 0.8 = 80\%$

4 a $\frac{7}{9}$ b $\frac{5}{12}$

5 a $21\% = \frac{21}{100}$ b $66\% = \frac{66}{100} = \frac{33}{50}$

  c $44\% = \frac{44}{100} = \frac{22}{50} = \frac{11}{25}$ d $\frac{35}{100} = 35\%$

  e $\frac{3}{10} = \frac{30}{100} = 30\%$ f $\frac{9}{20} = \frac{45}{100} = 45\%$

  g $\frac{7}{50} = \frac{14}{100} = 14\%$ h $\frac{3}{25} = \frac{12}{100} = 12\%$

  i $0.71 = \frac{71}{100} = 71\%$ j $0.07 = \frac{7}{100} = 7\%$

  k $0.7 = \frac{7}{10} = \frac{70}{100} = 70\%$ l $0.01 = \frac{1}{100} = 1\%$

  m $59\% = \frac{59}{100} = 0.59$ n $31\% = \frac{31}{100} = 0.31$

  o $5\% = \frac{5}{100} = 0.05$

5 Strengthen: Percentages

1 a 10   b £6
2 a 3   b 50   c 42   d 8.8
3 a 2   b 17
4 a £30   b £160   c £225   d £8.50
5 a $25\% = \frac{1}{2}$ of 50%   b 75% = 50% + 25%
6 a i £10   ii £5   iii £15
  b i 15kg   ii 7.5kg   iii 22.5kg
  c i 25ml   ii 12.5ml   iii 37.5ml
  d i 40m   ii 20m   iii 60m
  e i £60   ii £30   iii £90
7 a 9kg   b 13.5kg   c 27kg   d 2.25kg
8 $\frac{40}{100}$ of children like broccoli.

$\frac{40}{100} = 40\%$ of children like broccoli.

9 a 80% 0of students like chocolate cake.
  b 28% of people go to the gym.
  c 60% of people have a pet.
  d 80% of children like fruit.
  e 40% of students play sport regularly.

Challenge

No, because they are not all equivalent fractions:
$\frac{10}{15} = \frac{2}{3}, \frac{6}{9} = \frac{2}{3}$ but $\frac{9}{12} = \frac{3}{4}$

5 Extend

1 a i $\frac{1}{4}$ ii 25%

  b i $\frac{7}{10}$ ii 70%

  c i $\frac{3}{5}$ ii 60%

  d i $\frac{9}{20}$ ii 45%

2 a $\frac{1}{11}, \frac{1}{7}, \frac{1}{2}$ b $\frac{1}{3}, \frac{1}{8}, \frac{1}{9}$

3   0.1, 10%, $\frac{1}{10}$ 0.2, 20%, $\frac{1}{5}$ 0.25, 25%, $\frac{1}{4}$

  0.4, 40%, $\frac{2}{5}$ 0.5, 50%, $\frac{1}{2}$ 0.6, 60%, $\frac{3}{5}$

  0.75, 75%, $\frac{3}{4}$ 0.8, 80%, $\frac{4}{5}$

  5% does not belong in any of the groups.

4 a $1\frac{1}{2}$ b $1\frac{2}{3}$ c $1\frac{1}{2}$ d $1\frac{1}{2}$

  e $2\frac{2}{9}$ f $2\frac{1}{5}$

5   No; he has not cancelled the fraction to its simplest form. He can divide numerator and denominator by 3 to get $\frac{2}{3}$

6 a i $\frac{11}{20}$ ii 55%

  b 45% (100% − 55%)

7   48 adult size T-shirts

8 a Nicki 65%, Wei Yen 68%, Robin 80%, Alex 84%
  b Alex

9   30 000 m²

Challenge

a

Temperature (°F)	59	77	95	167	212
− 32	27	45	63	135	180
× $\frac{5}{9}$	15	25	35	75	100
Temperature (°C)	15	25	35	75	100

b   10°C, 59°F (15°C), 20°C, 77°F (25°C), 167°F (75°C), 84°C, 212°F (100°C)

5 Unit test

1 a 6m   b 40kg   c £18
2 a $\frac{1}{3}$ b $\frac{5}{6}$
3 1 more triangle
4 a $\frac{1}{8}$ b $\frac{2}{7}, \frac{4}{7}, \frac{5}{7}$
5 a $47\% = \frac{47}{100}$ b $3\% = \frac{3}{100}$

  c $70\% = \frac{70}{100} = \frac{7}{10}$ d $\frac{38}{100} = 38\%$

  e $\frac{9}{100} = 9\%$ f $\frac{3}{10} = 30\%$

6 a 0.75 = 75%   b 0.4 = 40%   c 0.05 = 5%
  d 0.5 = 50%   e 0.04 = 4%   f 0.25 = 25%
7 a 0.1   b 0.2, 0.3
8 45%, 0.45
9 a $1\frac{1}{5}$ b $4\frac{3}{4}$

10 $\frac{7}{9} = \frac{21}{27}$

11 a B   b C

12 a $\frac{4}{7}$ b $\frac{1}{5}$

13 $\frac{3}{5}$

14 $\frac{3}{4}$

15 a 7cm   b 36km
16 90%
17 20%
18 a T: $30\% = \frac{30}{100} = 0.3$

  b F: $1\% = \frac{1}{100} = 0.01$

Challenge

a   IS A POLYGON A DEAD PARROT?
b   Students' own answers

UNIT 6 Probability

6.1 The language of probability

1. Students' own answers, e.g.
   possible: something that can happen
   impossible: something that can never happen
   certain: something that will definitely happen
   predict: what you think will happen
   likely: something that is more likely to happen than not happen
   unlikely: something that is more likely to not happen than to happen

2. **a** 0.1 **b** 0.5 **c** 0.7
   d 0.9 **e** 10% **f** 50%
   g 80% **h** $\frac{1}{2}$ **i** $\frac{3}{4}$

3. **a** unlikely **b** certain **c** likely
   d very unlikely **e** even chance **f** certain
   g impossible

4.

5. **a–e** Students' own answers

6. **a** Spinner C
   b Spinner A
   c Spinner A
   d Spinner B landing on white is more likely, because the white section is bigger than the red section on Spinner A.
   e Students' own answers, e.g.

7. **a**

   b The non-identical twin, because 6% is greater than 1%.

8. **a** A $\frac{1}{8}$, B $\frac{3}{8}$, C $\frac{5}{8}$, D $\frac{7}{8}$
   b **i** C **ii** B
      iii D **iv** A

9. Students' own answers, e.g.
   Yes, because an event with a probability of 0.07 is very unlikely to occur.

10. Students' own answers, e.g.
    No, because 45% is less than an even chance.

11. Students' own answers, e.g.
    If I thought Ferrari were going to lose, I would have given them a 5% chance. Instead I gave them the much greater chance of 40% and they won.

Challenge

Students' own answers

6.2 Calculating probability

1. **a** **i** unlikely **ii** even chance
   b white

2. $\frac{1}{5}$

3. 50%

4. **a** 6-sided dice: 1, 2, 3, 4, 5, 6
      4-sided dice: 1, 2, 3, 4
      10-sided dice: 1, 2, 3, 4, 5, 6, 7, 8, 9, 10
   b 6-sided dice: 6 outcomes
      4-sided dice: 4 outcomes
      10-sided dice: 10 outcomes
   c the 4-sided dice

5. **a** 1, 2, 2, 3, 6 **b** 2, 2, 6
   c 3, 6 **d** 2, 2, 3

6. **a** $\frac{1}{6}$ **b** $\frac{3}{6}$ or $\frac{1}{2}$
   c $\frac{2}{6}$ or $\frac{1}{3}$ **d** 0

7. **a** 10
   b 1, 4, 9
   c $\frac{3}{10}$ or equivalent

8. **a** $\frac{4}{10}$ or equivalent
   b $\frac{4}{10}$ or equivalent

9. **a** **i** $\frac{2}{10}$ or $\frac{1}{5}$ **ii** $\frac{5}{10}$ or $\frac{1}{2}$
      iii $\frac{8}{10}$ or $\frac{4}{5}$ **iv** $\frac{6}{10}$ or $\frac{3}{5}$
   b **i** 0.2 **ii** 0.5
      iii 0.8 **iv** 0.6
   c **i** 20% **ii** 50%
      iii 80% **iv** 60%

10. **a** $\frac{3}{6}$ or $\frac{1}{2}$
    b $\frac{40}{100}$ or $\frac{4}{10}$ or $\frac{2}{5}$
    c $\frac{1}{7}$
    d $\frac{12}{52}$ or $\frac{3}{13}$

11. **b**

12. **a** $\frac{1}{4}$ **b** $\frac{1}{2}$

13. Spinner **b** is most likely to land on red.
    Probabilities are **a**: 50%, **b**: 60%, **c**: 50%

14. **a** A $\frac{6}{12}$ or $\frac{1}{2}$ or 0.5 or 50%
       B 1
       C $\frac{2}{12}$ or $\frac{1}{6}$
       D $\frac{9}{12}$ or $\frac{3}{4}$ or 0.75 or 75%
       E $\frac{5}{12}$
       F $\frac{3}{12}$ or $\frac{1}{4}$ or 0.25 or 25%
    b

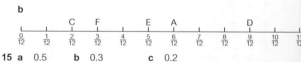

15. **a** 0.5 **b** 0.3 **c** 0.2

16. **a** $\frac{1}{10}$ or 0.1 or 10%
    b **i** 100 **ii** $\frac{1}{100}$ or 0.01 or 1%
    c $\frac{1}{1000}$ or 0.001 or 0.1%

17. **a** 28 **b** 21

Challenge

a 20
b 5 green, 10 red, 4 yellow, 1 black

6.3 More probability calculations

1. **a** $\frac{3}{4}$ **b** 0.4 **c** 80%
   d $\frac{3}{5}$ **e** 0.1 **f** $\frac{7}{10}$

2. **a** $\frac{3}{10}$ **b** 0.6 **c** 40%

3 **a** $\frac{6}{8}$ or equivalent

b Yes. Both 'blue or red' and 'not yellow' have the same number of outcomes.

4 **a** 6

b **i** $\frac{1}{6}$ **ii** $\frac{2}{3}$ **iii** $\frac{5}{6}$

iv 1 **v** $\frac{1}{2}$ **vi** $\frac{1}{3}$

5 **a** $\frac{2}{52}$ or $\frac{1}{26}$

b $\frac{8}{52}$ or $\frac{2}{13}$

c $\frac{39}{52}$ or $\frac{3}{4}$ or 0.75 or 75%

d $\frac{24}{52}$ or $\frac{6}{13}$

6 **a** **i** 60% **ii** 40%
iii 70% **iv** 50%

b 25%

c 0.25

7 0.7

8 $\frac{5}{12}$

9 0.73

10 95%

11 The probability our software will not crash is 98%.

12 0.2

Challenge

a $\frac{1}{2}$ **b** $\frac{1}{2}$ **c** $\frac{1}{2} + \frac{1}{2} = 1$

d $\frac{1}{6}$ **e** $\frac{5}{6}$ **f** $\frac{1}{6} + \frac{5}{6} = 1$

g P(event happening) + P(event not happening) = 1

6.4 Experimental probability

1 **a** $\frac{5}{20}$ or equivalent

b 25%

2 **a**

Colour	Tally	Frequency
Red	ℍℍ ℍℍ ‖	12
Blue	ℍℍ ‖	7
White	ℍℍ	6

b 25

c $\frac{12}{25}$

3 **a** unlikely **b** likely **c** even chance
d unlikely **e** likely **f** impossible

4

Colour	Frequency	Experimental probability
red	17	$\frac{17}{100} = 17\%$
green	28	$\frac{28}{100} = 28\%$
yellow	55	$\frac{55}{100} = 55\%$
Total frequency	100	

5 **a**

Number of leaves	Frequency	Experimental probability
3	156	$\frac{156}{200}$ or equivalent
4	26	$\frac{26}{200}$ or equivalent
5	18	$\frac{18}{200}$ or equivalent
Total frequency	200	

b Students' own answers, e.g.
Yes, because a probability of $\frac{18}{200}$ is very unlikely.

6 **a, b**

Outcome	Frequency	Experimental probability
symptom free	60	$\frac{60}{80}$
some improvement	15	$\frac{15}{80}$
no improvement	5	$\frac{5}{80}$
Total frequency	80	

c This is a valid claim, as the probability of improvement is $\frac{75}{80}$

7 $\frac{17}{500}$

8 **a** $\frac{38}{100}$ or 0.38 or 38%

b $\frac{62}{100}$ or 0.62 or 62%

9 **a**

Time (months)	Frequency	Experimental probability
5	7	7%
6	14	14%
7	35	35%
8	25	25%
9	15	15%
10	3	3%
11	1	1%
	Total = 100	

b 19%

10 **a** $\frac{415}{1000}$ or 0.415 or 41.5%

b No. This model assumes that rainfall is equally likely in all months and ignores any seasonal variations.

11 **a** B

b Students' own answers, e.g.
A Find out the arrival times of a random sample of 100 trains and the number that were late
C Drop a piece of toast on the floor 100 times and count the number of times it lands butter-side down
D Find out the price of the magazine each year for the last 10 years and how many price rises there were during that time.

12 **a** $\frac{2}{8}$ or equivalent

b Play more games

13 **a** Paul: $\frac{5}{10}$ or equivalent

Surinder: $\frac{41}{50}$ or equivalent

Amy: $\frac{7}{20}$ or equivalent

b Surinder, because she conducted the most trials.

c **i** 80
ii 66

iii $\frac{66}{80}$ or equivalent

14 **a** no
b a large number of batteries tested

Challenge

Students' own answers

6.5 Expected outcomes

1 **a** $\frac{1}{2}$ **b** $\frac{1}{6}$

2 **a** $\frac{1}{2}$ or 0.5 or 50%

b $\frac{1}{2}$ or 0.5 or 50%

3 $\frac{3}{10}$ or 0.3 or 30% (accept equivalent fractions)

4 **a** $\frac{5}{12}$

b $\frac{7}{12}$

c Game 2, because the chance of winning is greater.

5 40

6 a $\frac{3}{8}$　　**b** 9

7 a $\frac{3}{50}$ or 0.06 or 6%

b 15

8 about 25　　**b** about 5　　**c** about 40
d about 5

9 a 80p　　**b** £2　　**c** no

10 a $\frac{13}{50}$　　**b** lose £1.50　　**c** no

11 a 9　　**b** lose £4

12 a i $\frac{1}{6}$　ii $\frac{1}{6}$

b 30p

13 a $\frac{2}{5}$　　**b** 40 wins　　**c** £20

d any amount less than 50p (game would break even) and more than 20p (cost of a go)

Challenge

Students' own answers

6 Check up

1 A likely
B even chance
C impossible
D unlikely
E certain

2

C D — impossible; B — even chance; A; E — certain

3 a likely　**b** impossible　**c** unlikely
d certain　**e** even chance

4 unlikely

5 a white, white, blue, blue, red

b $\frac{2}{5}$　**c** $\frac{3}{5}$

6 a 0.1　　**b** 0.5　　**c** 0.6

7 a 20%　　**b** 40%　　**c** 90%
d 50%　　**e** 80%　　**f** 0%

8 0.9

9 a Yes, because 98% > 50%
b 2%

10 $\frac{2}{3}$

11 a red = $\frac{19}{50}$, amber = $\frac{7}{50}$, green = $\frac{24}{50}$

b Yes. Amber has an experimental probability of $\frac{7}{50}$, which is unlikely.

12 a Isabella: $\frac{6}{10}$　Tim: $\frac{24}{50}$

b Tim's, because he had a greater number of spins/trials.

13 25

14 a 0.1　　**b** 50

Challenge

a Students' own answers
b Students' own answers, e.g.

6 Strengthen: The language of probability

1 A4, B2, C1, D5, E3

2 a blue　　**b** red

3 C

4 a 2　　**b** 3　　**c** 1

5 a

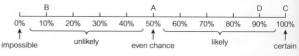

impossible — unlikely — even chance — likely — certain

b A even chance
B unlikely
C certain
D likely

6 Strengthen: Calculating probability

1 a 4
b A 2
　B 1
　C 3

c A $\frac{2}{4}$ or $\frac{1}{2}$ or 0.5 or 50%

　B $\frac{1}{4}$ or 0.25 or 25%

　C $\frac{3}{4}$ or 0.75 or 75%

2 a 10

b i $\frac{5}{10}$　ii $\frac{4}{10}$　　iii $\frac{1}{10}$

　iv $\frac{6}{10}$　v $\frac{9}{10}$

c

impossible — even chance — cer

d i 0.5　ii 0.4　　iii 0.1
　iv 0.6　v 0.9
e i 50%　ii 40%　　iii 10%
　iv 60%　v 90%
f i $\frac{3}{10}$ or 0.3 or 30%　　ii $\frac{1}{10}$ or 0.1 or 10%

　iii $\frac{7}{10}$ or 0.7 or 70%　　iv $\frac{6}{10}$ or 0.6 or 60%

　v $\frac{6}{10}$ or 0.6 or 60%

3 $\frac{9}{10}$

4 82%

5 55%

6 Strengthen: Experimental probability

1 a, b, c

Colour	Tally	Frequency	Experimental probability
blue	ＩＩＩＩ ＩＩＩＩ ＩＩＩ	13	$\frac{13}{20}$
yellow	ＩＩＩＩ	5	$\frac{5}{20}$
black	ＩＩ	2	$\frac{2}{20}$
	Total frequency	20	

d yellow
e blue likely, yellow unlikely, black very unlikely

2 a

Cashier desk	Tally	Frequency	Estimated probability
1	ＩＩＩＩ ＩＩＩＩ ＩＩＩＩ	14	$\frac{14}{100}$ = 14%
2	ＩＩＩＩ ＩＩＩＩ ＩＩＩＩ Ｉ	16	$\frac{16}{100}$ = 16%
3	ＩＩＩＩ ＩＩＩＩ ＩＩＩＩ ＩＩＩＩ ＩＩＩＩ	25	$\frac{25}{100}$ = 25%
4	ＩＩＩＩ ＩＩＩＩ ＩＩＩＩ ＩＩＩＩ ＩＩＩＩ ＩＩＩＩ	30	$\frac{30}{100}$ = 30%
5	ＩＩＩＩ ＩＩＩＩ ＩＩＩＩ	15	$\frac{15}{100}$ = 15%
	Total frequency	100	

b about 25%

c e.g. different cashier, fewer customers on Mondays

3 a 6   **b** 4   **c** $\frac{4}{6}$ or equivalent

d small number of matches/trials

Challenge

a Students' own answers, e.g. 1, 1, 1, 1, 1, 2

b Students' own answers, e.g. 1, 1, 1, 1, 1, 4, 4, 4, 4, 4

6 Extend

1 Students' own answers, e.g.

a 9   **b** even number

c 2   **d** 1, 2, 3, 4 or 5

e less than 7   **f** 1 or 2

2 60%

3 a $\frac{1}{240}$ **b** very unlikely   **c** $\frac{239}{240}$

4 a 80 nights   **b** $\frac{65}{80}$ or $\frac{13}{16}$

5 about 50%

6 a $\frac{22}{30}$ or equivalent

b $\frac{28}{40}$ or equivalent

c e.g. Probability describes the chance that something might happen. It cannot be used to predict exactly what will happen.

d 125

7 0.24

8 a

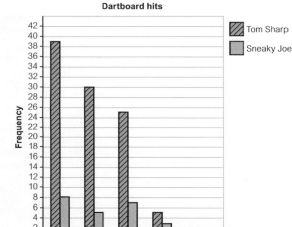

Dartboard hits

Tom Sharp
Sneaky Joe

b Tom Sharp $\frac{25}{100}$ or 25%

Sneaky Joe $\frac{7}{25}$ or $\frac{28}{100}$ or 28%

c Sneaky Joe, because 28% is greater than 25%.

d Tom Sharp's, because it is based on more throws.

e about 56

9 a **i** $\frac{1}{2}$ or 0.5 or 50%   **ii** about 15 times

b **i** $\frac{1}{4}$ or 0.25 or 25%   **ii** about 15 times

c $\frac{150}{510}$ or $\frac{5}{17}$ or 0.29 (2 d.p.)

Challenge

Students' own answers

6 Unit test

1 a **A** unlikely
B impossible
C likely

D impossible
E even chance

b

B, D   A   E   C

impossible   unlikely   even   likely   certain
                        chance

2 a likely   **b** unlikely   **c** even chance
d impossible   **e** certain

3 a **A** $\frac{1}{4}$ or 0.25 or 25%

B 0

C $\frac{3}{4}$ or 0.75 or 75%

D $\frac{1}{2}$ or 0.5 or 50%

E 1

b

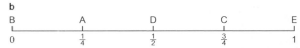

B   A   D   C   E

0         $\frac{1}{4}$ $\frac{1}{2}$ $\frac{3}{4}$ 1

4 a **A** 0.2
B 0.5
C 0.4
D 0.3
E 0.7

b

A D C B   E

0   0.5   1

5 a 10%   **b** 90%   **c** 40%   **d** 40%

6 0.7

7 80%

8 a 20 times

b

Outcome	Tally	Frequency	Experimental probability
curved surface	ⅢⅠ ⅢⅠ	8	$\frac{8}{20}$
flat surface	ⅢⅠ ⅢⅠ ⅠⅠ	12	$\frac{12}{20}$
	Total frequency	20	

c Yes, because $\frac{8}{20}$ is less than $\frac{10}{20}$ or 0.5.

d Drop the ball more times.

9 15

10 a 150 students
b 400 students
c $\frac{150}{400}$ or $\frac{3}{8}$

Challenge

Students' own answers

UNIT 7 Ratio and proportion

7.1 Direct proportion

1   a   6   b   8   c   9   d   £2.50

2   a   50 g   b   £0.20 or 20p
   c   5 kg   d   7 m

3   a   24   b   30   c   10   d   60

4   £42

5   £15

6   a   25 g   b   175 g

7   70

8   24

9   £15

10   a   96p   b   £1.44   c   £1.60

11   a   2.5 kg   b   12.5 kg   c   17.5 kg

12   a   387   b   645   c   903

13   a   16.5 m   b   25.5 m   c   42 m

14   175 g

15   2 hours 40 minutes

16   £105

17   a   240 g   b   180 g

18   £3

19   £2.50

20   £11

21   a   8 eggs   b   2 eggs   c   6 eggs   d   10 eggs

22   1 costs £140 ÷ 4 = £35, so 14 cost 14 × £35 = £490
   or 2 cost £140 ÷ 2 = £70, so 14 cost 7 × £70 = £490
   Students' own answers

23   a   direct   b   not direct   c   direct   d   not direct

Challenge

Students' own answers

7.2 Writing ratios

1   a   4   b   6   c   7   d   7

2   a   3   b   2   c   4   d   4

3   a   4   b   5

4   a   2 : 3   b   4 : 1   c   3 : 4   d   5 : 2

5   a   5 blue beads and 1 yellow bead in any order
   b   1 blue bead and 5 yellow beads in any order

6   No; students' own explanation, e.g. referring to the two different necklaces drawn in Q5

7   a   5 : 2   b   1250 : 1   c   1 : 3

8   a   1 : 3   b   1 : 3   c   1 : 3   d   3

9   a   4 : 2 (2 : 1)   b   6 : 3 (2 : 1)   c   4 : 4 (1 : 1)

10   a   1 : 10   b   5 : 1   c   1 : 6   d   1 : 5
   e   1 : 3   f   1 : 11

11   2 : 1

12   1 : 4

13   a   6 : 5   b   12 : 5   c   4 : 3   d   8 : 3
   e   3 : 10   f   3 : 4

14   2 : 5

15   Yes

16   Yes; 80 : 45 = 16 : 9 (divide by 5)

17   6 : 4 : 2

18   a   3 : 2 : 1   b   4 : 1 : 2   c   2 : 1 : 3
   d   2 : 6 : 3   e   5 : 3 : 4   f   4 : 5 : 3
   g   2 : 3 : 6   h   5 : 3 : 7   i   1 : 4 : 7

19   8 : 5 : 2

20   15 : 4 : 1

Challenge

Students' own answers, e.g.

a   8 : 10, 12 : 15, 40 : 50

b   2 : 4 : 6, 3 : 6 : 9, 10 : 20 : 30

7.3 Using ratios

1   3 : 2

2   a   7   b   6   c   12   d   4

3   a   100 cm   b   1000 m   c   1000 g   d   1000 ml

4   a   1 : 2 = 2 : 4   b   2 : 3 = 6 : 9

5   a   5 : 2 = 10 : 4   b   1 : 2 = 3 : 6
   c   6 : 1 = 18 : 3   d   7 : 3 = 14 : 6
   e   4 : 5 = 16 : 20   f   11 : 2 = 22 : 4
   g   6 : 7 = 42 : 49   h   15 : 8 = 30 : 16

6   12

7   12 teaspoons

8   2 tablespoons

9   £6000

10   a   3   b   12
   c   Students' own answers, e.g.
     Does 3 plus 12 equal 15? Is 3 : 12 in the ratio 1 : 4?

11   6 white and 12 black counters (Check: 6 + 12 = 18)

12   a   £14 : £7 (Check: £14 + £7 = £21)
   b   £18 : £27 (Check: £18 + £27 = £45)
   c   £56 : £40 (Check: £56 + £40 = £96)
   d   £16 : £12 (Check: £16 + £12 = £28)
   e   £27 : £45 (Check: £27 + £45 = £72)
   f   £44 : £16 (Check: £44 + £16 = £60)

13   a   6 g   b   18 g

14   a   about 40
   b   Students' own answers, e.g.
     The ratio is approximately 4 : 7.

15   Every 1 cm is the same as 10 mm.
   The ratio cm : mm is 1 : 10.

16   a   10 : 1   b   100 : 1   c   1 : 1000   d   1 : 1000
   e   1000 : 1   f   1 : 100

17   a   900 cm   b   20 mm   c   7000 ml   d   5 km
   e   2 m   f   3 cm   g   12 litres   h   100 mm
   i   0.1 km   j   1.5 litres

18   a   360 cm   b   2800 g   c   31 mm   d   8900 g
   e   3.9 km   f   6.3 m   g   8.4 cm   h   8.6 litres
   i   7000 cm

19   a   120 cm   b   48 cm

20   10.8 m

21   1.2 litres

22   a   i   £180   ii   £280
   b   i   £270   ii   £350
   c   Wilsons, £100

Challenge

Students' own answers, e.g.: No. They invested in the business in the ratio 2 : 3 so they should share the profits in the same ratio.

7.4 Ratios, proportions and fractions

1   a   i $\frac{1}{8}$ ii $\frac{7}{8}$

   b   i $\frac{1}{5}$ ii $\frac{1}{2}$

   c   i $\frac{4}{15}$ ii $\frac{11}{15}$

2   a $\frac{3}{4}$ b $\frac{4}{5}$

3   a   1 : 3
   b   i $\frac{1}{4}$ ii $\frac{3}{4}$

4   $\frac{3}{8}$

5   $\frac{3}{5}$

6   11 years old: $\frac{7}{15}$

   12 years old: $\frac{8}{15}$

7   a   2 : 3
   b   No, the proportion of boys out of the total number of children = $\frac{12}{30} = \frac{2}{5}$, which is not the same as $\frac{2}{3}$

8   a   i $\frac{4}{5}$ ii $\frac{1}{5}$
   b $\frac{4}{5} + \frac{1}{5} = 1$

9 a i $\frac{7}{10}$ ii $\frac{3}{10}$

 b $\frac{7}{10} + \frac{3}{10} = 1$

10 a i $\frac{5}{8}$ ii $\frac{3}{8}$

 b $\frac{5}{8} + \frac{3}{8} = 1$

11 a i $\frac{4}{11}$ ii $\frac{7}{11}$

 b $\frac{4}{11} + \frac{7}{11} = 1$

 c 8 litres of blue paint, 14 litres of yellow paint

12 $\frac{6}{42} = \frac{1}{7}$, $\frac{12}{60} = \frac{1}{5}$, $\frac{1}{5} > \frac{1}{7}$ so the second shop sold the greater proportion

13 Class 7H, $\frac{5}{8} > \frac{3}{8}$

14 Saturday, $\frac{1}{6} > \frac{1}{8}$

15 Neither, both have $\frac{3}{4}$ of seats for season-ticket holders

16 Bus 12, $\frac{6}{15} > \frac{1}{3}$

17 The first website, $\frac{5}{14} > \frac{4}{14}$

18 a $\frac{1}{13}$ b 1 : 12

19 Peter, $\frac{1}{8} > \frac{1}{10}$

Challenge

a ear : face = 1 : 3
 arm span : height = 1 : 1
 head length : height = 1 : 8
 elbow to hand : height = 1 : 5
 foot : height = 1 : 7
 hand : height = 1 : 10

b Students' own answers.

7.5 Proportions and percentages

1 a 10   b 20   c 25

2 a $\frac{23}{50} = \frac{46}{100}$ b $\frac{7}{25} = \frac{28}{100}$

 c $\frac{9}{10} = \frac{90}{100}$ d $\frac{13}{20} = \frac{65}{100}$

 e $\frac{126}{200} = \frac{63}{100}$

3 a 7 : 3   b $\frac{7}{10}$ c $\frac{3}{10}$

 d $\frac{7}{10} = \frac{70}{100} = 70\%$ e 30%

4 90%

5 26%

6 50%

7 a 70%   b 69%
 c the high street bookshop, 70% > 69%

8 penalties, 90% > 84%

9 the rectangle, $\frac{6}{25} > \frac{5}{25}$ (24% > 20%)

10 oatmeal bread, 0.58 > 0.55

11 a i 80% ii 20%
 b 80% + 20% = 100%

12 a i 70% ii 30%
 b 70% + 30% = 100%

13 a i 35% ii 65%
 b 35% + 65% = 100%

14 a 25%   b 80%
 c No, because the proportions say nothing about the overall numbers and 20% of a larger total may be more than 25% of a smaller total.

15 No, she is not correct; $\frac{16}{25} = 64\%$ of the skaters are children, and $\frac{9}{10} = 90\%$ of the people watching are adults, so there is a bigger proportion of adults watching than of children skating.

16 a 8 : 17
 b i 32% ii 68%
 c 32% + 68% = 100%

17 No, 70% > 60%

18 No; Jeff should work out for 75% of the time and rest for 25% of the time.

Challenge

a Students' own bar models

b 2 : 3

7 Check up

1 a £2   b £4   c £10

2 a 400 g   b 100 g   c 300 g

3 Yes, 28 = 7 × 4 so she needs 7 times as long; from 8 am to 11 30 am is $3\frac{1}{2}$ hours and $3\frac{1}{2} = 7 \times \frac{1}{2}$

4 £200

5 a 3 : 4   b 5 : 3

6 a 1 : 6   b 7 : 1   c 2 : 3   d 4 : 9 : 10

7 C

8 6 teaspoons

9 a 9   b 21

10 a 5   b 25

11 a £12 and £18
 b £12 + £18 = £30

12 2 km

13 $\frac{4}{11}$

14 a $\frac{1}{3}$ b $\frac{2}{3}$

15 a i 35% ii 36%
 b their second season, 36% > 35%

16 Swifts, $\frac{9}{20} > \frac{8}{20}$

17 a $\frac{3}{4}$ b $\frac{1}{4}$
 c 225 g of white flour and 75 g of wholemeal flour

18 a i 60% ii 40%
 b 40% + 60% = 100%

Challenge

1 a 8 squares shaded   b 15 squares shaded

2 Students' own answers, e.g. 4 boys, 5 girls (total 9);
 8 boys, 10 girls (total 18); 40 boys, 50 girls (total 90)

7 Strengthen: Direct proportion

1 a 2 × £3 = £6   b 5 × £3 = £15
 c 7 × £3 = £21

2 a 20 ÷ 5 = £4   b 3 × £4 = £12
 c 4 × £4 = £16

3 a £10   b £20   c £40   d £80

4 a £1   b £2   c £2.50   d £5

5 160 calories

7 Strengthen: Ratio

1 a 1 part shaded   b 3 parts shaded
 c 2 parts shaded

2 a i 5 parts shaded   ii 2 parts shaded
 iii 2 parts shaded
 b The numbers of shaded parts are the same.

3 1 : 3

4 a 1 : 2   b 2 : 1
 c 1 : 4   d 6 : 1
 e 3 : 4   f 2 : 9   g 2 : 3

5 No, he can divide both 6 and 9 by 3 giving 2 : 3

6 4 : 7

7 a 2 : 4   b 3 : 6   c 4 : 8   d 5 : 10
 e 8 : 10   f 12 : 15   g 20 : 25   h 40 : 50

8 Students' own answers, e.g. 4 : 6 and 6 : 9

9 4 teaspoons

10 a 9   **b** 15

11 28 kg

12 Students' own answers, e.g.
    a 4 parts total, 1 part shaded
    b 3 parts total, 1 part shaded
    c 5 parts total, 4 parts shaded
    d 5 parts total, 2 parts shaded

13 a £3 and £9   **b** £5 and £10
    c £16 and £4   **d** £8 and £12

14 a 20 ml   **b** 40 ml   **c** 20 + 40 = 60

7 Strengthen: Comparing proportions

1 $\frac{7}{10}$

2 $\frac{9}{10}$

3 a $\frac{3}{10}$ **b** $\frac{7}{10}$

4 $\frac{2}{5}$

5 $\frac{2}{3}$

6 a $\frac{40}{120}$ **b** $\frac{1}{3}$

7 a $\frac{3}{10}$ **b** $\frac{3}{5}$ **c** box B, $\frac{3}{5} > \frac{3}{10}$

8 a $\frac{5}{8}$ **b** $\frac{6}{16} = \frac{3}{8}$

    c the first group, $\frac{5}{8} > \frac{3}{8}$

9 73%

10 a $\frac{3}{5}$ **b** 60%   **c** 48%   **d** Ellie's

11 a 40%   **b** 45%   **c** Sarah

12 a Seals: 6 girls, 10 children
    Dolphins: 11 girls, 20 children
    b Seals: 60%
    Dolphins: 55%
    c Seals

13 a $\frac{2}{3}$ **b** $\frac{1}{3}$

14 a $\frac{3}{7}$ **b** $\frac{4}{7}$ **c** 3 : 4

15 a $\frac{11}{20}$ **b** $\frac{9}{20}$ **c** 11 : 9

16 a $\frac{2}{7}$ **b** $\frac{5}{7}$

    c Harry has written the number of gold rings as the denominator; he should have written the total number of rings.

Challenge

1 Lily; $\frac{1}{3}$ of the group are girls, and the ratio $\frac{2}{3} : \frac{1}{3}$ is equivalent to 2 : 1.

2 a 6 taxis
    b Students' own answers, e.g.
    4, 4, 4, 4, 4, 1;   4, 4, 4, 4, 3, 2;   4, 4, 4, 3, 3, 3
    c Students' own answers, e.g.
    The third way would be fairest; 12 people would pay £3.75 and 9 people would pay £5.

7 Extend

1 4 : 1

2 9 kg (Check: 9 + 6 = 15, 2 × 15 = 30)

3 66.96 milligrams

4 a Sunday   **b** Saturday, $\frac{7}{12} > \frac{5}{12}$

5 the riding club, $\frac{2}{5}\left(\text{or } \frac{4}{10}\right) > \frac{3}{10}$

6 Tony, $\frac{1}{5} < \frac{1}{4}$

7 a 15 : 4 : 1   **b** 6 g

8 2 : 3

9 3 : 1 : 4

10 a 12   **b** 9   **c** 20   **d** 10

11 a i £750   **ii** £3750   **iii** £1500
    b £750 + £3750 + £1500 = £6000

Challenge

a A 1.8 m, B 1.5 m, C 1.9 m
b Students' own answers

7 Unit test

1 a 6   **b** 18

2 £33

3 a 8   **b** 2   **c** 20

4 No, can only make 30 slices as 4 : 6 = 20 : 30

5 a 12 kg and 15 kg   **b** 12 + 15 = 27

6 a 1 : 5   **b** 6 : 1   **c** 3 : 5   **d** 4 : 3

7 a $\frac{2}{5}$ **b** $\frac{7}{20}$ **c** first set, $\frac{8}{20} > \frac{7}{20}$

8 a i 30% **ii** 20%
    b the first time, 30% > 20%

9 Sharks have greater proportion (Sharks 60%; Dolphins 40%)

10 a £4   **b** £8   **c** £32

11 £350

12 30 ml

13 a 12   **b** 20

14 a i $\frac{1}{5}$ **ii** $\frac{4}{5}$
    b 50 g of cherries and 200 g of sultanas

15 a i 70% **ii** 30%
    b 70% + 30% = 100%

16 a 4 : 1
    b i 80% **ii** 20%

17 Debbie, $\frac{5}{50} > \frac{4}{50}$

Challenge

1–3 Students' own answers.

4 a Using the cards in the order 1 : 4, 3 : 7, 1 : 3, 4 : 5, 3 : 5, 2 : 3 reaches £20
    b Students' own answers

UNIT 8 Lines and angles

8.1 Measuring and drawing angles

1   B and C
2   a   90°   b   right angle
3   a   60°   b   20°
4   a   Both angles are smaller than 90°.
    b   80°, 65°
5   a   140°   b   155°
6   a   acute   b   obtuse
7   360°
8   a   330°   b   275°   c   200°
9   a   Smaller angle = 120°,
        so marked angle = 360° − 120° = 240°
    b   Smaller angle = 50°, so marked angle = 310°
    c   Smaller angle = 160°, so marked angle = 200°
10 80°, 37°, 100° and 145° angles drawn accurately
11 Drawing of ladder at 75° from the horizontal
12 Drawing of two lines at 90°
13 a   140° angle drawn accurately
    b   60° angle drawn accurately; 300°
14 320°, 280°, 250° and 200° angles drawn accurately

Challenge

    a   9:00 or 3:00   b   10:00 or 2:00   c   8:00 or 4:00

8.2 Lines, angles and triangles

General note: alternative notation for labelling angles may be used throughout this section.

1   3
2   a   right angle   b   acute
    c   obtuse   d   acute
3   a   equilateral   b   isosceles   c   scalene
4   a   equilateral   b   isosceles
    c   right-angled   d   scalene
    e   c – scalene and right-angled
5   a   BC   b   CD   c   AD   d   BD
6   a   ∠ABC   b   ∠RQP   c   ∠LMN
    d   Yes, e.g. angle ABC or AB̂C , or angle CBA, ∠CBA or
        CB̂A in part a
7   a   ∠ABD   b   ∠ADB   c   ∠DAB
8   a   i   FG or GH or HI or IJ   ii   FG or GH or CD or DE
    b   No, because lines CI and CG meet at point C, and parallel
        lines never meet.
    c   acute   d   acute   e   e.g. angle BAJ
9   a   A or B or C
    b   AB (or BA), BC (or CB), AC (or CA)
    c   equilateral
    d   ∠BAC (or ∠CAB), ∠ACB (or ∠BCA), ∠ABC (or ∠CBA)
10 a   ∠BAC (or ∠CAB)
    b   ∠BAC, ∠ABC, ∠BCA, ∠DEF and ∠DFE (or equivalents in
        alternative notation)
    c   About 110°   d   About 60°
    e   ∠DEF (or ∠FED)
11 Students' own drawings, e.g.

12 a   isosceles
    b   6
    c   Any two of BG, BC, CD and DG
    d   Any pair from: ∠BGC, ∠BCG, ∠DCG and ∠CGD; ∠AHB,
        ∠BHG, ∠DFG, ∠DFE, ∠AGC and ∠CGE; ∠BGH and
        ∠DGF; ∠GBH and ∠GDF; ∠ABH and ∠EDF; ∠BAH,
        ∠CAG, ∠DEF and ∠CEG

Challenge

Students' own answers

8.3 Drawing triangles accurately

General note: alternative notation for labelling angles may be used throughout this section.

1   a   All three sides are equal length.
    b   Two sides are equal length.
    c   All three sides are different lengths.
2   7 cm, 6.2 cm and 50 mm lines drawn accurately
3   25° and 120° angles drawn accurately
4   6 cm line drawn accurately
5   Triangles drawn accurately
6   a   triangle drawn accurately
    b   59 mm
    c   angle CAB = angle CBA = 65°
    d   isosceles
7   Triangles drawn accurately
8   a   triangle drawn accurately
    b   AC = AB = 6 cm
    c   angle CAB = 60°
    d   equilateral
9   a   Accurate drawing of the truss
    b   No; this truss's height is less than 3 m.
10 a   C because the angles are shown approximately the correct
        size – angle 120° is obtuse.
    b   Triangle drawn accurately
11 Triangle ABC sketched, labelled and drawn accurately
12 Triangle DEF sketched and drawn accurately
13 Triangle sketched and drawn accurately
14 Triangle drawn accurately

Challenge

a–c   Student's own accurate drawing
d   4 cm   e   60°
f   All the small triangles are equilateral.   g   5
h   All smaller triangles will have sides 2 cm and angles 60°, 21
    triangles in total.

8.4 Calculating angles

1   a   140   b   115   c   80   d   185
2   a   120   b   40
3   a   360   b   2
4   a   a = 50°, b = 130°, c = 115°, d = 65°
    b   a + b = 180°, c + d = 180°
    c   Both answers are 180°.
5   a   100°   b   75°   c   90°
6   a   65°   b   143°   c   65°   d   25°
7   60°
8   a   60°   b   60°
9   angle A = 150°, angle B = 30°
10 a   100°   b   70°   c   360°
11 a   150°   b   195°   c   270°
12 a   90°   b   50°
13 A, E and F; B, H and D; C and G
14 a–d   Students' own drawings and measurements
    e   Vertically opposite angles are equal.
15 a   a = 40° (vertically opposite angles are equal)
    b   a = 30° (vertically opposite angles are equal)
        b = 150° (angles on a straight line add up to 180°)
    c   a = 20° (angles on a straight line add up to 180°)
        b = 160° (vertically opposite angles are equal)
        c = 20° (angles around a point add up to 360°)
    d   a = 135° (angles on a straight line add up to 180° or
        vertically opposite angles are equal)
    e   a = 30° (vertically opposite angles are equal)
    f   a = 40° (angles on a straight line add up to 180° or
        vertically opposite angles are equal)
16 m = 45°, n = 32°

Challenge

a   i   180° ii   120°   iii   72°   iv   45°
b   360° ÷ n
c   i   6   ii   4   iii   10   iv   12

8.5 Angles in a triangle

1 a 120 **b** 35 **c** 105 **d** 65
e 95

2 a 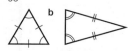 **b**

3 a 110° **b** 95° **c** 35°
4 a i 60° **ii** 120°
b i 90° **ii** 90°
c i 100° **ii** 80°
d Angles on a straight line add up to 180°.
5 a–c Students' own drawing
d The angles in a triangle add up to 180°.
The result is true for all triangles (but the demonstration here is not a proof).
6 a 120° **b** 60° **c** 60°
7 a 90° **b** 25° **c** 40°
8 Yes, 125° + 37° + 18° = 180°
9 60°
10 a Angles in a triangle add up to 180°, so $p + q = 80°$;
base angles in an isosceles triangle are equal, so
$p = q = 80°/2 = 40°$
b Angles in a triangle add up to 180°, so $r + s = 106°$;
base angles in an isosceles triangle are equal, so
$r = s = 106°/2 = 53°$
c Base angles in an isosceles triangle are equal, so $t = 45°$;
angles in a triangle add up to 180°, so $u = (180° - 45° - t)$
$= 180° - 90° = 90°$
11 a $a = 80°$ (angles in a triangle add up to 180°)
$b = 100°$ (angles on a straight line add up to 180°)
b $c = 50°$ (angles in a triangle add up to 180°)
$d = 130°$ (angles on a straight line add up to 180°)
c $e = 60°$ (angles on a straight line add up to 180°)
$f = 80°$ (angles in a triangle add up to 180°)
d *Unlabelled angle* = 65° (angles in a triangle add up to 180°)
so $g = 115°$ (angles on a straight line add up to 180°)

Challenge

 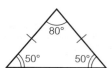

8.6 Quadrilaterals

1 4
2 a 270 **b** 335 **c** 80
3 A square, B kite, C rectangle, D parallelogram, E arrowhead, F trapezium, G rhombus, H trapezium
4 a square **b** rectangle
c trapezium **d** parallelogram
5 a Triangle and trapezium
b Triangle: 2 equal sides, right angle
Trapezium: no equal sides, 2 right angles
6 a Students' own drawing, e.g.

b Students' own drawing, e.g.

7 a 180° **b** 180° **c** 360° **d** Yes
8 a $a = 100°$ **b** $b = 100°$ **c** $c = 110°$
d $d = 103°, e = 77°$ **e** $f = 125°$ **f** $g = 70°$
9 a Base angles of isosceles triangles are equal; a–d are all the same, 30°.

b Opposite angles are equal (the marked pair are both 120°, and the other pair are both 60°).
10 a 130° **b** AD and BC **c** trapezium
11 $x = 60°$

Challenge

A rectangle, a square, a rhombus (as a square) and a right-angled isosceles triangle can be made.

8 Check up

General note: alternative notation for labelling angles may be used throughout this chapter.

1 a i acute **ii** obtuse
b i 80° **ii** 150°
2 a 70° **b** 170°
3 115° and 310° angles drawn accurately
4 a triangle ABC drawn accurately
b angle ACB = 60°
c equilateral
5 a 110° **b** 45°
6 $a = 120°$ ($a = 360° - 150° - 90°$)
7 a $x = 70°$ (angles in a triangle add up to 180°)
b $y = 30°$ (angles in a triangle add up to 180°)
8 a 108° (vertically opposite angles are equal)
b 72° (angles on a straight line add up to 180°)
9 $m = 70°$ (angles in a quadrilateral add up to 360°)
10 a rhombus **b** kite
11 a $a = 40°$ (angles in a triangle add up to 180°)
$b = 140°$ (angles on a straight line add up to 180°)
b $p = 120°$ (angles in a triangle add up to 180° so the missing angle in the triangle is 60°; angles on a straight line add up to 180° so $p = 180° - 60°$)
12 $x = 70°$ ($2x = 360° - 220°$)
13 a $x = 70°$ (base angles in an isosceles triangle are equal, and angles in a triangle add up to 180°)
b $y = 20°$ (base angles in an isosceles triangle are equal, and angles in a triangle add up to 180°)
c $z = 57°$ (angles in a quadrilateral add up to 360° and angles on a straight line add up to 180°)

Challenge

Students' own drawings and names

Strengthen: Measuring and drawing angles

1 a 70° **b** 40°
2 She has read the wrong scale.
3 a 50° **b** 15° **c** 63° **d** 105°
e 137° **f** 156°
4 a 120° **b** 120°
5 a 80° **b** 130° **c** 30° **d** 10°
6 a Juan
b–d

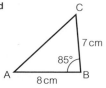

Strengthen: Calculating angles

1 a 180° **b** 90°
2 a 120° **b** 100° **c** 40° **d** 30°
3 360°
4 a 160° **b** 90° **c** 90°
5 a 80° **b** 20° **c** 130°
6 a $a = 110°, b = 40°, c = 30°, a + b + c = 180°$
b $d = 140°, e = 70°, f = 150°$
c i 180° **ii** 180° **iii** 180°
They all add up to 180°.
7 $z = 60°$
8 B Vertically opposite angles are equal.
9 a $a = 50°$ Vertically opposite angles are equal.

$b = 180° - 50° = 130°$ Angles on a straight line add up to 180°.

b **i** $c = 100°$ Vertically opposite angles are equal.
 ii $d = 30°$ Vertically opposite angles are equal.
  $e = 180° - 30° = 150°$ Angles on a straight line add up to 180°.

10 The angles in a quadrilateral add up to 360°.

11 a $a = 160°$ **b** $b = 30°$

12 a 120°
 b $c = 120°$, $d = 50°$, $e = 70°$

8 Strengthen: Solving angle problems

1 a

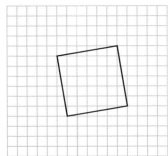

 b Students' own drawings, e.g. square, rectangle, parallelogram, rhombus

2 4 slices

3 a Students' own drawing, e.g.

 b Students' own drawing, e.g.

4 a Dashes on the lines show they are equal length.
 b isosceles
 c 80°

5 a $e = 40°$ (base angles in an isosceles triangle are equal)
 b $f = 100°$ (angles in a triangle add up to 180°)

6 a $a = 55°$ **b** $b = 70°$ **c** $c = d = 65°$

7 a i 60° **ii** $a = 120°$
 b i 80° **ii** $b = 100°$
 c i 50° **ii** $c = 130°$

8 $x = 60°$

9 a 150°   **b** 30°

Challenge

a 50°   **b** 110°
c Angle y increases to 115°.

8 Extend

1 45°

2 No, not necessarily; it depends on the sizes of the two acute angles.

3 Triangle ABC drawn accurately

4 Students' own drawings

5 No; Kevin has measured the smaller angle, but the marked angle is $360° - 120° = 240°$.

6 $a = 45°$ $b = 135°$ $c = 10°$

7 a 50°   **b** 150°   **c** 30°   **d** 140°
 e 210°

8 Students' own answers, e.g.
A reflex angle is between 180° and 360°, and the three angles in a triangle add up to 180°; one angle in a triangle cannot be greater than the sum of all three angles. But the angles in a quadrilateral add up to 360° so one of them can be reflex.

9 a Angle ABC = 120°, angle BCD = 60°, angle BAD = 60°
 b Opposite angles are equal.

10 He is partly correct: a rhombus is a special sort of parallelogram in which both pairs of parallel sides are the same length. So, all rhombuses are parallelograms, and some parallelograms are rhombuses.

11 a Triangle drawn accurately to scale
 b About 60°

Challenge

a

		Number of pairs of parallel sides		
		0	1	2
Number of pairs of equal sides	0		trapezium	
	1		trapezium (isosceles)	
	2	kite arrowhead		rectangle rhombus parallelogram

b A trapezium does not need to have a pair of equal sides, but it may have (if it is an isosceles trapezium).

c Students' own drawings

8 Unit test

1 130°

2 Angle D

3 70° angle drawn accurately

4 No, because the angles don't add up to 180°.

5 a 70° (360 − 190 − 100)   **b** 150° (360 − 60 − 60 − 90)

6 9 slices

7 a 60°   **b** 90°, 45°, 45°

8 $z = 75°$ $y = 105°$

9 Triangle drawn accurately

10 a Triangle ABC drawn accurately
 b Right-angled

11 a Parallelogram, rectangle, rhombus, square
 b Kite, parallelogram, rectangle, rhombus, square

12 $a = 60°$, $b = 120°$, $c = 20°$

13 $c = 140°$, $d = 40°$

Challenge

a Students' own shapes, e.g.

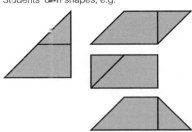

b Properties of the shapes drawn in part **a** (e.g. for those shown in part **a**, triangle, parallelogram, rectangle, trapezium)

UNIT 9

9.1 Sequences

1 14, 17, 20, 23, 26

2 6

3 24, 12, 6, 3, $1\frac{1}{2}$

4 8, 13, 18, 23, 28, 33; the numbers all end in 8 or 3.

5
  a 5, 8, 11, 14, 17
  b 10 000, 1000, 100, 10, 1
  c 3, 6, 12, 24, 48
  d 20.5, 20, 19.5, 19, 18.5
  e 3.2, 3.6, 4, 4.4, 4.8
  f 10, 5, 0, −5, −10
  g −7, −5, −3, −1, 1

6
  a 26, 31, 36   **b** 36, 28, 20
  c 3.6, 3.8, 4.0   **d** −4, −2, 0

7
  a First term 2, 'add 4', 18, 22, 26
  b First term 2, 'multiply by 3', 162, 486, 1458
  c First term 2, 'add 2', 10, 12, 14
  d First term 2, 'multiply by 2', 32, 64, 128
  e First term 52, 'subtract 4', 40, 36, 32
  f First term 32, 'add 4', 44, 48, 52
  g First term −15, 'add 3', −6, −3, 0
  h First term 4, 'subtract 3', −5, −8, −11
  i First term 2, 'subtract 3', −7, −10, −13
  j First term 1, 'multiply by 2', 16, 32, 64
  k First term 160, 'divide by 2', 20, 10, 5
  l First term 250 000, 'divide by 10', 250, 25, 2.5

8
  a Ascending   **b** Ascending
  c Descending   **d** Descending
  e Ascending   **f** Ascending
  g Descending   **h** Descending
  i Ascending   **j** Descending

9
  a Finite   **b** Infinite   **c** Finite
  d Infinite   **e** Finite   **f** Infinite

10
  a 2, 7, 12, 17, 22, 27, 32, 37, 42
  b Ascending, finite

11
  a 7, 7.6, 8.2, 8.8, 9.4, 10, 10.6, 11.2, 11.8, 12.4, 13
  b e.g. start at 0.76, multiply by 10

12
  a £250   **b** 9 months

13
  a 139 cm, 147 cm, 155 cm, 163 cm
  b 5 years
  c No, because you don't continue to grow 8 cm a year for the rest of your life.

14
  a 4 cells   **b** 80 minutes

15 Yes

16
  a **i** 1.8, 2, 2.2, 2.4   **ii** $3\frac{1}{2}$, 4, $4\frac{1}{2}$, 5
     iii −3, −1, 1, 3   **iv** 59.2, 59.9, 60.6, 61.3
     v $\frac{1}{32}$, $\frac{1}{64}$, $\frac{1}{128}$, $\frac{1}{256}$
     vi −4, −6, −8, −10
  b sequences v and vi

17
  a <u>9</u>, <u>12</u>, <u>15</u>, 18, 21, 24   **b** <u>46</u>, <u>49</u>, <u>52</u>, 55, 58, 61
  c <u>4000</u>, <u>2000</u>, <u>1000</u>, 500, 250, 125
  d <u>0.1</u>, <u>1</u>, <u>10</u>, 100, 1000, 10 000

18 Students' own answers

19
  a 23, <u>19</u>, 15, <u>11</u>, 7   **b** 23, <u>27</u>, 31, <u>35</u>, 39
  c −5, <u>−10</u>, −15, <u>−20</u>, −25
  d −5, <u>0</u>, 5, <u>10</u>, 15
  e 7.9, 8.3, <u>8.7</u>, 9.1, <u>9.5</u>, 9.9, <u>10.3</u>, <u>10.7</u>
  f 0.45, <u>0.5</u>, <u>0.55</u>, 0.6, 0.65, <u>0.7</u>, 0.75, <u>0.8</u>
  g 3.8, 4.0, 4.2, <u>4.4</u>, <u>4.6</u>, <u>4.8</u>, <u>5</u>, 5.2
  h −1.5, <u>−1.0</u>, −0.5, <u>0</u>, <u>0.5</u>, 1, <u>1.5</u>

Challenge

If you have included negative numbers in the sequence, you will have written all the whole numbers that exist. The list will not include any decimal numbers so is not a list of all numbers.

9.2 Pattern sequences

1 **a** 'add 2'   **b** 'add 6'   **c** 'add 10'

2 21, 25, 29, 33

3 4, 11, 18, 25

4 **a** 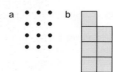 **b**

5 **a**
  b

Pattern number	1	2	3	4	5
Number of counters	5	9	13	17	21

  c Add 4 counters each time, one on each arm.

6 **a**
  b

Pattern number	1	2	3	4	5
Number of counters	3	5	7	9	11

  c Add 2 squares each time, one on each arm.

7 **a**

Week number	1	2	3	4	5
Number of cards	15	25	35	45	55

  b Week 7

8 **a** 3 × 4 = 12, 4 × 5 = 20, 5 × 6 = 30
  b e.g. Add 1 to the width and add 1 to the height.
  c 8 × 9

9 **a** Ben   **b** Students' own answers

Challenge
  a Students' diagrams showing: 25 rotten potatoes (day 3), 49 rotten potatoes (day 4), etc.
  b 1, 8, 16, 24, 32
  c Students' own answers, e.g. No, human interactions are far more complicated.

9.3 Coordinates and midpoints

1 9

2 7

3 I(7, 3), J(3, 7), K(5, 3), L(0, 3), M(8, 0), N(3, 8)

4 A(4, 5), B(−4, 5), C(4, −5), D(−4, −5), E(−4, 3), F(−1, 0), G(3, −4), H(0, −2), I(4, −3), J(−2, −5), K(−4, −1), L(−1, −4), M(−3, 0)

5 **a, b, c**

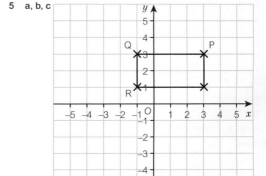

  d (3, 1)

6 **a**

x	1	2	3	4	5
y	8	16	24	32	40

  b (1, 8) (2, 16) (3, 24) (4, 32) (5, 40)
  c 'add 8' or '+8'

7 **a** 2, 4, 6

b

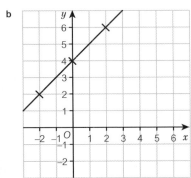

8 **a, b, c, d**

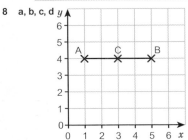

e (3, 4)

9

Line segment	Beginning point	Endpoint	Midpoint
CD	(5, 6)	(5, 4)	(5, 5)
EF	(−1, −1)	(3, −1)	(1, −1)
GH	(1, 4)	(5, 0)	(3, 2)
IJ	(−1, 5)	(−3, 1)	(−2, 3)

10 Students' own answers

11 **a** (1, 2) **b** (3, 0) **c** (3, 1) **d** (4, 4)
e (4, 0) **f** (2, 1) **g** (0, 0) **h** (1, −3)
i (1.5, 1) **j** (2, 1.5) **k** (−1, 0.5) **l** (−1.5, 0)

Challenge
(0, 1)

9.4 Extending sequences

1 **a** 3, $1\frac{1}{2}$, 4 **b** 4.5, 1.7, 3.3

2 **a** 10 **b** 46 **c** 6 **d** 4

3 **a** +4 **b** −5 **c** ×4

4 a, c, f, g, h

5 **a**

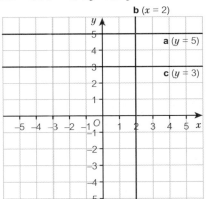

b 16, 25, 36
c square numbers
d No, because the difference between terms isn't the same each time.

6 **a** Number of dots: 1, 3, 6, 10, 15, 21
b 2, 3, 4, 5, 6, …
c 28
d No

7 **a** Not arithmetic **b** 0.5, +1
c Not arithmetic **d** Not arithmetic
e 25, −5 **f** 98, −9

8 13, 21, 34, 55

9 Month 12

10 **a** 5, 11, 23, 47, 95 **b** 7, 11, 19, 35, 67
c 127, 63, 31, 15, 7 **d** 2, 2, 2, 2, 2
e 32, 20, 14, 11, 9.5

11 **a** Students' own answers, e.g. multiply by 2 then add 3 gives 2, 7, 17, 37, 77, …
b Students' own answers

12 **a** add 5 sticks each time
b 6 + 5 = 11 sticks, 6 + 5 + 5 = 16 sticks
Yes, she is correct.
c 51 sticks

13 **a** 100 000, 1 000 000 **b** 625, 3125

14 **a** 3, 30, 300, 3000 **b** 10, 20, 40, 80
c 500, 100, 20, 4 **d** 800, 400, 200, 100

15 **a** Geometric **b** Arithmetic **c** Neither
d Neither **e** Arithmetic **f** Arithmetic

16 3

17 5 terms

Challenge
a Week 4
b Week 6
c Yes, the term-to-term rule is 'subtract 12'.

9.5 Straight-line graphs

1 10

2 **a** 30 **b** 5 **c** 20

3 **a** 7 **b** 10 **c** 22

4 **a** (4, 4), (1, 4), (−1, 4), (0, 4)
b

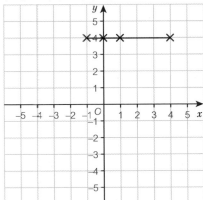

e.g. They all line up.
c e.g. All in the top half, in a line.

5 **a** (−4, 3), (−3, 3), (−2, 3), (−1, 3), (0, 3), (1, 3), (2, 3), (3, 3), (4, 3), (5, 3)
b The y-coordinate is always 3.
c **i** y = 3 **ii** x = 4 **iii** y = −2

6 A x = 1; B x = −3; C y = 2; D y = −1

7

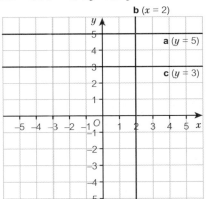

b (x = 2)
a (y = 5)
c (y = 3)

8 **a**

x	0	1	2	3	4
y	0	3	6	9	12

b (0,0), (1, 3), (2, 6), (3, 9), (4, 12)

c

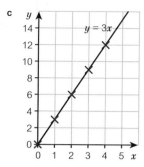

d 18

9

x	0	1	2	3	4
y	2	3	4	5	6

10 a

x	0	1	2	3	4	5
y	4	7	10	13	16	19

b smallest $x = 0$, largest $x = 5$; smallest $y = 4$, largest $y = 19$

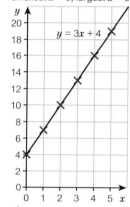

c $5\frac{1}{2}$

11 a

x	0	1	2	3	4	5
y	−1	2	5	8	11	14

b

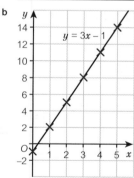

c They are all parallel.

12 a

x	−3	−2	−1	0	1	2	3
y	−3	−2	−1	0	1	2	3

b

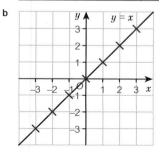

13 a e.g. (4, −4), (−1, 1) and (0, 0)

   b e.g. (1, 2), (4, 4) and (3, −1)

Challenge

a

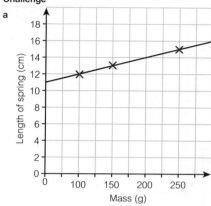

b 14 cm

c 50 g

9.6 Position-to-term rules

1 a '×3'   **b** '−5'   **c** '×5'

2 a 3   **b** 6   **c** 9   **d** 24

3 a 50   **b** 100

4 a 7, 8, 9, 10, 11, 12   **b** 2, 4, 6, 8, 10, 12

   c 3, 5, 7, 9, 11, 13   **d** 1, 4, 7, 10, 13, 16

5 a 3, 4, 5, 6, 7   **b** 2, 4, 6, 8, 10

   c −2, −1, 0 1 2   **d** 5, 10, 15, 20, 25

6 a 30   **b** 22   **c** 50   **d** 2

   e −5   **f** 10.5   **g** 9.3   **h** 5

7 6 terms

8 4 terms

9 a $8n$ **b** $11n$ **c** $10n$ **d** $9n$

   e n

10 a $2n$ **b** $12n$

11 a term = position number + 10 nth term = $n + 10$

   b $n + 4$ **c** $n + 11$ **d** $n + 20$

   e $n − 1$ **f** $n − 4$ **g** $n − 10$

12 a **i** $7n$ **ii** 7

   b **i** $25n$ **ii** 25

   c The multiple in the nth term is the same as the common difference.

13 a 12, 24, 36, 48, 60, 72   **b** $12n$

   c 2 hours

   d No. After 50 days, she will have to run for 600 minutes; this is 10 hours, which is unrealistic. She will have to stop adding 12 minutes each time at some point.

Challenge

   a $9n$ **b** 900

9 Check up

1 a 40, 50, 60   **b** 7, 9, 11   **c** 13, 16, 19

2 a 0, 2, 4, 6, 8   **b** even numbers

3 a 9, 5, 1, −3, −7

   b 0, 8, 16, 24, 32

   c 2, 6, 18, 54, 162

   d 48, 24, 12, 6, 3

4 a

○
○
○
○○○○○○○

   b Add 3 counters each time, one on each arm.

   c 1, 4, 7

   d 10, 13, 16, 19

5 a 34, 41, 48, 55   **b** −16, −18, −20, −22

   c $\frac{1}{11}, \frac{1}{13}, \frac{1}{15}, \frac{1}{17}$ **d** 64, 128, 256, 512

   e Sequences **a** and **b** are arithmetic. Sequence **d** is geometric. Sequence **c** is neither.

6 a 6, <u>3</u>, 0, −3, <u>−6</u>
b 1.5, <u>1</u>, 0.5, <u>0</u>, <u>−0.5</u>
c <u>11</u>, 4, <u>−3</u>, −10, −17
d <u>−4.1</u>, −3.2, −2.3, <u>−1.4</u>, −0.5, <u>0.4</u>

7 a 8, 9, 10, 11, 12 **b** 4, 8, 12, 16, 20

8 a 14 **b** 60 **c** 9.5 **d** 2.5

9 a position number × 6
b 10th term = 60, 50th term = 300
c $6n$

10 $n + 2$

11 a He has used the term-to-term rule, not the position-to-term rule.
b $3n$

12 a, b

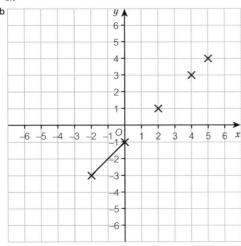

c (−1, −2)

13 A $x = 1$; B $y = 3$; C $y = −3$; D $x = 4$

14

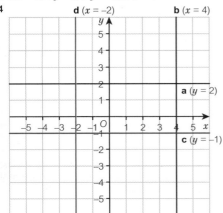

15 a

x	0	1	2	3	4	5
y	0	6	10	14	18	22

b

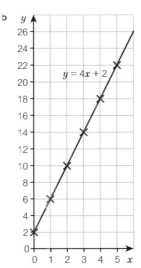

c 4

16

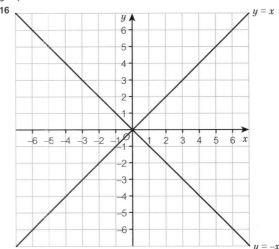

Challenge
a Students' correctly drawn lines for $x = 5$, $x = 1$, $y = −2$, $y = 2$
b The lines make the four sides of a square.
c Students' own answers
d Students' own answers

9 Strengthen: Sequences

1 a 3, 5, 7, 9, 11 **b** 10, 16, 22, 28, 34
c 20, 16, 12, 8, 4 **d** 15, 9, 3, −3, −9
e 3, 30, 300, 3000, 30 000 **f** 36, 18, 9, 4.5, 2.25

2 a i +2 ii ×3 iii −2 iv ÷10
b i and iii are arithmetic; ii and iv are geometric.
c i first term = 3, term-to-term rule = +2
ii first term = 1, term-to-term rule = ×3
iii first term = 10, term-to-term rule = −2
iv first term = 10 000, term-to-term rule = ÷10

3 a 6, 10, 14 **b** 4 hexagons at each stage
c 18 hexagons
d e.g. add 4 hexagons, draw the next pattern.
e 18, 22, 26, 30, 34

4 a 4, 11, 25, 53, 109 **b** 0, 5, 15, 35, 75
c 5, 8, 14, 26, 50

5 a 21 **b** 'add 3' **c** 33, 36, 39

6 a 60, 65, 70; '+5' **b** 36, 44, 52; '+8'
c 88, 84, 80; '−4' **d** 9, 17, 25; '+8'
e 1, 8, 15; '+7'

7 a <u>10</u>, 13, <u>16</u>, 19, 22 **b** 30, <u>34</u>, <u>38</u>, 42, 46
c <u>11</u>, 4, <u>−3</u>, −10, <u>−17</u> **d** <u>−1.5</u>, −1, <u>−0.5</u>, 0, <u>0.5</u>

9 Strengthen: The nth term

1 a −9, −8, −7, −6, −5

  b 100, 200, 300, 400, 500

2 a +3   b $n + 3$

3 a $n + 9$ b $n − 3$ c $n − 1$ d $n + 100$

4 a '+4'   b '+10'   c '×5'   d '×7'

5 b 10th term = 20, 50th term = 60

  c 10th term = 50, 50th term = 250

  d 10th term = 70, 50th term = 350

6 b $n + 10$ c $5n$ d $7n$

7 a $1 + 2 + 3 = 6$

  b

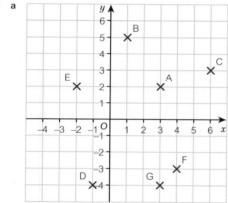

$1 + 2 + 3 + 4$ $1 + 2 + 3 + 4 + 5$
     10   15

  c The triangle numbers

  d The term-to-term rule

  e The position-to-term rule

  f $1 + 2 + 3 + 4 + 5 + 6 + 7 + 8 + 9 + 10 = 55$

9 Strengthen: Graphs

1 A(1, 3), B(5, 2), C(5, 0), D(1, −3), E(−2, −3), F(−3, 2)

2 a

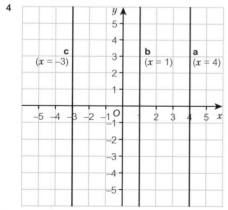

  b (3, −1) and (5, 0)

3 a Any four points with y-coordinate 2

  b The line crosses the y-axis at 2.

  c (0, 2), (5, 2), (−1, 2) and (2, 2)

  d e.g. The y-coordinates of the points that don't lie on the line are not equal to 2.

4

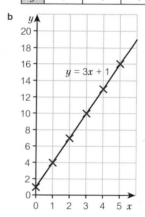

5 a She has plotted $y = 3$ not $x = 3$.

  b e.g. The line $x = 3$ is a straight line that crosses the x-axis. All the x-coordinates on the line are 3.

6 a

x	0	1	2	3	4
y	0	4	8	12	16

  b (0, 0), (1, 4), (2, 8), (3, 12), (4, 16)

  c, d

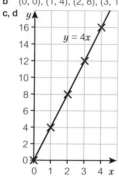

7 a

x	0	1	2	3	4	5
y	1	4	7	10	13	16

  b

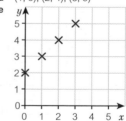

  d (0, 1)

Challenge

a 3   b '+1'

c

d (1, 3), (2, 4), (3, 5)

e

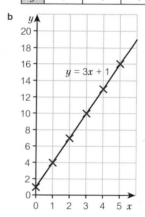

f +2

9 Extend

1 a First term 16, term-to-term rule '×10'

  b First term 0.25, term-to-term rule '×2'

  c First term 1, term-to-term rule 'add the two previous terms'

2 No, because the sequence will stop at 11 since 5.5 isn't a whole number.

3 Students' own answers

4 a 16   b square numbers

  c 1   d 1

  e No. The unshaded sequence is always 1, so it isn't increasing or decreasing.
    The shaded sequence does not go up or down in equal steps.

  f e.g. $6 × 6 + 1$ g 2, 5, 10, 17, …

5 **a** '×2' **b** $y = 2x$

c, d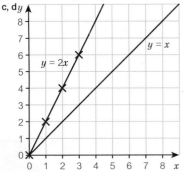

e e.g. Both graphs go through (0, 0), but $y = 2x$ is steeper.

6 **a** A(5, 15), B(5, −25), C(−15, −25)

b (−15, 15)

7 **a** $3n$

b $6n$

c $2n$

8 **a** −3, −2, −1, 0, 1; 100th term 96

b 2, 5, 10, 17, 26; 100th term 10 001

c −4, −2, 0, 2, 4; 100th term 194

d 199, 196, 191, 184, 175; 100th term −9800

9 **a** A, C

b Students' graphs

c e.g. They are all straight lines.

d Students' own answers

10 **a** **i** 0, 1, 2, 3, 4

ii 0, 2, 4, 6, 8

b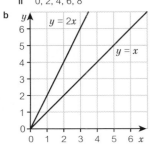

c (0, 0)

9 Unit test

1 **a** 200 **b** 9 **c** 10 000

2 **a** term-to-term rule '−10'; 60, 50, 40

b term-to-term rule '+12'; 88, 100, 112

c term-to-term rule '+1.5'; 14, 15.5, 17

d term-to-term rule '÷2' (or 'double denominator'); $\frac{1}{32}, \frac{1}{64}, \frac{1}{128}$

3 **a** 2, 7, <u>12</u>, 17, 22, <u>27</u> **b** <u>1.5</u>, 3, 6, 12, 24, <u>48</u>

4 **a**

Number of flowers	1	2	3	4	5
Number of beads	5	9	13	17	21

b 'add 4' or '+4'

5 **a** A(3, 1), B(−1, 3), C(1, 0), D(0, −2), E(1, −3)

b (0, 1.5)

6 **a**

(graph showing points plotted at y = 3 for x = 1 to 5)

b $y = 3$

c When $x = 50$, the y-coordinate is 3.

7 **a**

x	0	1	2	3	4	5
y	1	3	5	7	9	11

b 21

8 **a**

x	0	1	2	3	4
y	−3	0	3	6	9

b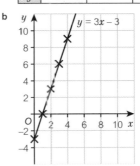

9 **a** '÷2' **b** 10th term 5; 50th term 25

c $\frac{n}{2}$

10 **a** 7, 10, 13, 16, 19 **b** 5, 4, 3, 2, 1

11 **a** Ascending **b** Descending

12 **a** Geometric **b** Arithmetic

13 **a** 20, 24, 28 **b** $4n$

Challenge

Students may interpret this in different ways, so produce many different answers. One way of looking at it is to view all the different squares of any size.

a 1; 4 + 1 = 5; 9 + 4 + 1 = 14; 16 + 9 + 4 + 1 = 30

b 25 + 16 + 9 + 4 + 1 = 55

c Sum of the square numbers up to 6^2, so 55 + 36 = 91

d Students' own answers

UNIT 10 Transformations

10.1 Congruency and enlargements

1 a 18   b 10   c 4   d 4
2 a C   b A
3 A and D
4 b

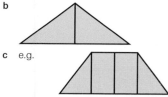

   c

   d

5 A and D, B and J, C and G, E and F, H and I
6 a Yes   b Yes
   c If two shapes are congruent, the lengths of matching
      sides are the same.
   d Yes   e Yes
   f If two shapes are congruent, the matching angles
      are the same size.
7 b

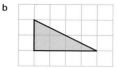

   c e.g.

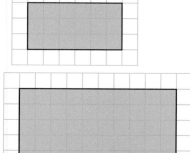

8 a Side x and side u are corresponding sides.
   b Side y and side v are corresponding sides.
   c Angle A and angle D are corresponding angles.
   d Angle C and angle E are corresponding angles.
9 B and C; the lengths of all the sides and the sizes of the
   angles are equal.
10 a Students' copies of diagram
    b

11 a i

    ii

b i

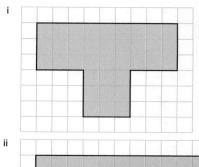

   ii

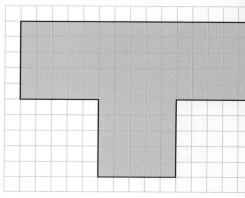

c i

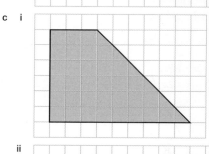

   ii

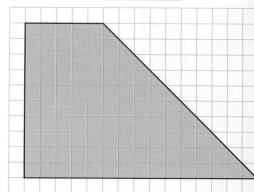

d i

   ii

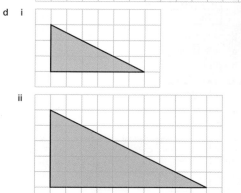

12 45 cm by 30 cm

13 a 15 cm   **b** 1.4 m

14 a i 1 : 2 **ii** 2
  b i 1 : 4 **ii** 4

15 a C   **b** D and F   **c** B and E

16 a Yes, the ratios of the side lengths are the same.
    16 : 8 = 2 : 1 and 12 : 6 = 2 : 1
  b No, the ratios of the side lengths are different.
    8 : 7 = 1.142... : 1 but 6 : 5 = 1.2 : 1
  c Yes, the ratios of the side lengths are the same.
    8 : 5 = 1.6 : 1 for height and width

17 Yes, every length will be 3 times as long as on the object.

Challenge

England: 4 congruent rectangles

Scotland: 2 pairs of congruent triangles

Wales: no congruent shapes

United Kingdom: 2 sets of 4 congruent triangles

10.2 Symmetry

1 a Cuboid **b** Cube   **c** Cylinder
  d Square-based pyramid

2 a Isosceles, 2 equal angles (or sides)
  b Scalene, no equal angles (or sides)
  c Equilateral, 3 equal angles (or sides)

3 a Students' own tracing and folding
  b An equilateral triangle has 3 lines of symmetry.
    An isosceles triangle has 1 line of symmetry.
    A scalene triangle has 0 lines of symmetry.

4 b

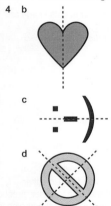

  c

  d

5 A 2, B 1, C 0, D 4 E 1, F 1, G 2, H 1

6 a, b

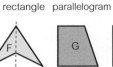

square   rectangle parallelogram   isosceles
                     trapezium

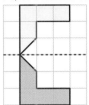

kite   arrowhead   trapezium   scalene
                               quadrilateral

7 a 25°, 5 cm   **b** 70°, 5 cm
  c 60°, 15 cm

8 Circle; *any* diameter of a circle is a line of symmetry so it has an infinite number of them, but a square has only 4 lines of symmetry.

9 a 2   **b** 8   **c** 2   **d** 1
  e 4

10 a 3   **b** 4   **c** 5   **d** 6
  e 2   **f** 8   **g** 1   **h** 1

11

Shape	Number of lines of symmetry	Order of rotational symmetry
square	4	4
rectangle	2	2
parallelogram	0	2
isosceles trapezium	1	1
kite	1	1
rhombus	2	2
sosceles triangle	1	1
equilateral triangle	3	3

12 A, B and D

13 a Yes; for a cylinder standing on end, there is 1 horizontal plane of symmetry and an infinite number of vertical planes.
  b No

14 a 2   **b** 3

Challenge

1 a 2   **b** 2

2 a Students' own answers
  b Students' own answers

3 Students' own answers, e.g. BOX, COOK

4 Students' own answers, e.g. 609, 818

5 Students' own answers, e.g. pod, MOW

10.3 Reflection

1 a The x-axis is horizontal and the y-axis is vertical.
  b (3, 4)

2

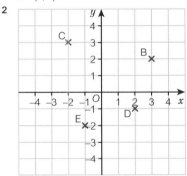

3 A $x = 3$ B $x = -2$ C $y = 1$ D $y = -2$
  E $y = x$

4 C

5 (Assuming yellow shapes are the object and green shapes the reflection; allow alternative correct answers if students have assumed green is the object.)
  a Correct reflection
  b No, it's a rotation. The correct reflection is

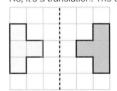

  c No, it's a translation. The correct reflection is

  d Correct reflection

6

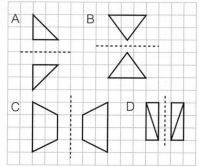

b Yes, all the side lengths and angles are equal.

7 A (3, 1), B (1, 3), C (3, 3), D (3, 6), E (4, 8)

8

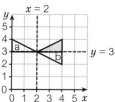

9

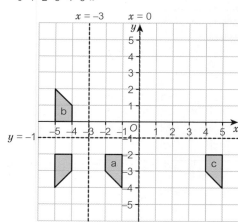

10 a–c

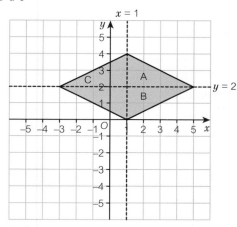

d Rhombus

11 b $y = 1$ **c** $x = -2$ **d** $x = 2$ **e** $y = -1$

12 b–d

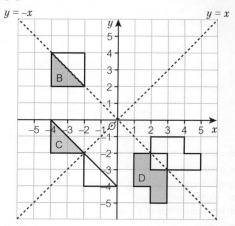

Challenge

1 a

	Coordinates of vertices					
Object ABC	A	(1, 4)	B	(0, 3)	C	(2, 3)
Image DEF	D	(4, 1)	E	(3, 0)	F	(3, 2)

b The x- and y-coordinates in the object and its image are swapped.

c Students' own shapes and their reflections in $y = x$

d The x- and y-coordinates in each object and its image are swapped.

2 a The x- and y-coordinates in the object and its image are swapped, and their signs reversed.

b Students' own shapes and their reflections in $y = -x$

10.4 Rotation

1 a i Quarter   **ii** Half
   iii Half   **iv** Quarter
   b 90°   **c** 180°

2 A anticlockwise, B clockwise

3 b 90° turn anticlockwise or 270° turn clockwise
   c 180° turn
   d 90° turn anticlockwise or 270° turn clockwise
   e 180° turn
   f 90° turn clockwise or 270° turn anticlockwise

4 A 180° rotation is half a turn, and two half turns in either direction make a full turn, so the first half turn in either direction gives the same result; it does not matter whether the turning is clockwise or anticlockwise.

5 a 180° turn
   b 90° turn clockwise or 270° turn anticlockwise
   c 90° turn clockwise or 270° turn anticlockwise

6 b 90° rotation anticlockwise about (1, 0)
   c 180° rotation about (1, 0)
   d 90° rotation clockwise about (1, 0)
   e 180° rotation about (1, 0)
   f 180° rotation about (1, 0)

7 a (0, 1), (2, 1) and (0, 0)
   b (−1 ,0), (−3, 0) and (−1, 1)
   c (0, −1), (0, −3) and (−1, −1)

8 a

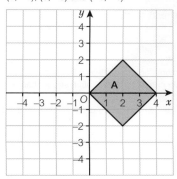

b Square

c

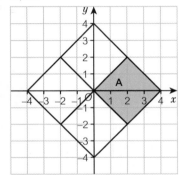

d Square

9 a 90° rotation clockwise about (2, −1)

   b 90° rotation anticlockwise about (2, −2)

   c 180° rotation about (1, 0)

   d 180° rotation about (4, 1)

Challenge

a Successive rotations of 90° (or 270°) clockwise (or anticlockwise) about (2, 0)

b Students' own logos

10.5 Translations and combined transformations

1 A right, B left

2 a, b

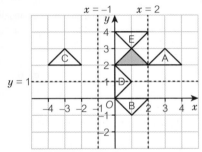

3 a 5 squares right

   b 5 squares right, 4 squares down

   c 3 squares left, 4 squares up

   d 5 squares left, 2 squares up

   e 3 squares left, 2 squares down

   f 1 square right, 4 squares down

   g 3 squares right, 4 squares up

4 a 1 square left, 2 squares down

   b 4 squares right, 2 squares down

   c 2 squares right, 1 square up

   d 7 squares right, 1 square up

5

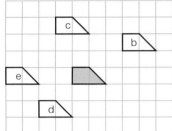

6

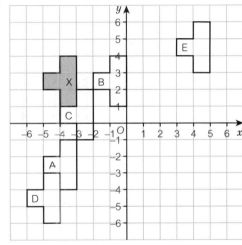

7 No

8 a

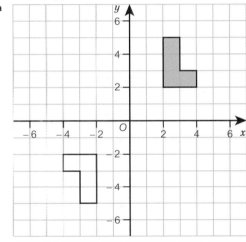

b

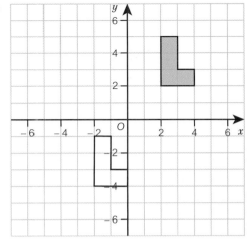

c

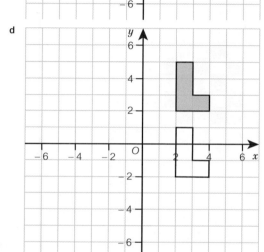

d

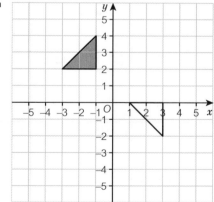

9 Yes, Larry is correct.
   a 180° rotation about (0, 0)
   b Translation 4 squares left, 6 squares down
   c Reflection in the x-axis
   d Translation 4 squares down

10 Yes, 4 squares left and 3 squares right is the same as 1 square left, and 2 squares down and 4 squares up is the same as 2 squares up.

11 a

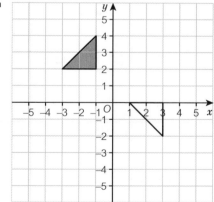

b

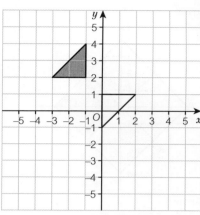

12 a T
   b F, different sizes
   c T
   d T
   e F, different orientations
   f T
   g F, different sizes
   h F, different sizes
   i T
   j F, different orientations

Challenge

a 90° clockwise (or 270° anticlockwise) about (2, 2)
b e.g. reflection in $x = 2$ followed by translation 4 squares up
c e.g. translation 4 squares up followed by reflection in $x = 2$

10 Check up

1 A and D, B and F, C and E
2 a 1   b 4   c 2   d 0
   e 2
3

4 a 4   b 2   c 3   d 1
5 a 8   b 8
6

7 a 5 squares right, 1 square up
   b 1 square left, 4 squares down
8 D
9 a

   b 1 : 2

329

10 a, b

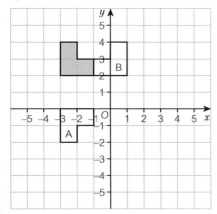

11 The direction (clockwise)

12 a, b

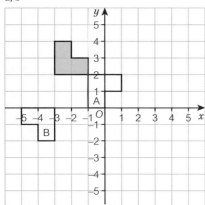

13 a, b

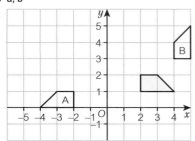

Challenge

a Yes, with suitable explanation.

b No, for example a parallelogram has rotational symmetry of order 2 but no lines of symmetry.

10 Strengthen: Shapes and symmetry

1 C, D, G

2 a 3 and 5 **b** 3 **c** 3

3 A, B, C and D

4 a–c
Rectangle: 2 lines
Square: 4 lines
Parallelogram: no lines
Rhombus: 2 lines
Isosceles trapezium: 1 line

5 a–d

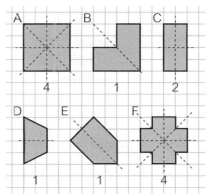

6 b, c A 2, B 5, C 3, D 1, E 2, F 4

7 A order 1, B order 2, C order 4, D order 3, E order 5

10 Strengthen: Translations, reflections and enlargements

1 a 5 squares right, 2 squares down
  b 3 squares right, 1 square up
  c 5 squares left, 3 squares down

2 a 2 squares right, 3 squares up
  b 2 squares right, 1 square down
  c 2 squares left, 4 squares down
  d 3 squares right
  e 5 squares left, 2 squares up
  f 5 squares up

3

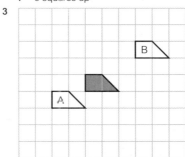

4 a Correct
  b Incorrect, it should be half a square to the left.
  c Incorrect, it should be one square down.

5

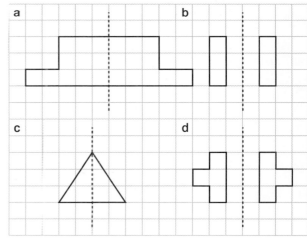

6 a

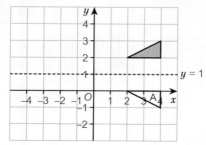

b

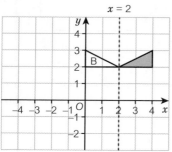

c

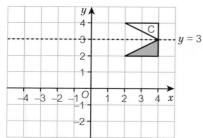

7 a i 1   **ii** 2   **iii** 1
   iv 2
  b i 3   **ii** 6   **iii** 3
   iv 6
  c i

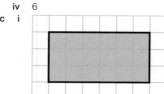

   ii 1 : 3
  d i

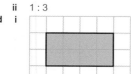

   ii 1 : 2

8 a

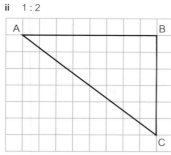

b 5 cm on original shape, 10 cm on enlarged shape
10 cm is 2 times 5 cm

10 Strengthen: Rotations and combined transformations

1 a 90° clockwise
  b 180° clockwise
  c 90° anticlockwise
  d 180° anticlockwise

2 a–c

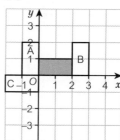

3 a

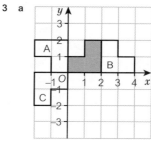

  b

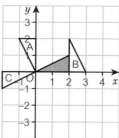

4 a No, need to know direction (clockwise or anticlockwise)
  b Yes
  c No, need to know the centre of rotation
  d No, need to know the angle of rotation
  e Yes
  f Yes

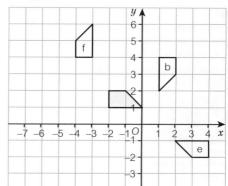

5 a–d

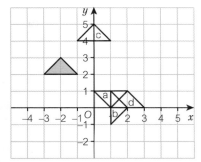

Challenge

a

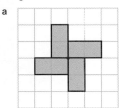

b 4

c No, it only has rotational symmetry.

10 Extend

1 a Rotation 180° about (2, 2.5)

b Rotation 180° about (–0.5, 1)

c Rotation 180° about (–4, 1.5)

2 a i 2   ii 3   iii 1.5

b 6 cm, 12 cm, 18 cm

c 2 cm², 8 cm², 18 cm²

d

Rectangles	Ratio of lengths	Ratio of perimeters	Ratio of areas
A : B	1 : 2	1 : 2	1 : 4
A : C	1 : 3	1 : 3	1 : 9

e Ratio of lengths and perimeters is the same.
Ratio of areas is the square of ratio of lengths,
$2^2 = 4$ and $3^2 = 9$

3 a i 40 cm   ii 75 cm²

b Enlargement drawn accurately

4 a, b

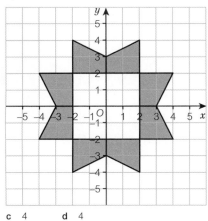

c 4   d 4

5 a $x = 3$

b

Triangle ABC	A (2, 1)	B (2, 4)	C (1, 2)
Triangle DEF	D (4, 1)	E (4, 4)	F (5, 2)
Triangle GHI	G (8, 1)	H (8, 4)	I (7, 2)

c J (10, 1), K(10, 4), L(11, 2)
To find J and K, add 2 to the x-coordinates of G and H,
y-coordinates stay the same.
To find L, add 4 to the x-coordinate of I, y-coordinate
stays the same.

6 a True   b False

c False   d True   e False

7 a 6   b Infinite number   c 2

Challenge

There are 12 different pentominoes (18 including reflections)

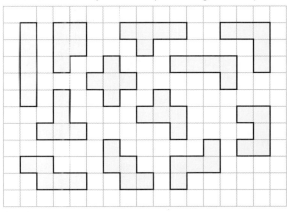

10 Unit test

1 a A and C

b D, G, F and H

2 a 1   b 0   c 5   d 2

e 0

3 No, it only has 1 (vertical) line of symmetry.

4

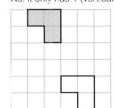

5

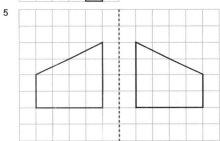

6 a 1, b 2, c 5, d 2, e 2

7 The centre of rotation (3, 1)

8 a–d

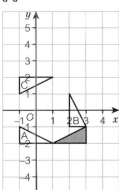

9 He has reflected the shape in the line $x = 2$.

10 a, b

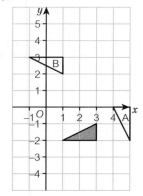

11 a

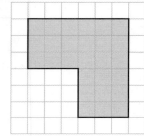

b 1 : 3

12 e.g. rotation 90° anticlockwise about (−4, −2) followed by a
translation 5 squares right and 2 squares up

13 a 3 **b** 4 **c** 2

Challenge

a

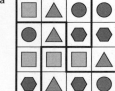

b

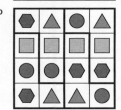

Index